Advanced Coatings for Buildings

Advanced Coatings for Buildings

Editors

Anibal C. Maury-Ramirez
Inês Flores-Colen
Hideyuki Kanematsu

MDPI • Basel • Beijing • Wuhan • Barcelona • Belgrade • Manchester • Tokyo • Cluj • Tianjin

Editors
Anibal C. Maury-Ramirez
Universidad El Bosque,
Decanatura Facultad de
Ingeniería
Colombia

Inês Flores-Colen
CERIS (Civil Engineering
Research and Innovation for
Sustainability), DECivil
(Department of Civil
Engineering, Architecture and
Georesources), IST (Instituto
Superior Técnico),
University of Lisbon
Portugal

Hideyuki Kanematsu
National Institute of Technology,
Suzuka College
Japan

Editorial Office
MDPI
St. Alban-Anlage 66
4052 Basel, Switzerland

This is a reprint of articles from the Special Issue published online in the open access journal *Coatings* (ISSN 2079-6412) (available at: https://www.mdpi.com/journal/coatings/special_issues/Prot_Coat_Build).

For citation purposes, cite each article independently as indicated on the article page online and as indicated below:

LastName, A.A.; LastName, B.B.; LastName, C.C. Article Title. *Journal Name* **Year**, *Article Number*, Page Range.

ISBN 978-3-03943-078-9 (Hbk)
ISBN 978-3-03943-079-6 (PDF)

Contents

About the Editors

Aníbal Maury-Ramírez is the current Dean of the Engineering Faculty of Universidad El Bosque (Bogotá, Colombia). He is a civil engineer from Universidad del Norte (Barranquilla, Colombia), where he worked his thesis on the "Development of a Neural-Fuzzy Model to Estimate the In-Let Flows to Waste Treatment Plants". In recognition of his academic performance he received the ALFA Scholarship on Materials Science from the European Union. This time, working at Tampere University (Tampere, Finland), he researched the "TiO2 Impregnation of Concrete and Plaster Surfaces". Based on the obtained results, he developed his doctorate on "Cementitious Materials with Air-Purifying and Self-Cleaning Properties Using Titanium Dioxide Photocatalysis" at Ghent University (Ghent, Belgium). Afterwards, with the aim of developing urban infrastructure with air-purifying and self-cleaning properties, he conducted a post-doctorate at the Hong Kong Polytechnique University (Hong Kong, China). Today, Dr. Maury Ramírez is an associate researcher (Colciencias), the author of more than 15 high-impact articles, 50 articles in the proceedings of international conferences as well as several books. In addition, he has been acknowledged as an Editorial Board Member and Guest Editor of Coatings (MDPI, Switzerland), a Member of the American Concrete Institute, a Senior Member of the International Union of Laboratories and Experts in Construction Materials, Systems and Structures (RILEM), and a Member of the Hong Kong Concrete Institute. His main scientific contributions lies in the development of sustainable building materials, particularly of green walls and roofs, photocatalytic materials, nanomaterials, and sustainable concrete (i.e., supplementary cementitious materials, recycled aggregates, permeable concrete). Finally, he has been involved in the design of sustainable buildings and communities. He was a Professor at Universidad del Atlántico (Barranquilla, Colombia), a Director of Post-Graduate Studies on Civil Engineering at Pontificia Universidad Javeriana Cali (Cali, Colombia) and the Dean of the Engineering Faculty at Universidad de La Salle (Bogotá, Colombia).

Inês Flores-Colen is an Associate Professor of the Construction Section, at the Department of Civil Engineering, Architecture and Georesources, Instituto Superior Técnico, University of Lisbon, and a research member at the research unit CERIS (also vice-president since January 2019). She is a civil engineer (1996), with an MSc (2002) and PhD in Civil Engineering (2009). She has been published in the field of construction, civil engineering and coatings: 2 books; 7 book chapters (3 national and 4 international); over 50 papers in international journals; 60 reports (technical-scientific dissemination and institutional consultancy); and 35 international congress papers. Her supervision activities include 2 post-doctorate researchers (1 finished); 9 PhD students (1 finished); and 80 Masters students. She is the referee of 20 international journals. She participates (ed) in eight competitively financed research programs (1 international and ongoing—CCS-CCU—PPI/APM/2019/1/00042/U/00001), two of which as a Principal Investigator (one national and ongoing - WGB_Shield PTDC/ECI-EGC/30681/2017; Nanorender—PTDC/ECM/118262/2010) and one as the responsible of the subcontracted IST team in a P2020 project (PEP -POCI-01-0247-FEDER-017417). She is a member of CIB W70 and W86 commissions (in this commission, she has been the secretary since June 2019). Her Scopus h-index is 17. Her research interests are coatings; innovative and sustainable solutions; performance; durability; pathology and in situ testing; maintenance and rehabilitation; thermal retrofitting (mortars and

ETICS); and aerogel-based renders. She is the first author of a national patent PT (108864B) published in July 2019 related to the development of an aerogel-based thermal render.

Hideyuki Kanematsu is a professor of materials science and engineering at the National Institute of Technology at Suzuka College in Japan. His areas of interest include biofilm engineering, surface engineering, corrosion and STEM education (active learning by e-learning). He holds a B.Eng. (1981), a M.Eng (1983) and a Ph.D in Materials Science and Engineering (1989), all from Nagoya University. He is an active member, board member, and editorial member of ASM International, the National Association for Surface Fishing, USA (NASF), the Institute of Metal Finishing, UK (IMF), as well as of the Minerals, Metals & Materials Society, USA (TMS), American Chemical Society (ACS), Electrochemical Society (ECS), the Japan Institute of Metals (JIM) and the Iron & Steel Institute of Japan (ISIJ), among others.

 *coatings*

Editorial

Advanced Coatings for Buildings

Aníbal Maury-Ramírez [1,*]**, Inês Flores-Colen** [2,*] **and Hideyuki Kanematsu** [3,*]

1   Engineering Faculty, Universidad El Bosque, Edificio Fundadores 3er Piso, Av. Cra 9 No. 131 A-02, Bogotá 111711, Colombia
2   CERIS, DECivil, IST, Universidade de Lisboa, Av. Rovisco Pais, 1049-001 Lisboa, Portugal
3   Department of Material Science and Engineering, National Institute of Technology (KOSEN), Suzuka College, Suzuka 510-0294, Japan
*   Correspondence: amaury@unbosque.edu.co (A.M.-R.); ines.flores.colen@tecnico.ulisboa.pt (I.F.-C.); kanemats@mse.suzuka-ct.ac.jp (H.K.)

Received: 22 July 2020; Accepted: 24 July 2020; Published: 24 July 2020

Abstract: Based on five Special Issues in *Coatings*, this e-book contains a series of fifteen articles demonstrating actual perspectives and new trends in advanced coatings in buildings. Innovative materials and multiperformance solutions provide a basis, contributing also to better protection of buildings' surfaces during the service life, and users' wellbeing.

Keywords: protective coatings; air-purifying coatings; anti-fouling coatings; ultra-low biofouling coatings; biological coatings

1. Introduction

Based on the current need to achieve sustainable development, multiperformance coatings have been developed for building applications (new and existent ones) to improve durability and to achieve novel, sustainable and health functions. Although there are already some commercially available coating products on the market, their application in the construction sector is still not very common. With some exceptions, this situation is most probably due to the lack of reported scientific evidence and the fact that there are still numerous unanswered questions about the concept and application. Therefore, this e-book is aimed to provide recent research and works on the application potential of advanced coatings for buildings and intends to give new scientific insights into the interrelation between science, engineering, architecture and other disciplines by means of advanced techniques for coatings' characterization and their application.

2. Multiperformance, Durable and Sustainable Coatings

This e-book, entitled "Advanced Coatings for Buildings", contains a collection of five Special Issues consisting of fifteen articles which cover fundamental studies and applications of multiperformance coatings designed to improve durability and to achieve novel, sustainable and health functions in buildings. COVID-19 redefined the concept of sustainability, making crucial the inclusion of human health [1]. Therefore, solutions proposed by engineers, architects and scientists should be addressed, giving buildings innovative functions such as: air-purifying and fire-resistant properties, long-lasting aesthetical appearance, water resistance, anti-fouling and anti-microbial surfaces. The current threat of interacting with microorganisms also generates new challenges that must be acknowledged to improve human quality of life, for example, by developing biological coatings for buildings which can act as a thermal, moisture, noise, and electromagnetic barrier. On the urban scale, they might reduce the heat island effect and sewage system load, improve runoff water and air quality, and reconstruct natural landscapes including wildlife. In summary, this e-book on coatings compiles the following Special Issues demonstrating the bright future of advanced coatings for buildings: "Advanced Protective Coatings for

Buildings" [2,3], "Nanocoatings with Air-Purifying Properties" [4–6], "Antifouling Coatings" [7–10], "Ultra-Low Biofouling Materials and Coatings" [11] and "Biological Coatings for Buildings" [12–16]. Going into detail, the Special Issues include the following topics:

- "Advanced Protective Coatings for Buildings": nanomaterial-based protective coatings, multifunctional coatings, bio-based protective coatings, smart protective coatings, sustainable protective coatings, low-embodied energy protective coatings, ecotoxicity on protective coatings, durable protective coatings and recycled protective coatings.
- "Nanocoatings with Air-Purifying Properties": comparison and assessment of the application potentials of novel nanocoatings for indoor and outdoor air purification.
- "Antifouling Coatings": coatings for commercial facilities, residential buildings, food processing, kitchens, baths and toilets, oil and water pipes.
- "Ultra-Low Biofouling Materials and Coatings": fundamentals and new concepts on ultra-low biofouling surface chemistries, novel nanostructured materials as coatings and coating additives, advancement in coating deposition technologies to improve coating antifouling performance, the self-assembly of chemistries and/or nanomaterials to fabricate antifouling thin films and coatings, advancements in understanding interfacial interactions between antifouling surfaces and microbial organisms.
- "Biological Coatings for Buildings": mechanical, thermal and hydraulic performance of biological coatings, modeling of biological coatings, life cycle assessment of biological coatings, construction of biological coatings and novel biological coatings.

Based on these interesting topics, we encourage you to read through this e-book and use the valuable information provided therein to help us move forward in the important area of multiperformance coatings with potential application in buildings for improving durability and to achieve innovative, sustainable and health functions. This approach certainly is a breakthrough to solve the current human threats for sustainable development during buildings' service life.

Funding: This research received no external funding.

Conflicts of Interest: The authors declare no conflict of interest.

References

1. Hakovirta, M.; Denuwara, N. How COVID-19 Redefines the Concept of Sustainability. *Sustainability* **2020**, *12*, 3727. [CrossRef]
2. Borsoi, G.; Esteves, C.; Flores-Colen, I.; Veiga, R. Effect of Hygrothermal Aging on Hydrophobic Treatments Applied to Building Exterior Claddings. *Coatings* **2020**, *10*, 363. [CrossRef]
3. Beh, J.H.; Yew, M.C.; Yew, M.K.; Saw, L.H. Fire Protection Performance and Thermal Behavior of Thin Film Intumescent Coating. *Coatings* **2019**, *9*, 483. [CrossRef]
4. Cruz-Pacheco, A.F.; Muñoz-Castiblanco, D.T.; Gómez Cuaspud, J.A.; Paredes-Madrid, L.; Parra Vargas, C.A.; Martínez Zambrano, J.J.; Palacio Gómez, C.A. Coating of Polyetheretherketone Films with Silver Nanoparticles by a Simple Chemical Reduction Method and Their Antibacterial Activity. *Coatings* **2019**, *9*, 91. [CrossRef]
5. Rosales, A.; Maury-Ramírez, A.; Gutiérrez, R.-D.; Guzmán, C.; Esquivel, K. SiO$_2$@TiO$_2$ Coating: Synthesis, Physical Characterization and Photocatalytic Evaluation. *Coatings* **2018**, *8*, 120. [CrossRef]
6. Guzmán-Aponte, L.A.; Mejía de Gutiérrez, R.; Maury-Ramírez, A. Metakaolin-Based Geopolymer with Added TiO$_2$ Particles: Physicomechanical Characteristics. *Coatings* **2017**, *7*, 233. [CrossRef]
7. Kanematsu, H.; Oizumi, A.; Sato, T.; Kamijo, T.; Honma, S.; Barry, D.M.; Hirai, N.; Ogawa, A.; Kogo, T.; Kuroda, D.; et al. Biofilm Formation of a Polymer Brush Coating with Ionic Liquids Compared to a Polymer Brush Coating with a Non-Ionic Liquid. *Coatings* **2018**, *8*, 398. [CrossRef]
8. Nguyen, M.D.; Bang, J.W.; Kim, Y.H.; Bin, A.S.; Hwang, K.H.; Pham, V.-H.; Kwon, W.-T. Anti-Fouling Ceramic Coating for Improving the Energy Efficiency of Steel Boiler Systems. *Coatings* **2018**, *8*, 353. [CrossRef]
9. Zhang, M.; Wang, R.; Li, L.; Jiang, Y. Size Distribution of Contamination Particulate on Porcelain Insulators. *Coatings* **2018**, *8*, 339. [CrossRef]

10. Ba, M.; Zhang, Z.; Qi, Y. Fouling Release Coatings Based on Polydimethylsiloxane with the Incorporation of Phenylmethylsilicone Oil. *Coatings* **2018**, *8*, 153. [CrossRef]
11. Hecker, M.; Ting, M.S.H.; Malmström, J. Simple Coatings to Render Polystyrene Protein Resistant. *Coatings* **2018**, *8*, 55. [CrossRef]
12. Naranjo, A.; Colonia, A.; Mesa, J.; Maury-Ramírez, A. Evaluation of Semi-Intensive Green Roofs with Drainage Layers Made Out of Recycled and Reused Materials. *Coatings* **2020**, *10*, 525. [CrossRef]
13. Gutiérrez, R.-D.; Villaquirán-Caicedo, M.; Ramírez-Benavides, S.; Astudillo, M.; Mejía, D. Evaluation of the Antibacterial Activity of a Geopolymer Mortar Based on Metakaolin Supplemented with TiO_2 and CuO Particles Using Glass Waste as Fine Aggregate. *Coatings* **2020**, *10*, 157. [CrossRef]
14. Verdier, T.; Bertron, A.; Erable, B.; Roques, C. Bacterial Biofilm Characterization and Microscopic Evaluation of the Antibacterial Properties of a Photocatalytic Coating Protecting Building Material. *Coatings* **2018**, *8*, 93. [CrossRef]
15. Naranjo, A.; Colonia, A.; Mesa, J.; Maury, H.; Maury-Ramírez, A. State-of-the-Art Green Roofs: Technical Performance and Certifications for Sustainable Construction. *Coatings* **2020**, *10*, 69. [CrossRef]
16. Illera, D.; Mesa, J.; Gomez, H.; Maury, H. Cellulose Aerogels for Thermal Insulation in Buildings: Trends and Challenges. *Coatings* **2018**, *8*, 345. [CrossRef]

Article

Evaluation of Semi-Intensive Green Roofs with Drainage Layers Made Out of Recycled and Reused Materials

Alejandra Naranjo [1], Andrés Colonia [2], Jaime Mesa [3] and Aníbal Maury-Ramírez [4,*]

[1]   Department of Civil and Industrial Engineering, Engineering and Sciences Faculty, Pontificia Universidad Javeriana Cali, Cali 760031, Colombia; alejanaranjo@gmail.com

[2]   Gerencia, Dos Mundos, Cali 760050, Colombia; gerencia@dosmundos.org

[3]   Department of Mechanical Engineering, Faculty of Engineering, Campus Tecnológico, Universidad Tecnológica de Bolívar, Cartagena de Indias 130001, Colombia; jmesa@utb.edu.co

[4]   Engineering Faculty, Universidad De La Salle, Sede Candelaria, Bogotá 111711, Colombia

*   Correspondence: amaury@lasalle.edu.co; Tel.: +57-134-17-900

Received: 10 April 2020; Accepted: 27 May 2020; Published: 29 May 2020

Abstract: Green roof systems represent an opportunity to mitigate the effect of natural soil loss due to the development of urban infrastructure, which significantly affects natural processes such as the hydrological water cycle. This technology also has the potential to reduce the indoor building temperature and increase the durability of waterproof membranes, reduce run-off water and heat island effects, create meeting places, and allow the development of biological species. However, despite the described benefits, the use of this technology is still limited due to the costs and the environmental impact from using non-renewable building materials. Therefore, this article presents the hydraulic and thermal analysis of different semi-intensive green roofs using recycled (rubber and high density polyethylene (HDPE) trays) and reused materials (polyethylene (PET) bottles) in their drainage layers. Then, three roof systems were evaluated and compared to traditional drainage systems made with natural stone aggregates. Results showed that some systems are more useful when the goal is to reduce temperature, while others are more effective for water retention. Additionally, this study presents evidence of the potential of reducing the dead loads and costs of green roofs by using recycled and reused materials in drainage systems.

Keywords: green roofs; recycled and reused materials; drainage; thermal insulation; dead load

---

1. Introduction

In recent years, the use of green roofs has increased, not only for aesthetic reasons but also to improve the environmental quality of the environment [1]. Plants can reduce heat through the reflection of solar radiation and the generation of shade. They can also decrease heat through the evapotranspiration process, which reduces the temperature inside and outside buildings [2]. In this way, green roofs are considered a modern and ecological technology to face climate change and the most common environmental problems in the urban environment [3]. Green roofs represent the opportunity to expand the presence of vegetated surfaces by replacing conventional waterproofed surfaces, generating thermal insulation, improving the internal environment of buildings, and consequently causing a reduction in the energy consumption of such buildings [4,5].

Though green roofs have the benefits described above, their use in many countries is limited due to costs and environmental impacts of using non-renewable materials, as well as the lack of formal methods focused on their design, construction, and maintenance [6]. A growing trend regarding the construction of green roof systems has been the use of recycled and reused materials in their drainage and substrate layers, which can generate environmental, technical, economic, and aesthetic benefits while also providing the possibility of incorporating waste into the construction production chain. Most of the existing works in the current literature regarding the use of recycled materials in green roofs have sought to evaluate their hydraulic performance (drainage capacity and water retention) and the response of vegetation growth compared to conventional green roof systems. Table 1 summarizes the most relevant works concerning the use of recycled materials in green roofs. Recycled materials such as rubber, glass, and construction waste have been widely studied in the last decade, demonstrating that replacing conventional green roof materials is a feasible solution.

Table 1. Summary of works related to the use of recycled materials in green roofs.

Author	Materials	Scope	Results
Chi-Feng et al. 2018 [7]	Recycled glass	Vegetation growth	Good results depend on controlling the mixing ration between the recycled glass and organic substances, as well as the location of the glass in the substrate.
Eksi and Rowe 2016 [8]	Recycled crushed porcelain	Vegetation growth and water retention	Improved water retention and plant growth when reducing the number of large particles. Additionally, porcelain provides less embodied energy to construct a green roof.
Bates et al. 2015 [9]	Recycled construction waste	Vegetation growth	Pure crushed bricks or mixtures primarily based on crushed bricks enabled the most diverse ruderal plant assemblages. Ash from solid waste provided the worst response.
Rincón et al. 2014 [10]	Recycled rubber	Environmental performance	The material provided a similar response compared to pozzolana, in addition to a lower environmental impact.
Mickovski et al. 2013 [11]	Recycled construction waste	Vegetation growth and drainage properties	Good response and properties compared to conventional materials.
Pérez et al. 2012 [12]	Recycled rubber from tires as the drainage layer	Potential Energy Savings	The material provided a similar response to pozzolana in terms of hydraulic behavior. The use of recycled rubber can be useful as a potential energy-saving material for the continental Mediterranean climate during summer.

To contribute to these research efforts, the present study aimed to evaluate the performance of green roofs with drainage layers made out of recycled and reused materials for reducing the impact on virgin materials and assuring technical benefits such as temperature reduction, water retention, and the regulation of rainwater. In addition, this study analyzed the reduction of dead load compared to traditional green roofs, which is relevant in terms of building structures. The subsequent sections of this paper are organized as follows: Section 2 summarizes the materials and methods employed to conduct the experimentation and to analyze the hydraulic and thermal performance of the proposed green roof systems. Results are presented in Section 3, while Sections 4 and 5 correspond to discussion and conclusions, respectively.

2. Materials and Methods

This section describes the materials used to build the green roof prototypes and the methodology for monitoring and measuring hydraulic and thermal performances. Measurement instruments, constructive characteristics of each green roof prototype, and the rain simulation system are described in detail as well. The dead loads of the green roof systems were determined as the approximate cost per square meter.

Based on the characteristics, amount, and local availability, four materials were selected to develop the drainage layer of green roof systems: (a) basalt gravel (typical material), (b) recycled rubber from tires, (c) recycled polyethylene (PET) bottles, and (d) recycled high density polyethylene (HDPE) trays.

In this case, the basalt gravel was considered as a reference material, since it is commonly used as a typical drainage layer of green roofs. The other three materials are conventionally considered waste from other human activities.

2.1. Green Roof Materials

This study considered five layers to construct the roof prototypes: (i) the roof structure, (ii) the geotextile, (iii) the drainage layer (using materials aforementioned), (iv) the substratum layer, and (v) the vegetation layer. Figure 1 describes the distribution of layers in the proposed green roof prototypes. Each layer is described as follows:

1. Roof structure: This consists of a box made of galvanized steel (gauge 20) and thermally isolated through polyethylene foam.
2. Geotextile membrane: This is a high resistance geotextile membrane. Reference T2400 PAVCO, manufactured by Mexichem®.
3. Drainage layer: This is designed to facilitate water drainage during and after rain conditions. In this study, four different drainage layers were analyzed: (a) basalt gravel, (b) recycled rubber from tires, (c) recycled PET bottles, and (d) recycled HDPE trays.
4. Substratum: This consists of organic and inorganic materials, with the former consisting of organic soil (70 v/v percent) and rice husks (30 v/v percent). In the case of PET bottles, the substratum was added into each cell (bottle section) as well as the vegetal layer.
5. Vegetal layer: This consists of a high resistance grass denominated *Stenotaphrum Secundatum*, also named San Agustín grass. This vegetation provides resistance to high temperatures and human traffic.

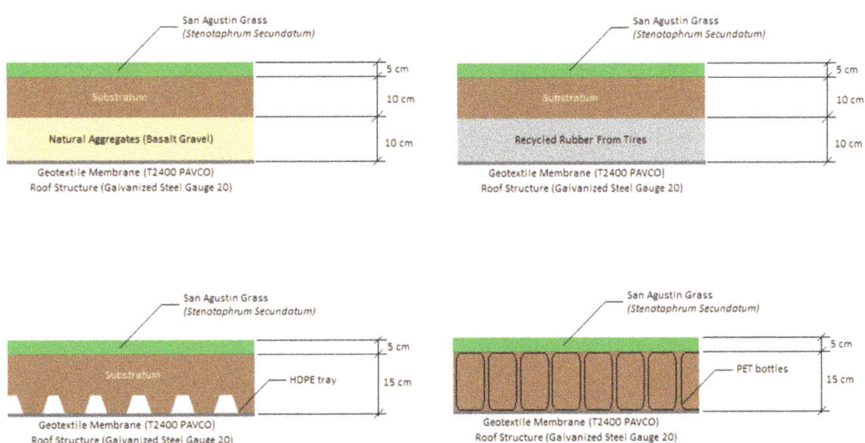

Figure 1. Cross sections of the green roof prototypes made out of recycled or reused materials.

A detailed description of each one of the selected materials for the drain layer is summarized in Table 2, including details of appearance, shape, and typical composition.

Table 2. Description of materials selected for drainage layers.

Material	Sample Photo	Appearance and Shape	Composition
Basalt Gravel: natural stone, traditionally used as a drainage layer in green roofs.		Large, irregularly shaped particles with a rough, porous, and vesicular surfaces, preserving traces of bubbles produced by expanding water vapor generated during the cooling and solidification of lava.	45%–54% silica and generally rich in iron and magnesium [13].
Recycled Rubber: recycled elastic polymer used to replace basalt stone in the drainage layer in the green roof systems.		Coarse particles in the form of rectangular prisms and granules of fine angular black particles. Obtained from recycled car tires.	Rubber 45%, carbon black 21%, metal 20%, textiles 4%, additives 8%, zinc oxide 1%, and sulfur 1% [14].
Recycled PET Bottles: polymer denominated as polyethylene terephthalate PET or PETE.		Cells of 4 × 4 bottles split in half (eight halves with mouths and eight halves with bottoms), each with five perforations of 3 mm diameter, arranged in a reticular way, and joined by galvanized wire staples located at the height of 7.5 cm. Smooth texture, translucent light green.	Polymer molecules comprised of carbon, hydrogen, and oxygen. $(C_{10}H_8O_4)_n$
Recycled Polymer Trays HDPE: high-density polyethylene obtained from recycled product containers.		Dark material. 60 cm × 60 cm × 10 cm, consisting of nine containers or concavities of cylindrical, conical and round shapes that allow for the retention and storage of water. Additionally, trays have diffusion holes that allow for air and water vapor circulation, as well as drainage.	Polymer molecules comprised of carbon, hydrogen, and oxygen. $(C_2H_4)_n$

2.2. Green Roofing Prototypes

Figure 2 shows the full prototype and the cross-section of the four proposed roof systems. The dimensions of the four roof structures were 1.20 m × 0.60 m (W × D). Height varied according to the type of roof, using 0.15 and 0.20 m for drainage with recycled PET bottles and HDPE trays and for basalt gravel and recycled rubber, respectively. Inside the roof structure, a drainage point was installed using a $\frac{1}{2}''$ galvanized pipe towards the center of the structure using a 2% slope.

(a) (b) (c) (d)

Figure 2. Cross-section detail of roofing prototypes: (**a**) Prototype 1: basalt gravel. (**b**) Prototype 2: recycled rubber. (**c**) Prototype 3: recycled polyethylene (PET) bottles. (**d**) Prototype 4: recycled high density polyethylene (HDPE) trays.

2.3. Rain Simulator Installation

The rain simulator system was designed to provide controlled and homogeneous drip irrigation under the effective area of the roof prototypes, guaranteeing water flow and the optimization of water consumption. Figure 3 shows the rain simulation full assembly. Water retainers were installed on the roof prototypes to avoid water losses and to provide a better approximation of natural precipitation conditions. Figure 4 summarizes the instruments and materials used to calibrate the rain simulation system.

The rain simulator was calibrated using a pluviometer and by running several preliminary tests to verify the proper functionality of the system. According to the preliminary tests, it was possible to identify that the outer roof prototypes (1 and 4) received a flow rate of 0.58 L/m, while the inner roof prototypes (2 and 3) received a flow rate of 0.61 L/m. Due to restrictions of the experiment, each simulation was carried out on sets of two roofs, always in the same way, prototypes 1 and 2, and prototypes 3 and 4. Then, given the average flow rate per nozzle of 0.6 L/min, the total flow rate in two nozzles corresponded to 1.2 L/min. Based on these flows, it was found that 0.6 min was required to obtain 1 mm of precipitation on each roof. In this way, the time necessary to simulate each rain was calculated. Then, this value was multiplied by the number of millimeters.

Finally, the approximate cost for each prototype (USD per square meter) is summarized in the Appendix A at the end of this manuscript. The cost was directly estimated from the materials employed in the green roof construction (labor not included).

(a)

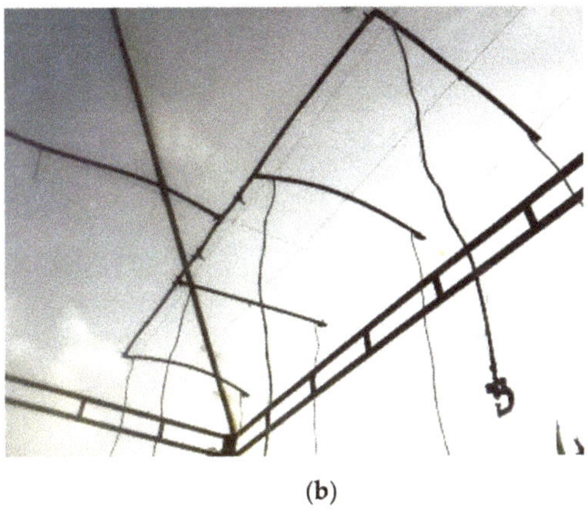

(b)

Figure 3. *Cont.*

(c)

(d)

Figure 3. Rain simulator system: (**a**) Overall picture of rain simulator and water supply tank. (**b**) Floating irrigation system. (**c**) Water retainers. (**d**) Spatial orientation of the rain simulation experiment with respect to cardinal points.

(a)   (b)   (c)

(d)   (e)   (f)

Figure 4. Calibration and rain simulation monitoring elements: (**a**) calibration test tube, (**b**) micro-spray nozzle, (**c**) timer, (**d**) submersible pump, (**e**) chronometer to measure drainage performance, and (**f**) tank for water supply for pumping.

2.4. Method

This subsection describes the definition of experimentation parameters, as well as the procedure followed to perform and monitor the rain simulation for each roof prototype.

2.4.1. Definition of Experimental Conditions

The conditions to simulate precipitations were established according to the geo-climatic characteristics (altitude, temperature, radiation, and geographic location) of Valle del Cauca, Colombia, more precisely in the city of Santiago de Cali. We used the official data record from the weather station of the Universidad del Valle—located at the latitude (3°22′39.66″), longitude (76°32′05.26″) coordinates of the WGS 84 system—because it is the closest station to the study area with official data [15].

A review of rainfall records was necessary to define the quantity and characteristics of precipitations in the study zone. Data from 1966 to 2015 were considered to define the most relevant precipitation cycles or scenarios. Based on these, three main types of precipitation were considered.

Typical rainfall: This consists of a cycle of several consecutive days of rain, the intensity of which corresponds to the wet season of the periodic cycles presented in the geographical study area. The criteria used for the selection were:

(a)   Minimum five days in a row of rainfall;
(b)   Not having days without precipitation between the chosen days;
(c)   The chosen cycle must start and end on a day without precipitation;
(d)   Incremental trend;
(e)   Rainy months of the city of Santiago de Cali (April, May, June, October, and November);

IFD (intensity–frequency–duration) curves were not considered in this rainfall scenario due to the fact that the capacity of the spray system per roof prototype was 1.2 L/min. Rainfall for the five days were simulated as follows: precipitation of 4 mm in a single period of 2.4 min (2 min and 24 s); 12 mm precipitations in two periods of 3.6 min (3 min and 36 s); 50 mm precipitation in three periods, the first two of 9 min (15 mm) and the last of 12 min (20 mm); 51 mm precipitation in three periods, the first two of 9 min (15 mm) and the last of 12.6 min (21 mm); and finally, 1 mm precipitation over 36 s. The time between each period was thirty minutes for each measurement, and the data employed to estimate typical conditions were taken from 23 to 29 October 1987 (See Appendix A—Figure A1a).

Intense rainfall: This was considered under the variables of intensity, frequency, and duration. It was selected the one with historical data higher than 30 mm/h that would have occurred in those atypical events and under the effects of La Niña phenomenon. It was also important as a requirement to have an IFD curve. The selected parameters were the total precipitation (49.70 mm) and a duration of 1 h (intensity: 49.7 mm/h).

Considering the need to simulate intense rain in the closest way to the IFD curve and taking into account the capacity and restriction of the irrigation system, this precipitation was simulated in four periods seeking to represent ten-minute intervals of distribution of the IFD curve. For this reason, the 17.9 mm precipitation that took a time of 10.74 min was joined with the second 10 mm precipitation, thus generating continuous precipitation of 16.74 min (16 min and 44 s). The second period corresponded to a 10 mm precipitation in a period of 6 min, the third to an 8 mm precipitation in 4.8 min (4 min and 48 s), and, finally, the fourth to a 3.8 mm precipitation in of 2.28 min (2 min and 16 s). Periods two, three, and four started every 10 min. Recorded data for intense rainfall conditions were taken from 2 June 1995 (see Appendix A—Figure A1b).

Saturation rainfall: This consisted of finding the number of precipitation millimeters required for simulating over the area of the roofs (average condition). In such a way, precipitation generated an effect close to the ground saturation. The way to find this was to drop the total capacity of the sprinkler system on the roof, taking a rain gauge as a measurement reference. Then, when the evacuation flow was similar in volume to the water entering to the system, the reference volume of the rain gauge was taken, giving a value of 10 mm over 6 min.

2.4.2. Rain Simulation

The rain simulations on the green roof prototypes were performed in three cycles that included the three types of rainfall chosen (typical, intense, and saturation). These considerations were included to obtain data with technical guidelines for the standardization and validation of the methods used and to guarantee the following aspects:

1. To get consistency between the independent results obtained with the same method, the same test material, the same conditions were used (same operator, same apparatus, and after short time intervals);
2. Measurement repeatability to provide very similar indications to each other for repeated applications and under the same measurement conditions;
3. To assure reproducibility of the experiment and the quality of the obtained data;
4. The cycles carried out in the experiment are described in Table 3 below:

Table 3. Simulated precipitation cycles.

CYCLE 1 (Post-Sowing)	CYCLE 2 (Post-Drought)	CYCLE 3 (Stable Cycle)
4 dry days Saturation rainfall 1 dry day Typical rainfall 4 dry days Intense rainfall 8 dry days (prolonged drought simulation period)	Saturation rainfall 2 dry days Typical rainfall 4 dry days Intense rainfall 4 dry days	Typical rainfall 4 dry days Saturation rainfall 4 dry days Intense rain 2 dry days

2.4.3. Monitoring of Performance Parameters

This subsection describes, in detail, the three main variables analyzed in this study: hydraulic performance, thermal performance, and dead load (Figure 5). Each one of these aspects is described in detail as follows:

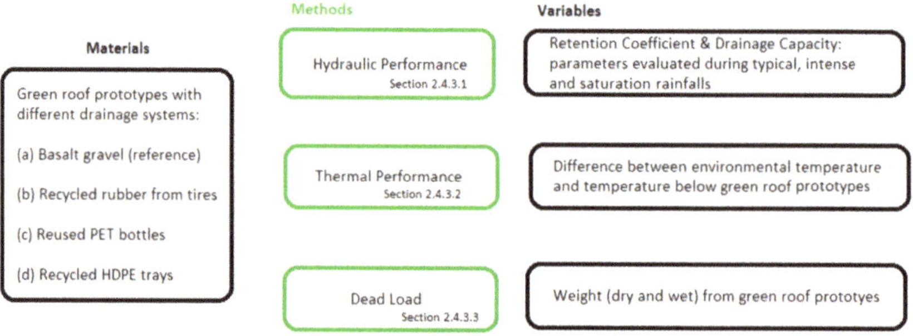

Figure 5. Main variables monitored on the green roof prototypes.

Hydraulic Performance: Retention Coefficient and Drainage Capacity

The benefits of green roofs related to hydraulic performance were evaluated by analyzing the retention coefficient and drainage efficiency. Retention is defined as the volume of water that a green roof can store during and after a rain event. Such a volume of water can be calculated through the equation $V = P \times A \times C$, where P is the precipitation (mm), A is the area of the green roof, and C is a measure of the water storage of the roof system which varies between 0 and 1.0, with being 1.0 the maximum retention value [16]. Since the objective was to compare the retention capacity (storage), variable C (retention coefficient) of such equation was solved to find C = V/(P × A). Considering that the post-sowing cycle is a cycle that only occurs once after the installation of the green roof, the post-drought and stable cycles were those taken for the analysis, they should occur permanently during the lifespan of green roofs.

In addition to the water retention capacity, there is a variable that is related to the delivery of water flows, ultimately used for the design of urban rainwater drainage systems, which on cities serve on a large scale to regulate the evacuation times of rainwater to public sewerage systems. Such capacity mitigates the risk of possible flooding or stream formation in extreme events, and reduces the demand for large conduction drainage systems.

The performance of the drainage from the green roofs made with recycled and reused materials is determined by the concept of efficiency, and it is interpreted in an inverse way when the time variable is associated. Usually, a green roof is considered more efficient than those which demand less resources. In the drainage study, on the contrary, the efficiency in the performance of a green roof was recognized as the one that delivers the water flow using the highest amount of time, demonstrating that no associated side effects, such as an intolerable increase in weight or excessive saturation of the systems due to the presence of water, were generated. This condition of efficient performance was identified by having higher flow delivery times, avoiding the collapse of city collection systems for the evacuation of rainwater. For this analysis, the accumulated in milliliters of the drained water was taken for each time interval, which was directly measured on the drainage of green roof prototypes.

The volume of drained water collected in plastic containers was measured and subsequently measured in a 1000 mL cylinder for each rainfall simulations. The collection of retention and drainage information was manually carried out at the same time when precipitation was simulated. These conditions varied for each type of rainfall. In the case of intense rainfall, data were collected every 10 min for one hour, making a consistent relationship with the precipitation intervals of the IFD curve. For typical rain, the frequency of data collection was every 30 min. A pluviometer, chronometer, plastic containers, and a test tube were employed to measure water retention and drainage capacity.

Temperature

Temperature data were taken from the environment or project site that corresponded to the space under the tent and at the bottom of the support that contained each green roof prototype, with constant measurement of 24 h every thirty minutes. It is important to clarify that the temperature under the tent was higher than the environment temperature due to the local greenhouse effect caused by the tent. However, all the green roof prototypes were exposed to the same temperature conditions. The temperature was monitored and recorded through an data logger (HOBO ONSET—H08-004-02). One day of temperature monitoring was divided in four periods (morning, noon, sunset, and early morning), and data were collected in the following dates (mm/dd/yy): 07/14/15 for morning and sunset periods and 07/24/15 for early morning period. Therefore, the obtained results were comparatively satisfactory and consistent.

Dead Load

The variable of dead load was determined by the weight per square meter of each roof system, and its interpretation was that the less weight it has for each area unit, the more efficient the system was because of the lower structural requirements. To consider the weight of the substratum in a water-saturated state, as suggested by County Flat Roofing (2005) [17] and according to what has been established by the NSR-10 (Known as "Norma Sismo Resistente de Colombia from 2010"or Colombian building code 2010), wet soil weighs approximately 1750 kgf/m³. Both saturated and unsaturated loads were determined in this study.

3. Results

3.1. Hydraulic Performance

The results obtained in the different types of simulated rainfall conditions are shown below.

3.1.1. Retention Coefficient C

Figure 6 shows the response of the four green roof prototypes under typical rainfall conditions. The rainfall simulation followed the pre-defined five-days sequence of precipitation (4, 12, 50, 51, and 1 mm). On the other hand, Figure 7 summarizes the behavior of the green roof prototypes for the intense rainfall condition (simulating a precipitation rate of 49 mm/h).

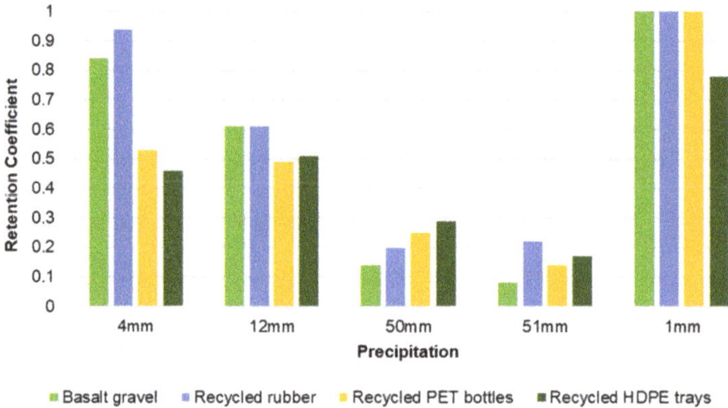

Figure 6. Average behavior of the retention coefficient (C) for typical rain conditions between post-drought and stable cycles.

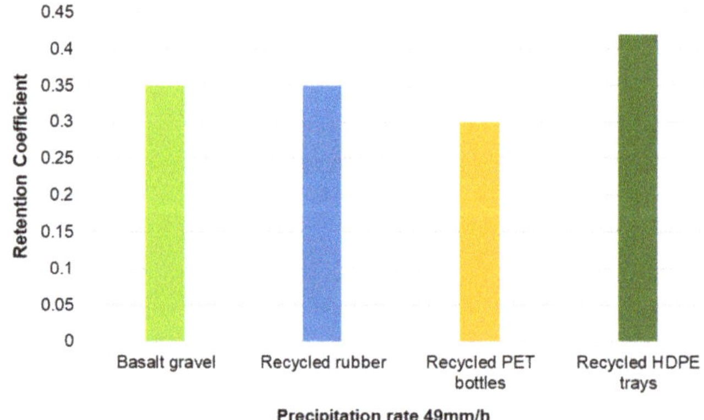

Figure 7. Average behavior of the retention coefficient for intense rain condition between post-drought and stable cycles.

3.1.2. Drainage Capacity

Drainage capacity was measured for both typical and intense rainfall conditions. Figures 8 and 9 show the graphical results according to the predefined simulation parameters.

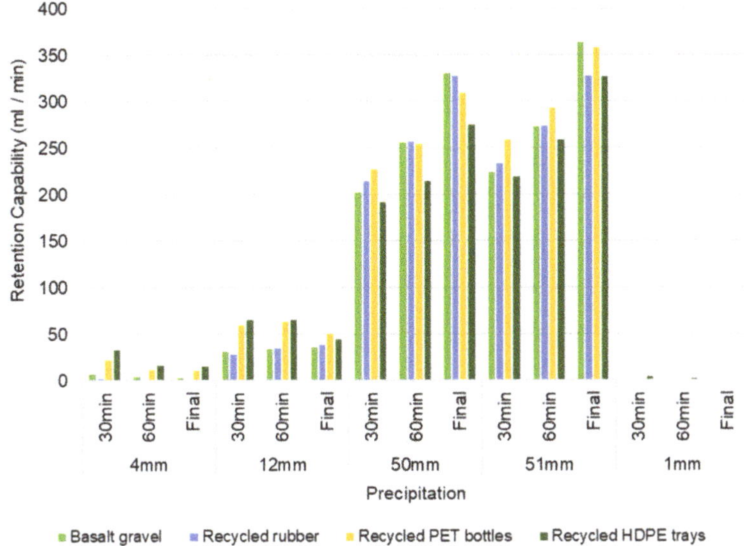

Figure 8. Average behavior of drained water (mL/min) for typical rain conditions between post-drought and stable cycles.

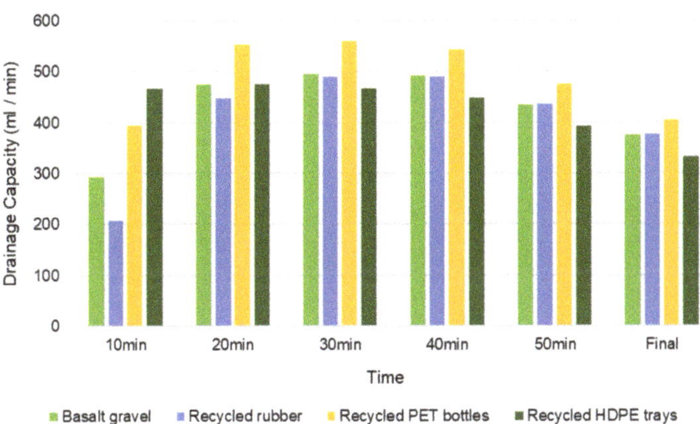

Figure 9. Average behavior of drained water (mL/min) for intense rain conditions between post-drought and stable cycles.

3.2. Temperature

Temperature data from the four moments of the day are summarized in Figure 10. From left to right: morning, noon, sunset, and early morning. In the first segment, a change between the environment and the sunrise effect was observed. In the second, a notable difference was observed due to the high-temperature condition that occurred in the city of Cali and the decreasing contribution of the roofs. In the third segment, again, a change was observed between the roofs and the environment due to the arrival of night. Finally, the condition of lower environment temperature in the night and early morning hours was observed.

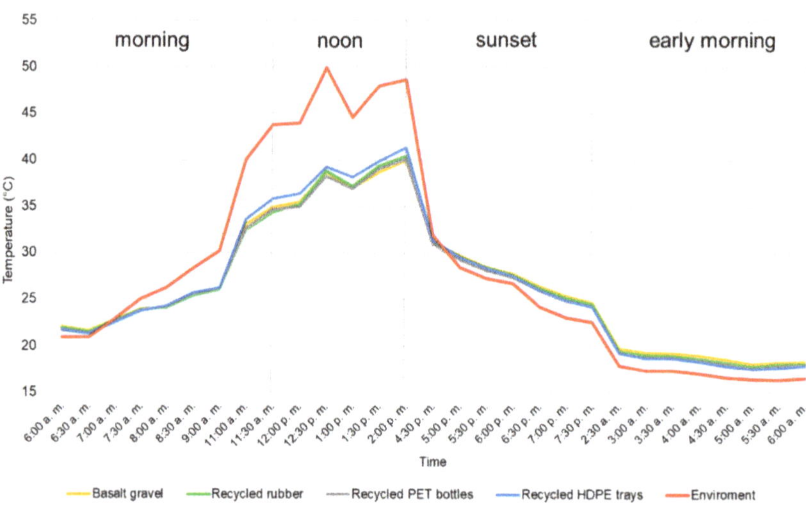

Figure 10. Temperature behavior during the day for roof prototypes and the environment.

3.3. Dead Loads

Weight was measured during the experimentation process for each green roof prototype (see Table 4). The results showed that both unsaturated and saturated conditions provided the following order of efficiency in terms of weight: (1) PET bottles, (2) HDPE trays, (3) rubber, and (4) gravel. Likewise, it was observed that in both conditions, the green roof with a traditional drainage layer (gravel) showed higher values than roofs made from recycled or reused materials, which provided weight reductions between 33% and 72%.

Table 4. Weight per area for each roofing system proposed. Value for 1 m^2.

Description	Roof 1 Basalt Gravel	Roof 2 Recycled Rubber	Roof 3 Recycled PET Bottles	Roof 4 Recycled HDPE Trays
No saturated system kgf/m^2	174.6	117.7	49.3	70.0
Percentage difference in the unsaturated state, of the weight of the systems, made with recycled and reused materials, concerning the conventional green roof system (gravel).	0.0%	32.6%	71.8%	60.0%
Saturated system kgf/m^2	183.3	126.5	53.3	75.3
Percentage difference in the saturated state of the weight of the systems made with recycled and reused materials concerning the conventional green roof system (gravel).	0.0%	31.0%	71.0%	58.9%

Table 5 summarizes the weight of each component per area for the four green roof systems analyzed in this study. Figure 11 illustrates the graphical distribution of layers through pie charts.

Table 5. Weights per area (m^2) for components of the studied green roof systems.

Description	Roof 1 Basalt Gravel kgf/m^2	Roof 2 Recycled Rubber kgf/m^2	Roof 3 Recycled PET Bottles kgf/m^2	Roof 4 Recycled HDPE Trays kgf/m^2
San Agustín grass	16.7	16.7	7.5	16.7
Substratum	39.2	39.2	38.3	48.8
Drainage layer	118.2	61.3	2.9	4.0
Geotextile	0.6	0.6	0.5	0.5
Total	174.6	117.6	49.3	70.0

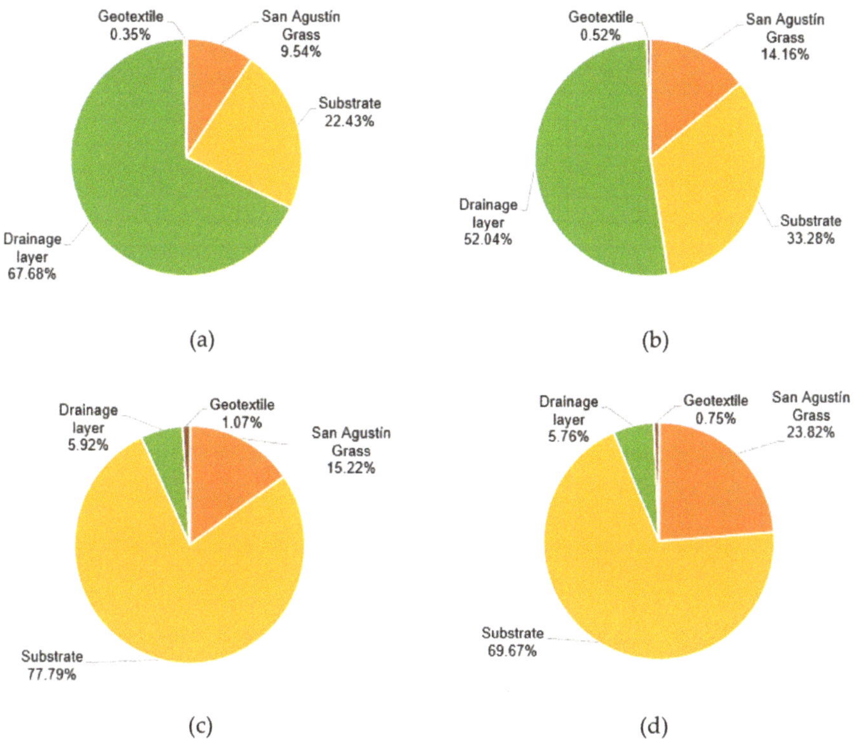

(a)

(b)

(c)

(d)

Figure 11. Weight distribution for each green roof prototype. (**a**) basalt gravel, (**b**) recycled rubber, (**c**) recycled PET bottles, and (**d**) recycled HDPE trays.

4. Discussion

4.1. Retention Coefficient—C

4.1.1. Typical Rainfall

In Figure 6, it can be seen that on the days when the precipitation was less, the roof prototypes had a higher retention capacity. In roofs with drainage layers of granular materials (gravel and rubber), the performance in low-intensity rainfall provided a retention coefficient of 1.0. In the case of 1 and 4 mm rainfalls, it ranged from 0.84 to 0.94, respectively. Meanwhile, the roofs whose drainage layers consisted of the container type system i.e., bottles and trays, showed retention coefficients of 1.0 and 0.78, respectively, for 1 mm rainfall. In the case of 4 mm rainfalls, they presented retention coefficient values that oscillated between 0.53 and 0.46.

In the case of intermediate intensity rain (12 mm), it was observed that the granular-type roofs presented a decrease in the coefficient concerning their performance in low-intensity rain and an inverse effect of the container-type roofs, which improved their performance with the same type of rain. Despite this effect in intermediate intensity rain, the best performing granular types continued with an average value of 0.61. When the intensity of the rain increased to 50 and 51 mm, the behavior of all roofs showed significant reductions, oscillating their retention coefficient in average values between 0.08 and 0.29, respectively.

The described above behaviors during the typical simulated rain cycles, with significant differences between the granular roof system (gravel and rubber) and the container-type roof systems (bottles and trays), showed that the first ones started with almost a total retention that later abruptly decreased with the increase in the intensity of the rain. In contrast, the second ones on less intense days did not present good behavior, but they later stabilized and provided acceptable retention performances. This was because container types with drainage systems that used bottles and trays had holes in their design that allowed for water flow, which in low-intensity rain made drainage faster and, therefore, higher compared to granular-type roofs in which, due to the size, distribution, and specific surface of its rubber and gravel particles, water had to travel longer distances and take more time to move than due to gravity. In some cases, it was not even possible to evacuate, showing a retention coefficient of 1.0. As simulated rain intensified (and due to the same design conditions), the effects were reversed due to the storage capacity of the container-type roofs—especially tray roofs, which conserved water up to a certain level, thus allowing for the improvement of the retention coefficient. The opposite was evident in the granular-type roofs, which suffered a saturation effect in the presence of intense rain, and the incoming water came out in similar amounts as a consequence of gravity.

4.1.2. Intense Rainfall

In terms of intense rainfalls, it was observed that in both cycles, the behavior of the green roof prototypes in regards to retention capacity was similar, with average values ranging between 0.30 and 0.42. The roof prototype with the highest efficiency in rain retention was the tray-type, with an average value of 0.42. In contrast, the least effective was the bottle-type, with an average value of 0.30. The lowest retention capacity of the roofs was obtained in the PET bottle arrangements due to the existing spaces inside the cells that comprised the system, allowing for the direct flow of water through the roof system. However, it was found that although the percentage of voids per square meter between the bottles was 6%, the impact on the retention coefficient was very low compared to the other systems studied.

To summarize, the results from the retention coefficient (C) were comparable to conventional intensive green roofs, showing even better performance for typical rainfall conditions. The literature has shown that C values for conventional green roofs can range, for example, from 55% to 75% [18], from 39% to 43% [19], and from 43% to 61% [20]. This study demonstrated that is possible to obtain values close to 1.0 in the case of low intensity rainfalls and close to 60% in the case of intermediate intensity rainfalls (for granular type-roofs). To provide better results using recycled materials, it is necessary to characterize all physical properties and preferable geometries to arrange materials in detail. It is important to clarify that this study was limited to use recycled materials without additional processing or modifications.

4.2. Drainage Capacity

4.2.1. Typical Rainfall

The behavior of the roofs in typical rain, related to the flow of evacuated water, was similar to that presented in the retention coefficient. It was observed that in the first two days of the cycle with rain of 4 and 12 mm (Figure 8), the roof prototype of recycled trays presented higher levels in milliliters per minute than the others. In the same way, it was observed that roofs of gravel and rubber presented deficient levels, showing better performance during the first days. This allowed us to conclude that for rain events equal to or less than 12 mm, the roof system that delivers lower flow rates was the gravel roof, followed by that made out of recycled rubber.

Figure 8 shows that on the third day, when the simulated rainfall was 50 mm, the condition of the roofs was the opposite, and the tray roof was the one with the highest efficiency because it showed the lowest levels of water evacuation. It was observed that the difference between the tray roof and the others was enlarged over time. This was a consequence of the maturity of the vegetal layer, which improved self-regulation and reduced flow delivery times.

When the simulated precipitation was 51 mm, the trend of less flow evacuated by the tray roof continued, and at the same time, it was observed that the gravel roof ended with the highest values.

This was because in the first two intervals (30 and 60 min), its evacuation was lower because in the first moments of rain, it retained better; at the end, however, it increased the evacuation quantities.

Finally, it was observed that for 1 mm precipitation, the only roof system that showed water delivery was the tray roof, which demonstrated a high sensitivity to the initial moments of rain, in which it did not demonstrate acceptable performance. It should be noted that due to the design of the drainage layer, it was possible to store water in such a way that it was able to self-regulate flow evacuation and improve retention capacity in more extended and demanding periods of rain.

4.2.2. Intense Rainfall

It was observed that the flow behavior was similar for the four green roofs, with the best performance being the tray roof and the lowest performance being the one made up of bottles. The key factor of the trays was their operation based on their modular design, while in systems with granular-type drainage, in addition to the size and distribution of particles, specific surface area and absorption of the material also had influence. The low performance of the bottle system roof was because it was made up of isolated units that formed cells and not a consolidated and consistent layer structure that facilitated the retention, self-regulation, and homogeneous life of the vegetation.

It was observed that tray roof delivered one of the highest flows at the beginning due to its design. However, this situation was later found to be the opposite, and it was the roof that, over time and in the presence of higher precipitation, behaved more efficiently in self-regulation because it evacuated the drained water more slowly. The above was reflected in lower volumes for each calculated time.

Between minute ten and minute twenty, the drainage curves showed a noticeable increase in the amount of water drained concerning the following intervals; this was because the IFD curve (see Appendix A—Figure A1b) of the selected intense rain (49.7 mm/h), as it was intended to perform in the most precise way. Therefore, and according to the conditions of such curve, in the interval of the first twenty minutes, 27.9 mm were simulated continuously over 16 min and 44 s. From minute twenty, the flows were more regulated according to the conditions of the selected IFD, which generated the stability of the drainage curves. The gravel and recycled rubber roofs behaved similarly, delivering little water at the beginning compared to the others, but they stabilized over time, and despite showing good performance, they were surpassed by the tray roof system.

The results showed that the proposed green roof systems provided good performance in terms of water drainage. In the case of bottle system and tray roofs, an intrinsic ability to self-regulate was evident, which is highly desirable. Responses in respect to the reference green roof made of gravel demonstrated that all considered recycled materials provided similar responses under the same rainfall conditions. Further modifications regarding the arrangement of layers and systems (e.g., bottles and trays) have to be studied to analyze the drainage response and optimize area and material distribution.

4.3. Temperature

In all the analyzed intervals, the behavior of the roofs tended to preserve the ideal conditions of thermal comfort, which ranged between 21 and 25 °C according to the Colombian Sustainable Building Code. However, it was not possible to reach this value in some analyzed segments. However, it was possible to mitigate the effect of external temperature, reducing the intense use of active cooling systems. The following observations were obtained from the temperature behavior obtained from the four different green roof systems:

(a)   In the first half-hour of the morning for all the analyzed days, it was generally observed, as shown in Figure 10, that environmental temperature was lower than the temperature of the four roof systems, presenting a change due to the sunrise, where the condition was contrary and the roofs had the lowest temperatures. Before reversing this condition, the roof systems with container type drains (trays and bottles) presented lower temperatures compared to those with a granular type drainage layer (gravel and rubber). The difference between the highest and lowest temperature roof did not exceed, in any case, 1°C.

(b) Between 11:00 a.m. and at 2:00 p.m. for all the analyzed days, it was generally observed that the temperature of the environment was notably higher than the temperature of the roofs. In this analysis in which the sun radiated more intensely, the benefits of lowering the temperatures of the green roof systems and the effectiveness of using recycled and reused materials in their drainage layer were demonstrated. The most extreme case of the difference between the environment value and the lowest roof was 11.7 °C. Likewise, it was observed that the smallest difference was 2.8 °C between the environment and the tray roof.

(c) At the end of the afternoon, it was observed that the environmental temperature was higher than the roof temperature until between 4:30 p.m. and 6:00 p.m., when the condition was the opposite and the green roofs were the ones with a higher temperature, demonstrating the benefits of lowering temperatures at critical moments that were opposite to the night. However, a temperature inversion was observed one hour earlier between the roofs and the environment.

(d) According to Figure 10, in the early morning hours, when the temperature was lower compared to other times of the day, the temperature of the environment was lower than the temperature of the roofs (in some cases the differences reached up to 2 °C). Moreover, the behavior between roofs showed that the gravel and trays roofs provided the highest and lowest temperatures, respectively.

Regarding temperature, the results showed the same behavior with minimum variations with respect to the reference green roof system (basalt gravel). Therefore, the use of recycled materials, as proposed in this study, is suitable from the perspective of environmental temperature reduction. The next step in this research direction is the analysis of different vegetation layers (growing and development) using recycled materials that can provide better responses with respect to temperature and CO_2 reduction.

4.4. Dead Load

According to Figure 11, the drainage layer was the most representative component when analyzing the variable weight. In green roofs with granular materials, values of 52% for rubber and 68% for gravel were found (these were related to the sum of all green roof components), while, on the contrary, the percentage of this layer only represented 6% for both cases of roof systems with a container-type drainage layer. In these, the most representative component was the substratum, with percentages of 78% for bottles and 70% for trays, in terms of the sum of all their components.

When analyzing the behavior of the drainage layers made with recycled and reused materials compared to the traditional gravel, the second one had a higher weight in all cases. For example, the recycled rubber drainage layer weighed 48% less than the gravel layer. In the case of the layers of bottles and trays, these weighed 98% and 97% less than that of gravel, respectively. Likewise, the low weight of the drainage layers of the PET bottle and HDPE tray systems must be highlighted, since, in addition to reducing the weight more than significantly, they facilitated handling at the time of construction, reducing mechanical risk factors associated with the installation work.

When analyzing the total weight of green roof systems, it was found that the system whose drainage layer was composed of gravel presented a higher percentage of weight than the systems consisting of rubber, bottles, and trays (33%, 72%, and 60% respectively), which made it a more demanding system when designing the structures of buildings due to the required increase of loads.

The composition of the weight inside the roofs showed that in the case of systems with a gravel and rubber drainage layer, the highest proportion was determined by the drainage layer, while the bottle and tray roofs showed the highest representative weight or load.

With respect to the overall sustainability of the proposed green roof prototypes from recycled material, it is important to clarify that this study did not cover the lifecycle assessment of materials involved in the construction of the systems. Therefore, the environmental impacts of materials were out of the scope of this article. Hydraulic and thermal performances were included and widely analyzed. Dead loads and initial cost were also calculated as additional parameters for the proposed green roof prototypes.

In terms of initial cost, it is important to mention that recycled materials provided a cost reduction per square meter that was greater than 10% (in the case of HDPE trays) compared to basalt gravel (Table A1). Maintenance and another lifecycle cost were not included in this article.

5. Conclusions

In this article, scale prototypes of three semi-intensive green roof systems with different types of drainage systems made out of recycled and reused materials (rubber, trays, and bottles) were successfully evaluated and compared to a traditional green roof system that used natural aggregates of (gravel) as its drainage layer.

In terms of hydraulic performance, the behavior of the systems using trays, bottles, rubber, and gravel (reference) was analyzed in different precipitation regimes (typical and intense). The results showed that for "typical precipitations," the granular drainage systems (gravel and rubber) were very efficient because they retained all the precipitation (retention coefficients close to 1.0), while the systems composed of module containers (bottles and trays) retained approximately half of the water supplied at that level of precipitation (retention coefficients close to 0.5). For the "intense precipitation," the coefficients reached values close to 0.3 for all roof systems, except for the gravel system, which reduced its water storage capacity to almost zero (0.08). In general, this study demonstrates the enormous potential of all the green roofs analyzed in this study to reduce the maximum flow of runoff water volumes, as they could increase retention time when they are implemented in large areas at the urban level.

With respect to the thermal behavior, it was possible to verify the effect of temperature reduction of all the roofing systems. During days when the ambient temperature was very high (approximately 50 °C), a reduction in temperature that ranged from 10.6 to 11.7 °C was found for the investigated green roof systems. This makes it evident that the use of green roof systems with drainage layers made out of recycled and reused materials have, like gravel roofs, the potential to reduce the consumption of electrical energy in buildings derived from artificial cooling.

Finally, a reduction from 33% to 72% in weight per area (dead load) of the green roof was observed when using recycled and reused materials compared to natural materials in the drainage layers. This is significantly important because the ease and costs of implementation of green roofs depends on the structural condition of the building. Therefore, for the load capacity of an existing building, the dead and live loads must be assessed in order to verify whether, with the increase in the dead load generated by the weight of the green roof, the building can withstand the loads added to it without affecting its resistance, as well as if it complies with the specifications of the building construction code. Otherwise, a structural reinforcement must be designed to guarantee safety.

For new buildings, a structural calculation must be made according to all the loads that act on the structural system and that come from the weight of all the permanent elements in the construction (dead loads), the occupants and their belongings (live loads), environmental effects, differential settlements, and dimensional change restriction following current regulations. The last indicates that due to the relatively lower density and lower absorption capacity of the recycled and reused materials evaluated in this research, the implementation of green roof systems, both in existing and new buildings, would be easier and cheaper.

Future research works will be oriented to evaluate the sustainability and economic performance of green roofs based on recycled or reused materials. Considering environmental impacts in terms of material lifecycle and lifecycle costs derived from installing, maintaining, upgrading, and disposing constructive components. All those parameters need to be included to generate holistic evaluations to facilitate the decision-making during green roof design and construction.

Author Contributions: Conceptualization, methodology, and editing, J.M. and A.M.-R.; components, classification, and technical performance, A.N. and A.C. All authors have read and agreed to the published version of the manuscript.

Funding: This research received no external funding.

Acknowledgments: The authors thank Dos Mundos (Cali, Colombia) and Pontificia Universidad Javeriana Cali (Colombia) for the technical support given during the experiments reported in this article.

Conflicts of Interest: The authors declare no conflict of interest. The funders had no role in the design of the study; in the collection, analyses, or interpretation of data; in the writing of the manuscript, or in the decision to publish the results.

Appendix A

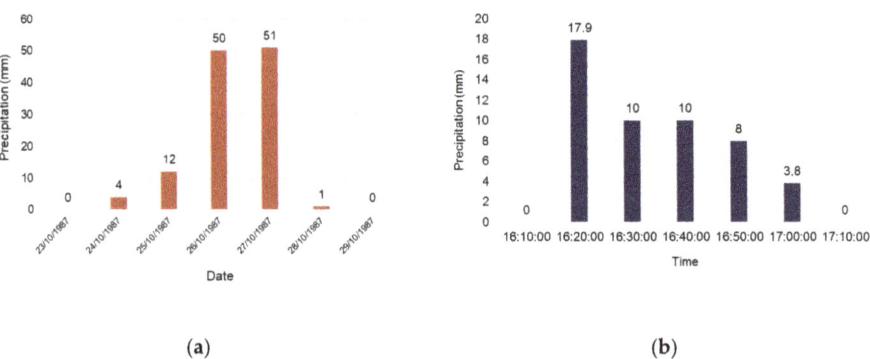

(a)   (b)

Figure A1. Data employed to estimate typical and intense rainfall conditions. (a) Typical rainfall conditions (from 23 to 29 October 1987). (b) Intensity–frequency–duration (IFD) curve for intense rainfall (data from 2 June 1995).

Table A1. Approximated costs for each roofing system proposed (Value for 1 m^2).

N°	Green Roof System	Approximated Costs USD
1	Recycled HDPE trays	$93.71
2	Recycled PET bottles	$97.59
3	Recycled rubber	$103.16
4	Basalt gravel	$104.49

References

1. Li, J.F.; Wai, O.W.H.; Li, Y.S.; Zhan, J.M.; Ho, Y.A.; Li, J.; Lam, E. Effect of green roof on ambient CO_2 concentration. *Build. Environ.* **2010**, *45*, 2644–2651. [CrossRef]
2. Fujii, S.; Cha, H.; Kagi, N.; Miyamura, H.; Kim, Y.S. Effects on air pollutant removal by plant absorption and adsorption. *Build. Environ.* **2005**, *40*, 105–112. [CrossRef]
3. Jim, C.Y.; Tsang, S.W. Biophysical properties and thermal performance of an intensive green roof. *Build. Environ.* **2011**, *46*, 1263–1274. [CrossRef]
4. Ahmad, S.; Hashim, N.M. Effects of Soil Moisture on Urban Heat Island Occurrences: Case of Selangor, Malaysia. *Humanit. Soc. Sci. J.* **2007**, *2*, 132–138.
5. Bliss, D.J. Stormwater Runoff Mitigation and Water Quality Improvements through the Use of a Green Roof in Pittsburgh, Pennsylvania. Master's Thesis, Faculty of the School of Engineering, University of Pittsburgh, Pittsburgh, PA, USA, 2010.
6. Naranjo, A.; Colonia, A.; Mesa, J.; Maury, H.; Maury-Ramírez, A. State-of-the-Art Green Roofs: Technical Performance and Certifications for Sustainable Construction. *Coatings* **2020**, *10*, 69. [CrossRef]
7. Chen, C.F.; Kang, S.F.; Lin, J.H. Effects of recycled glass and different substrate materials on the leachate quality and plant growth of green roofs. *Ecol. Eng.* **2018**, *112*, 10–20. [CrossRef]
8. Eksi, M.; Rowe, D.B. Green roof substrates: Effect of recycled crushed porcelain and foamed glass on plant growth and water retention. *Urban For. Urban Green.* **2016**, *20*, 81–88.
9. Bates, A.J.; Sadler, J.P.; Greswell, R.B.; Mackay, R. Effects of recycled aggregate growth substrate on green roof vegetation development: A six year experiment. *Landsc. Urban Plan.* **2015**, *135*, 22–31. [CrossRef]

10. Rincón, L.; Coma, J.; Pérez, G.; Castell, A.; Boer, D.; Cabeza, L.F. Environmental performance of recycled rubber as drainage layer in extensive green roofs. A comparative Life Cycle Assessment. *Build. Environ.* **2014**, *74*, 22–30. [CrossRef]

11. Mickovski, S.B.; Buss, K.; McKenzie, B.M.; Sökmener, B. Laboratory study on the potential use of recycled inert construction waste material in the substrate mix for extensive green roofs. *Ecol. Eng.* **2013**, *61*, 706–714. [CrossRef]

12. Pérez, G.; Vila, A.; Rincón, L.; Solé, C.; Cabeza, L.F. Use of rubber crumbs as drainage layer in green roofs as potential energy improvement material. *Appl. Energy* **2012**, *97*, 347–354. [CrossRef]

13. Gill, R. *Igneous Rocks and Processes: A Practical Guide*; Wiley: New York, NY, USA, 2010; ISBN 9781444330656.

14. Evans, A.; Evans, R. *The Composition of a Tyre: Typical Components*; Banbury: Oxon, UK, 2006.

15. Instituto de Hidrología, Meteorología y Estudios Ambientales [IDEAM]. *Información Climatológica de la Estación Univalle*; Variación de Parámetros Climáticos: Cali, Colombia, 2014.

16. Mihelcic, J.R.; Zimmerman, J.B.; Auer, M.T. *Environmental Engineering: Fundamentals, Sustainability, Design*, 2nd ed.; Wiley: New York, NY, USA, 2014; ISBN 9781118741498.

17. C County Flat Roofing. County Flat Roofing Main Webpage. 2020. Available online: https://countyflatroofing.co.uk/ (accessed on 20 December 2019).

18. Metselaar, K. Resources, Conservation and Recycling Water retention and evapotranspiration of green roofs and possible natural vegetation types. *Resourc. Conserv. Recycl.* **2012**, *64*, 49–55. [CrossRef]

19. Wong, G.K.L.; Jim, C.Y. Quantitative hydrologic performance of extensive green roof under humid-tropical rainfall regime. *Ecol. Eng.* **2014**, *70*, 366–378. [CrossRef]

20. Young, J.; Jung, M.; Han, M. A pilot study to evaluate runoff quantity from green roofs. *J. Environ. Manag.* **2015**, *152*, 171–176.

Article

Effect of Hygrothermal Aging on Hydrophobic Treatments Applied to Building Exterior Claddings

Giovanni Borsoi [1,*], Carlos Esteves [1], Inês Flores-Colen [1] and Rosário Veiga [2]

[1] CERIS, Civil Engineering Research and Innovation for Sustainability, Instituto Superior Técnico, University of Lisbon, 1000-147 Lisbon, Portugal; carlosdcesteves@gmail.com (C.E.); ines.flores.colen@tecnico.ulisboa.pt (I.F.-C.)

[2] LNEC, National Laboratory for Civil Engineering, 1000-147 Lisbon, Portugal; rveiga@lnec.pt

* Correspondence: giovanni.borsoi@tecnico.ulisboa.pt; Tel.: +351-218-443-975

Received: 26 February 2020; Accepted: 3 April 2020; Published: 7 April 2020

Abstract: Hydrophobic materials are among the most commonly used coatings for building exterior cladding. In fact, these products are easily applied to an existing surface, significantly reduce water absorption and have a minimal impact on the aesthetic properties. On the other hand, although these products have a proven effectiveness, their long-term durability to weathering has not yet been systematically studied and completely understood. For these reasons, this study aims to correlate the effect of artificial aging on the moisture transport properties of hydrophobic treatments when applied on building exterior claddings. Three hydrophobic products (an SiO_2-TiO_2 nanostructured dispersion; a silane/oligomeric siloxane; and a siloxane) were applied on samples of limestone and of a cement-based mortar. The moisture transport properties (water absorption, drying, water vapor permeability) of untreated and treated specimens were characterized. Furthermore, the long-term durability of the specimens was evaluated by artificial aging, that is, hygrothermal cycles (freeze–thaw and hot–cold). All treatments have significant hydrophobic effectiveness and improve the long term-durability of the treated specimens. However, the results showed that the three hydrophobic products have different effectiveness and durability, with the SiO_2-TiO_2 nanostructured dispersion being the most durable treatment on limestone, and the siloxane the most suitable for cementitious mortar.

Keywords: hydrophobic products; silicon-based compounds; claddings; durability; moisture transport properties

---

1. Introduction

The use of coatings on building façades is a fundamental action for the protection of external surfaces from weathering. Protective coatings should be considered for the conservation of historical façades, as well as for the maintenance of modern buildings. In fact, these products can help in the conservation of ancient materials by increasing the durability of the treated surface.

Water is among the most aggressive atmospheric agents, being the main physical-chemical degradation cause of porous building materials [1]. In fact, water can trigger several surface anomalies both on ancient and modern buildings, leading to several physical (e.g., loss of cohesion and material, condensation in the interior of the building, reduction of thermal conductivity, formation of salt efflorescence, hygrothermal aging in buildings) and aesthetic changes (e.g., stains, biofilm formation) on the affected surface [1–3].

For these reasons, excessive water penetration and retention should be avoided in order to balance the water content in the façade cladding. In fact, porous building materials can balance the moisture content according to their water transport properties, that is, water vapor adsorption, water capillary suction, and water vapor condensation [4,5]. Water transport mechanisms can act simultaneously, sequentially, or separately, and their action depends on the exposure conditions and on the moisture content of the material [6].

Through the formation of a thin protective coating, the application of hydrophobic treatments intends to reduce mostly water absorption, and thus, to protect the treated surface from the possible physical-chemical and biological alterations induced by the presence of water [7]. Additionally, surface coatings hinder the penetration of deleterious environmental particles (e.g., pollutants and salts) in the treated surface.

A wide majority of hydrophobic products present a surface tension lower than water and by modifying the contact angle of the treated surface help avoid wettability or limit water absorption within the surface [3,8,9]. In fact, hydrophobic products are generally non-polar materials, which repel water (which is polar), whereas they have an affinity with other non-polar materials, making them attractive to, for example, alkanes (fats and oils) and noble gasses [10].

Among the requirements of an ideal hydrophobic product, the protective coatings should be compatible with the treated material, and thus, not remarkably modify the physical-chemical properties of the treated surface. In fact, the hydrophobic materials should deeply penetrate the pore network and confer water-repellent properties to the treated material. However, they should not drastically alter the water vapor transport and the drying kinetics, and thus the breathability of the treated cladding [1,8].

The effectiveness of the hydrophobic products also depends on the chemical affinity between the product and the treated surface [11]. A lack of a proper knowledge of the hydrophobic products and of the moisture transport properties of the treated surface can lead to the formation of stains and byproducts (e.g., salts), alteration of the colour and brightness, and even an acceleration of the degradation of the treated surface. In addition to a proper hydrophobic effectiveness and compatibility with the treated surface, the hydrophobic product should have a suitable durability to weathering agents.

Organic silicon compounds, such as silanes, siloxanes, silicon resins and silicanates, are among the most commonly used hydrophobic products [12]. These materials have been widely used due to their high resistance to oxidation processes, UV radiation and extreme pH environment [13]. Additionally, these materials ensure ease of application and respect the aesthetic properties of the treated surface. When applied to the building material, the hydrolytically sensitive alkoxy groups of silicone compounds react with water or humidity, forming non-stable silanol intermediates, which spontaneously polycondensate to form stable covalent bonds. This hydrophobic film is irreversibly bonded to the mineral substrate [2,7].

Hydrophobic treatments are generally subjected to weathering and tend to alter their water-repellent properties (by physical-chemical degradation or leaching), and thus, protective action over time. The loss of hydro-repellency is attributed to the synergic effect of atmospheric agents (e.g., hygrothermal variations, solar radiation, rain, atmospheric pollutants) [1]. The accumulation of atmospheric particles with hydrophilic properties on the surface also has an important role in the surface degradation process [9,14]. Additionally, weathering can speed up the alteration and degradation of the polymeric structure of the hydrophobic material, inducing an increase of the polarity and a loss of water-repellency, as well as chromatic alteration and the formation of stains. In fact, photo-oxidation—induced by UV radiation and a degradation of the Si–O bonds due to the extreme pH environment—can lead to loss of adhesion, yellowing and a reduction of the surface gloss [8,9].

Although in some cases a long-term resistance of the hydrophobic treatment is reported [15], with an effective water repellence after 3–5 years of exposure to severe weathering [7,8], the durability of most commercially available hydrophobic products is not reported by manufacturers. Furthermore, the relationship between the physical-chemical properties of the hydrophobic product and of the treated material, and environmental factors, have not yet been completely understood.

For the reasons mentioned above, this paper aims to discuss the factors that influence the durability and effectiveness of hydrophobic products. Three commercially available products (a silicon and titanium dioxides-based nanostructured dispersion; a silane/ oligomeric siloxane; and a siloxane) were applied on limestone and on mortar specimens. The moisture transport properties (water absorption by capillarity and under low pressure, drying, water vapor permeability) of untreated and treated samples were characterized. Furthermore, with the intention of evaluating the durability of these treatments to weathering, treated and untreated specimens were subjected to artificial aging tests, that is, hygrothermal cycles (freeze–thaw and hot–cold) and were then tested.

This works ultimately intends to provide tools to enhance the effectiveness and durability of hydrophobic products in the construction sector.

2. Materials and Methods

2.1. Materials

2.1.1. Substrates

Two types of substrates were selected for the application of hydrophobic products: (a) Moleanos limestone and (b) a cement-based rendering mortar.

Moleanos is a dense, fine-grained, yellowish bioclastic limestone (>98% $CaCO_3$, density = 2.67 g/cm^3), quarried in central Portugal and used as a decorative flooring or finishing building material [16,17].

The rendering mortar was obtained by following the recommendation of the EN 1015-2 [18]. A pre-dosed cement-based mortar (weber.rev ip©) was used, by mixing three parts of pre-dosed mortar with one part of water (in volume); this was then applied on ceramic hollow bricks.

The characteristics of the limestone and of the mortar, as well as the substrate where the mortar was applied (ceramic brick), are presented in Table 1 [19]. Additionally, the cement-based rendering mortar has a higher surface roughness, if compared to the dense Moleanos limestone.

Table 1. Average results and standard deviation of capillary water absorption coefficient, open porosity and bulk density of the substrates [19].

Substrates	Capillary Absorption Coefficient (kg·m^{-2}·min$^{-0.5}$)	Open Porosity (%)	Bulk Density (kg/m^3)
Limestone	0.252 ± 0.031	10.96 ± 1.31	2385 ± 34
Mortar	0.431 ± 0.018	30.23 ± 2.04	1441 ± 101
Ceramic brick	0.074 ± 0.031	25.49 ± 0.49	1979 ± 12

2.1.2. Hydrophobic Products

The selection of hydrophobic products was based on market research. The main aim was the evaluation of products with various chemical compositions and, therefore, that possibly differentiated in terms of their effectiveness and durability.

Three silicon-based products were chosen, selecting different manufacturers:

- $H_{SILA/SIL}$: A solvent-based silane/oligomeric siloxane-based emulsion, which also contains a biocide additive ($\leq$ 0.01% in volume); the silane used is a triethoxyoctylsilane;
- H_{SIL}: A solvent-based siloxane product (polydimethylsiloxane);
- H_{NST}: A water emulsion of silicon (SiO_2) and titanium dioxides (TiO_2) nanoparticles, which contains a reduced concentration of silane (N-octyltriethoxysilane $\leq$ 2.5% in volume) and a biocide ($\leq$ 0.0015% in volume).

Silane-modified siloxane polymers are widely adopted as hydrophobic materials; siloxane guarantees improved adhesion properties with the treated substrate, whereas the silane is used as coupling agent and improves the hydrophobic properties of the treatment [20].

Silica nanoparticles, which present hydrophilic properties, can be modified during their synthesis by adding a coupling agent such as silane or sodium sulphate, which confers hydrophobic properties and prevents nanoparticle agglomeration [21,22]. These products are generally composed of hydrophobic hybrid crystalline SiO_2–TiO_2 nanoparticles with a crystallite size equal to 5–20 nm [23,24].

The use in coatings of high refractive index oxides, such as TiO_2, has significantly grown in recent years due to the advancements of nanoparticle manufacturing processes and because of their beneficial properties. Nanostructured titanium oxide has photocatalytic properties which can result in multifunctional self-cleaning and biocidal coatings [25].

The physical/chemical characteristics and application protocol of the hydrophobic products is reported in Table 2. It is worth noting that most solvent-based hydrophobic products can be harmful for both the operator and the environment, due to their high content of volatile organic compound (VOC) (e.g., polycyclic aromatic hydrocarbons and alkanes) [9]. Additionally, silane and biocide additives (isothiazol-3-one, 3-iodo-2-propynylbutyl carbamate, among others) are generally toxic for the operator and detrimental for the environment.

Table 2. Chemical and physical characteristics and amount of product used in the application of the hydrophobic products.

Hydrophobic Product	Color	Density * (g/cm^3) at $T = 20\,°C$ and RH% = 60	Drying Time (h)**	Number of Applications	Amount of Product Per Application (L/m^2)
$H_{SILA/SIL}$	Whitish	1.02	24	2	1.01 ± 0.06
H_{SIL}	Transparent	0.78	2	1	0.40 [a]
H_{NST}	Whitish	1.01	3	2	0.11 ± 0.01

* As referred in the product technical sheet; ** Time necessary to achieve constant mass after the product application; [a] 1 application per specimen.

2.2. Methods

2.2.1. Specimen Preparation

Cylindrical specimens, with 20 cm diameter and 2 cm thickness, were drilled and cut from Moleanos limestone blocks and used for capillary water absorption, drying and water vapor permeability tests. Prismatic specimens (30 cm × 30 cm, 2 cm thickness) were used for the analysis of water absorption under low pressure. Before the application of the hydrophobic product, in order to obtain a constant moisture content in all the specimens, limestone specimens were stored in a conditioned room at $T = 23 \pm 2\,°C$ and $RH = 50 \pm 5\%$.

Concerning the mortar specimens, cylindrical specimens with 20 cm diameter and 2 cm thickness were produced by using dedicated molds; these were used for water vapor permeability tests. Additionally, a 2 cm-thick layer of the mortar was applied on hollow ceramic bricks (29 cm × 17 cm × 4 cm). These latter specimens were used for all the other tests (capillary water absorption, drying kinetics and water absorption under low pressure).

All mortar specimens were cured for 2 days at $T = 23 \pm 2\,°C$ and $RH = 95 \pm 5\%$, and later stored in a conditioned room at $T = 23 \pm 2\,°C$ and $RH = 50 \pm 5\%$ for 26 days, before testing.

All specimens (limestone and mortar) were sealed along their side surface with liquid paraffin (applied multiple times by brushing, until obtaining a layer of approximately 1 mm).

2.2.2. Application Protocol

The three hydrophobic products were applied by brushing, following the recommendations of the products technical data sheets (Figure 1a). Two applications were carried (in orthogonal directions) in the case of $H_{SILA/SIL}$ and H_{NST}, whereas one application was performed in the case of H_{SIL}. The interval between applications was around 2 h for $H_{SILA/SIL}$ and 3 h for H_{NST}. The applications were performed under controlled conditions ($50 \pm 5\%$ RH, $T = 20 \pm 2\,°C$) and the treated specimens were stored at the same hygrothermal conditions for 7 days, with the aim of completing the polymerization of the silicon-based products. Untreated samples were stored in the same conditions.

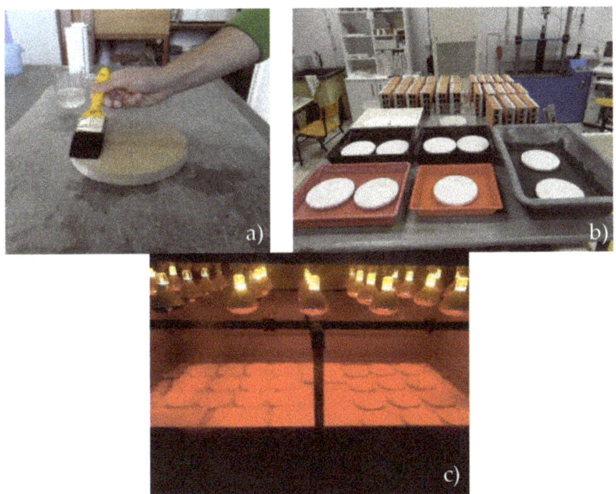

Figure 1. (**a**) Application of hydrophobic product on cylindrical limestone specimens; (**b**) capillarity water absorption on limestone and mortar specimens; (**c**) artificial aging test (hygrothermal cycles).

2.2.3. Moisture Transport Properties

The capillarity water absorption coefficient (C) was determined as the initial slope of the absorption curve using a protocol based on EN 1015-18 [26] (Figure 1b). The determination of water absorption by capillary action is achieved through the evolution of the amount of water absorbed by the solid unit surface area (kg/m^2), as a function of the square root of time (t$^{1/2}$).

Water absorption at low pressure was carried out with Karsten tubes applied on the specimens for 60 min, following RILEM recommendations [27]. The water absorption coefficient for 60 min (C^{60}, kg·m^{-2}·min$^{-1/2}$) was calculated according to the following Equation:

$$C^{60} = \frac{A_{bp} \times 10^{-3}}{A_c \times 10^{-4} \times \sqrt{60}}$$ (1)

where A_{bp} is the water mass absorbed after 60 min (kg), and A_c is the contact area of the pipe with the surface (assumed to be 5.7 cm^2, i.e., the contact area of the Karsten pipe).

Drying tests were carried out sequentially at the end of the capillary water absorption tests in order to allow a direct correlation between the water absorption and drying results. The drying kinetics of the specimens were verified according to RILEM [28] and UNI [29] recommendations, which considers the initial drying (based on the slope of the initial drying curve) and the drying index (DI). The latter is an empirical quantity that expresses the drying curve into a single quantitative parameter reflecting the global drying kinetics. The drying index, obtained from the average drying curve of those of the different specimens (in equivalent conditions), was calculated according to

$$DI = \frac{\int_{t_0}^{t_f} f\left(\frac{M_x - M_1}{M_1}\right) dt}{\left(\frac{M_3 - M_1}{M_1}\right) \times t_f}$$ (2)

where M_x is the specimen mass weighted during the drying process (g); M_1 is the specimen mass in dry state (g), M_3 the specimen mass in the saturated state (g), and t_f is the final time of the drying process (h).

Additionally, with the aim of understanding the influence of the hydrophobic products on the drying kinetics of the treated surface, the different steps of drying and the critical moisture content (i.e., the transition between the 1st and 2nd step of drying) were studied.

The water vapor permeability was determined as specified in EN1015-19 [30]. The water vapor diffusion resistance coefficient (μ) was calculated according to Equations (3) and (4):

$$\Lambda = \frac{m}{A \times \Delta_P} \tag{3}$$

$$\mu = \frac{1.94 \times 10^{-10}}{\Lambda \times e} \tag{4}$$

where Λ is the water vapor permeance (kg/m^2·s·Pa); m is the linear relationship slope of time versus mass change (kg/s); A is the specimen area (0.126 m^2); Δ_P is the difference between the outdoor and indoor vapor pressure (Pa); μ is the water vapor diffusion resistance coefficient; and e is the specimen thickness (m).

In all tests, three untreated specimens and nine treated specimens were analyzed, considering their average values and relative standard deviations.

2.2.4. Accelerated Aging Test

Accelerated aging tests were performed on untreated and treated limestone and mortar specimens to verify the durability of the hydrophobic treatments to weathering cycles [31]. Since temperature shock, rain and solar irradiation are the main degradation agents of porous materials [32], the specimens were subjected to hygrothermal cycles (hot–cold and freeze–thaw).

The test conditions, which represent extreme climate conditions, were adapted from EN 1015:21 [33]. This methodology was also validated in previous research by the authors [34,35]. Hot-cold cycles consist of storing the specimens firstly within a closed apparatus with infrared lamp (Figure 1c), which provides high temperature, and later within a deep-freeze cabinet with low temperature. Freeze-thaw cycles were carried out by exposing the specimens to a sprinkler system (simulating rain), followed again by a storage within a deep-freeze cabinet. Hot-cold and freeze-thaw cycles were carried out sequentially on the same specimens. Eight cycles of each type were performed, improving the indications (four cycles) of the norm previously mentioned (Figure 1c). Further details on the weathering cycles are provided in Table 3.

Table 3. Accelerated aging test: Hygrothermal cycles test conditions.

Hot-Cold Cycles	Freeze-Thaw Cycles	Exposure Time (h/mins)
Infrared lamps (60 ± 2 °C)	Sprinkler system, water at T = 20 ± 1 °C	8 h ± 15 min
Stabilization (20 ± 2 °C, 65 ± 5% RH)	Stabilization (T = 20 ± 2 °C, 65 ± 5% RH)	30 ± 2 min
Deep freeze cabinet (T = −15 ± 1 °C)	Deep freeze cabinet (−15 ± 1 °C)	15 h ± 15 min
Stabilization (T = 20 ± 2 °C, 65 ± 5% RH)	Stabilization (T = 20 °C, 65% RH)	30 ± 2 min

At the end of the hygrothermal cycles, specimens were stabilized for 48 h at T = 20 ± 2 °C, 50 ± 5% RH. All the tests mentioned in the previous section were repeated on artificially aged untreated and treated specimens.

3. Results

3.1. Capillary Water Absorption and Water Absorption with Karsten Tubes

The results show that, in the case of limestone specimens, the capillary water absorption coefficient (C) of treated specimens considerably reduced after the application of the hydrophobic products (33% in the case of H$_{Sila-Sil}$, 51% with H$_{sil}$, 88% with H$_{NST}$) (Table 4, Figure 2).

Table 4. Average results and relative standard deviation of the capillarity water absorption coefficient (C) of treated and untreated specimens, before and after artificial aging tests.

Substrates	Capillarity Water Absorption Coefficient ($kg \cdot m^{-2} \cdot min^{-0.5}$)							
	Before Artificial Aging				After Artificial Aging			
	Untreated	Treated			Untreated	Treated		
		$H_{Sila\text{-}Sil}$	H_{Sil}	H_{NST}		$H_{Sila\text{-}Sil}$	H_{Sil}	H_{NST}
Stone	0.234 ± 0.031	0.175 ± 0.012	0.116 ± 0.032	0.040 ± 0.001	0.131 ± 0.031	0.053 ± 0.013	0.028 ± 0.010	0.068 ± 0.041
Mortar	0.182 ± 0.031	0.005 ± 0.001	0.004 ± 0.001	0.008 ± 0.001	0.124 ± 0.013	0.010 ± 0.001	0.011 ± 0.003	0.012 ± 0.002

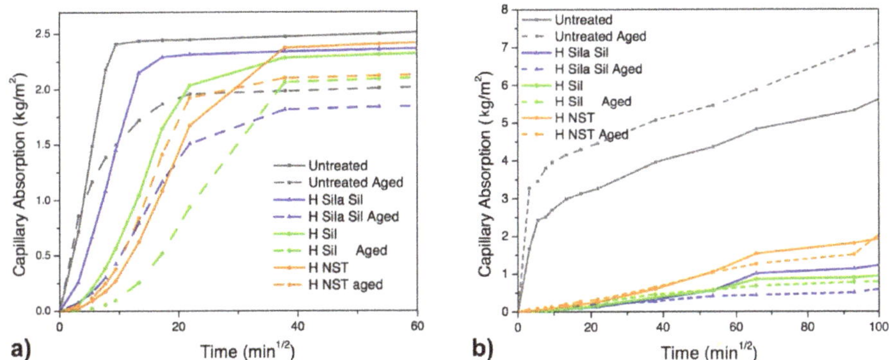

Figure 2. Capillary water absorption curves for the sound and treated substrates, before and after aging tests, of (**a**) limestone and (**b**) mortar, where solid lines are the unaged specimens, and dotted lines the aged specimens.

All hydrophobic products penetrated within the porous network of the substrate, reducing the wettability and providing a hydrophobic coating [9]. The higher reduction of the capillary water absorption coefficient of the specimens treated with H_{NST}, which can penetrate deeper within the treated substrate due to its nanosize [1], can be attributed to the chain arrangement (creation of Si-O-Ti bonds) of the TiO_2-SiO_2 nanoparticles. The copolymerization of the TiO_2 and silane within the silica network can give rise to the formation of a homogeneous organic–inorganic hybrid xerogel, with improved hydrophobic properties [23,24].

After accelerated aging tests, untreated specimens show a significant reduction of the capillary water absorption (around 44%), which can be attributed to the modification (destruction) of capillary pores (typically observed in altered and decayed materials) [36]. In fact, a significant increase of porosity is observed mostly between 5 to 10 freeze-thaw cycles [37]. Thus, artificial aging induces the generation of new pores and the expansion of existing pores in the specimens.

Additionally, the specimens with H_{NST} treatment, which show the highest reduction of capillary water absorption before artificial aging, show an opposite trend after artificial aging, with a decrease of up to 60% when compared to the unaged specimens. On the other hand, aged specimens treated with $H_{Sila/Sil}$ and H_{Sil} show a significant reduction of the capillarity water absorption (80% and 50%, respectively), when compared to the untreated aged specimens.

This can be justified by the lower durability of H_{NST} treatment to hygrothermal aging cycles. As a matter of fact, it is reported [38] that weathering can induce a weakening of the film adhesion and reduce the durability of TiO_2-SiO_2 protective coatings.

After aging, all specimens still maintain a significant reduction (40%, 21% and 52% for $H_{Sila/Sil}$, H_{Sil} and H_{NST}, respectively) of the capillary water absorption when compared to the untreated specimens.

Concerning mortar specimens, all treatments show a higher reduction of the capillarity water absorption when compared to treatments on limestone, although the total amount of absorbed water is higher (total saturation of the specimens is not completely achieved at the end of the test). A remarkable reduction of the capillarity water absorption was observed in all treated mortar specimens (>95%), when compared to untreated specimens. This behavior is attributed to the higher open porosity of rendering mortars (Table 1), allowing the easier penetration of hydrophobic products into the pores of the substrate coating [19]. The deeper penetration of the hydrophobic products leads to a reduction of the wettability of the substrate, resulting in an almost hydrophobic surface. All treatments show a similar behavior, almost waterproofing the mortar specimens, and the treatments show a good durability after artificial aging, with a minimal increase of the capillarity water absorption coefficient when compared to unaged specimens.

Additionally, all aged treated mortar specimens significantly reduce their capillary water absorption coefficient when compared to aged untreated specimens. However, it is worth noting that a slight decrease of the capillarity water absorption coefficient was observed if comparing unaged treated specimens to aged treated ones. This modification can be attributed to the possible alteration of the pore size distribution of the substrates.

A possible cause for the reduction of capillary water absorption after freeze-thaw cycles can be the reduction of capillary suction, resulting from an increase in the amount of bigger pores (above the capillary range) as a consequence of micro-cracking. Additionally, in the case of the cement-based mortar, the exposure to water with these cycles can induce a self-healing effect that promotes hydration reactions, further explaining the reduction of the capillary absorption rate with ageing.

When considering the results of water absorption by Karsten tube in the limestone specimens, a trend similar to that seen in the capillarity water absorption test was observed (75% in the case of H_{Sila}-Sil, 93% with H_{sil}, 92% with H_{NST}). In the case of the mortar specimens, the reduction was similar to that observed in capillary water absorption tests (96% in the case of H_{Sila}-Sil, 98% with H_{sil}, 99% with H_{NST}) (Table 5).

Table 5. Average results and relative standard deviation of the water absorption coefficient under pressure (C^{60}) of treated and untreated specimens, before and after artificial aging tests.

Substrates	Coefficient of Water Absorption at 60 min (C^{60}) ($kg \cdot m^{-2} \cdot min^{0.5}$)							
	Before Artificial Aging				After Artificial Aging			
	Untreated	Treated			Untreated	Treated		
		$H_{Sila\text{-}Sil}$	H_{Sil}	H_{NST}		$H_{Sila\text{-}Sil}$	H_{Sil}	H_{NST}
Stone	0.673 ± 0.178	0.167 ± 0.073	0.041 ± 0.022	0.053 ± 0.032	0.702 ± 0.131	0.083 ± 0.058	0.046 ± 0.021	0.147 ± 0.052
Mortar	0.893 ± 0.062	0.033 ± 0.012	0.023 ± 0.005	0.013 ± 0.005	0.906 ± 0.008	0.007 ± 0.009	0.013 ± 0.005	0.003 ± 0.002

If comparing the results of capillary water absorption and water absorption under low pressure of the untreated specimens, the opposite trend is seen. In fact, there is a reduction of capillary water absorption after aging; however, an increase of water absorption under low pressure is observed in equivalent conditions [39]. It is generally assumed that pores ranging from 1 to 10 μm act as capillary pores, whereas pores >10 μm contribute to the water permeability through gravity (e.g., percolation) or wind driven water ingress [40,41]. Thus, this confirms that artificial aging cycles possibly contribute to an increase in the amount of pores with dimensions greater than 10–20 μm.

Furthermore, $H_{Sila/Sil}$ e H_{Sil} treatments decrease the water absorption coefficient under pressure after artificial aging, both when applied on mortar or limestone. On the other hand, a decrease of 25% in the C^{60} was observed in the aged limestone specimens treated with H_{NST}, if compared to the unaged specimens, whereas an opposite trend is observed when considering treated mortar specimens.

These results point out that $H_{Sila/Sil}$ e H_{Sil} treatments have lower variation of the water absorption after artificial aging, compared to H_{NST} treatments. In fact, the latter shows both an increase of the capillary water absorption and water absorption under low pressure, possibly due to its physical-chemical alteration.

3.2. Drying Rate

When observing the drying curves, two stages can be observed (Figure 3). In the first stage of drying, called the constant drying period or initial drying rate, the drying front is at the surface and the drying rate is constant and controlled by the external conditions [42]. This first phase (initial drying rate) ends after 24 h in the case of all treated and untreated limestone and treated mortar specimens, whereas for untreated mortar specimens it ends after ≥72 h (Figure 3). When compared to the sound untreated specimens, all treated specimens decrease the initial drying rate products (first stage of drying), both in the limestone (3–12%) and mortar specimens (6–36%). More specifically, $H_{Sila/Sil}$ has an almost negligible influence on the initial drying rate of the treated specimens (3% reduction, when compared to untreated specimens), whereas H_{Sil} and H_{NST} induce a slightly higher reduction (up to 12%).

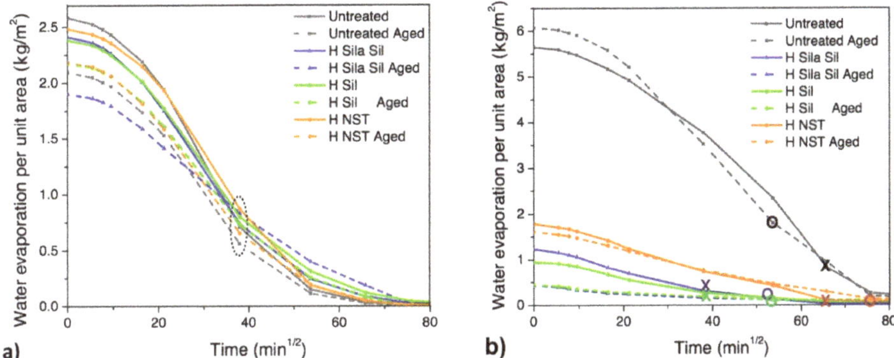

Figure 3. Drying curves for the sound, treated and aged substrate of (**a**) limestone and (**b**) mortar, where the dotted lines are the aged specimens, and solid lines the unaged specimens. The dotted ellipse (**a**) identifies the end of the 1st step of drying (critical moisture content) in (**a**), whereas this spot is highlighted as *X* (unaged specimens) or *O* (aged specimens) in (**b**).

In the second stage of drying, identified by the change in the slope of the drying curve, the moisture content can no longer support the demands of the evaporation flux, and, thus, the drying process occurs in the vapour phase. The transition between the first and second step of drying (i.e., the critical moisture content) occurs when the superficial moisture has evaporated. In this phase, the drying front progressively recedes into the material and the properties of the liquid and of the substrate control the rate of drying [42].

When considering the second stage of drying in limestone specimens, although the critical moisture content is identified at 24 h in all cases, it can be observed that (aged and unaged) untreated specimens almost achieve complete drying at 72 h, and a similar trend is observed in the case of the specimens treated with H_{NST} (Figure 3a). On the other hand, H_{Sil} and $H_{Sila/Sil}$ slightly delay the drying process (the second step of drying ends at 96 h), when compared to H_{NST} treatment.

It can be concluded that the hydrophobic treatments induce only a slight retarding effect on the drying behavior of the limestone.

When considering the mortar specimens, all the hydrophobic treatments remarkably reduce the total amount of water absorbed; however, the H_{NST} treatment takes a longer time to dry completely when compared to the H_{Sil} and $H_{Sila/Sil}$ treatments. Conversely, the H_{NST} treatment only slightly influences the initial drying rate (6% reduction) when compared to the $H_{Sila/Sil}$ treatment (13%) and, especially, the H_{Sil} treatment (36%). Additionally, in accordance with the results observed in the previous section, H_{NST} treatment also increases the drying time (8%) in the mortar specimens, whereas H_{Sil} and $H_{Sila/Sil}$ show a significant decrease (30–39%). The difference in the behavior of the hydrophobic products is also observed in the second step of drying, which starts at 24 h in the case of $H_{Sila/Sil}$ and H_{Sil} treatment, whereas the critical moisture content is identified at 72 h in the case of the H_{NST} treatment (as in the case of untreated specimens). Results obtained with contact angle measurements in a previous work confirm this trend and the drying index (Table 6), that is, a higher reduction of the wettability on both substrates in the case of $H_{Sila/Sil}$ treatment [19].

Table 6. Drying index (I_s) of aged treated and untreated specimens, before and after artificial aging tests.

Substrates	Drying Index (DI)							
	Before Artificial Aging				After Artificial Aging			
	Untreated	Treated			Untreated	Treated		
		$H_{Sila-Sil}$	H_{Sil}	H_{NST}		$H_{Sila-Sil}$	H_{Sil}	H_{NST}
Stone	0.089	0.095	0.101	0.096	0.074	0.093	0.085	0.079
Mortar	0.191	0.117	0.133	0.207	0.215	0.171	0.159	0.155

After hygrothermal aging, $H_{Sila/Sil}$ treatment show an improvement of the initial drying rate (20% and 27%, for limestone and mortar specimens, respectively), with worse results when compared to H_{NST} treatment (reduction of 5% and 24%, for limestone and mortar specimens, respectively) and H_{Sil} treatment (reduction of 9% and 12%, for limestone and mortar specimens, respectively). It can be seen that artificial aging slightly speeds up the drying process of the mortar specimens (the critical moisture content is identified at 48 h in the case of untreated aged specimens, and at 72 h with untreated unaged specimens). Additionally, in accordance with previous observations, the second step of drying of aged specimens treated with HSil and HSila/Sil starts at 48 h, and at 96 h in the case of the H_{NST} treatment (Figure 3b).

After artificial aging, $H_{Sila/Sil}$ treatment shows the highest variation of drying behavior, with an increase (26%) of the DI for limestone and decrease of 21% for mortar specimens (the slower the drying, the higher the DI). In accordance with previous observations, H_{Sil} treatment also induces an increase (15%) of the DI for limestone, and, conversely, a significant decrease (26%) for mortar specimens. On the other hand, H_{NST} treatment shows the best performance on limestone specimens, with only a slight DI increase (6%), and, however, a significant decrease for mortar specimen (28%).

3.3. Water Vapor Permeability

The results of the water vapor permeability test, as expressed by the water vapor diffusion resistance coefficient (μ), show a μ decrease for all hydrophobic treatments on limestone and mortar specimens, reducing the breathability of the substrates (Table 7). The reduction in water vapor permeability is an inevitable consequence of the water repellence properties of polymer film; however, the lowest possible decrease is pursued [43].

Table 7. Average results and relative standard deviation of the water vapor diffusion resistance coefficient (μ) of treated and untreated specimens, before and after artificial aging tests.

Substrates	Water Vapor Diffusion Resistance Coefficient (μ)							
	Before Artificial Aging				After Artificial Aging			
	Untreated	Treated			Untreated	Treated		
		$H_{Sila-Sil}$	H_{Sil}	H_{NST}		$H_{Sila-Sil}$	H_{Sil}	H_{NST}
Stone	3.20 ± 0.31	10.46 ± 1.24	7.30 ± 1.34	5.57 ± 0.75	4.37 ± 1.17	12.58 ± 0.92	6.39 ± 0.90	4.67 ± 0.52
Mortar	2.19 ± 0.05	2.15 ± 0.05	2.27 ± 0.05	2.26 ± 0.14	2.75 ± 0.07	3.02 ± 0.07	2.69 ± 0.04	2.71 ± 0.14

This reduction is more significant in limestone specimens, whereas it is almost negligible in mortar specimens. In fact, an increase of μ of 227% ($H_{Sila/Sil}$), 129% *(H_{Sil}) and 74% (H_{NST}) is observed on treated limestone specimens, when compared to untreated ones, whereas a minimal μ increase (<4%) is observed on the treated mortar specimens.

After artificial aging, only the H_{NST} treatment applied on limestone maintains reasonably higher μ values (increase of 7%) when compared to untreated specimens, whereas H_{Sil} and $H_{Sila/Sil}$ treatments still induce a drastic μ increase (46% and 188%, respectively). In the case of mortar specimens, the higher increase of μ was observed with $H_{Sila/Sil}$ treatment (9%), and only a slight μ decrease in the case of H_{Sil} and H_{NST} treatments (<2%). In general, the hydrophobic treatment that illustrates the most suitable behavior to water vapor permeability was H_{NST}, with a moderate reduction of the μ on both substrates, even after artificial aging.

4. Discussion

The variation of the moisture transport properties of the substrates treated with the hydrophobic products is presented in Figure 4. In general, it can be observed that the hydrophobic products induce a decrease of the capillary water absorption coefficient (C), this decrease being more relevant with H_{NST} treatment on limestone specimens. However, after artificial aging, $H_{Sila/Sil}$ and H_{Sil} treatments show a higher durability, with a higher decrease of the C, if compared to specimens treated with H_{NST}. Concerning mortar specimens, all treatments maintain a similar water absorption coefficient (considerably lower than untreated specimens), even after artificial aging.

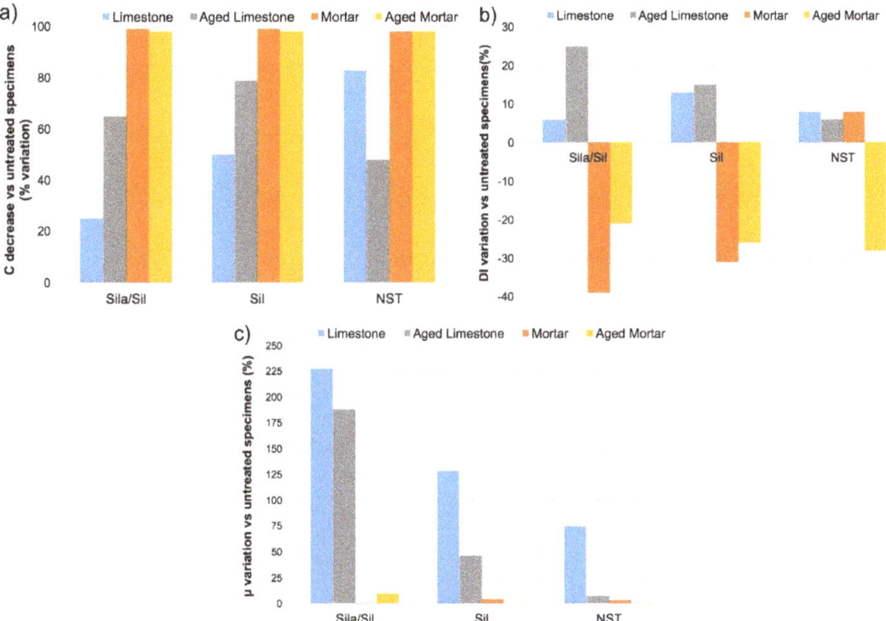

Figure 4. Percentage variation of (**a**) capillary water absorption coefficient (C), (**b**) drying index (DI) and (**c**) diffusion resistance of the water vapor coefficient (μ) of treated specimens, before and after artificial aging, when compared to untreated specimens.

The different moisture transport properties of mortar and limestone can be attributed to the higher open porosity (30%) of the mortar when compared to the studied limestone (11%). Additionally, mortar specimens generally have a higher volume of coarse pores (>100 μm) when compared to this type of compact Moleanos limestone [40].

Regarding the difference in the effectiveness and durability of the hydrophobic products, it is worth noting that (monomer) silane molecules are considerably smaller (10 to 15 Å) when compared to (oligomeric) siloxane molecules (25 to 75 Å) [44]. Thus, the longer Si–O chains of the siloxanes compared to the silane ones have a lower penetration depth in the compact limestone. Additionally, organo-modified siloxanes have an extremely high reactivity, which can hinder their in-depth penetration [2]. The silane has a potentially deeper penetration in the treated surface, with a higher reduction of the hydrophilicity and, thus, improved hydrophobic effectiveness. Furthermore, concerning the silane, it is generally assumed that the larger the molecule of the alkyl group linked to the silicon atom (which constitutes the structure of this compound), the higher the water repellency of the silane [8,45]. On the other hand, the long siloxane chains are more affected by environmental agents and weathering, undergoing degradation processes which can reduce their effectiveness as hydrophobic products [14].

The optimal performance obtained with the nanostructured product based on SiO_2 and TiO_2 (H_{NST}) on the limestone can probably be attributed to the chain arrangement of the TiO_2-SiO_2 nanoparticles, due to the creation of the Si–O–Ti bond [23]. In fact, the copolymerization of the TiO_2 and silane within the silica network can give rise to the formation of homogeneous organic–inorganic hybrid xerogel [24]. Additionally, the nanosize (<0.1 μm) of the silicon titanium oxide particles can match the dimension of pore network of the limestone. On the other hand, the low durability of H_{NST} treatment can be attribute to the photocatalytic oxidation (of the organic radicals) and thermal degradation of the SiO_2-TiO_2 composite, which can weaken the adhesion and thus, durability of the coating [38].

Concerning the drying index, the hydrophobic products induced an increase of the DI on limestone specimens, with higher variation in the case of the H_{Sil} and $H_{Sila/Sil}$ treatments. After aging, $H_{Sila/Sil}$ significantly increase the DI of the treated limestone specimens, whereas H_{Sil} and H_{NST} maintain values similar to unaged specimens. Greater variations are observed in mortar specimens; in fact, the H_{NST} treatment induces an increase of the DI, and a drastic decrease is observed after artificial aging, indicating the probable degradation of the H_{NST} treatment. On the other hand, the $H_{Sila/Sil}$ and H_{Sil} treatments decrease the DI on mortar specimens, with a lower decrease of the latter after artificial aging compared with the H_{NST} treatment. As reported in other works [46], the silane/siloxanes products can modify the pore netwok of the treated material by increasing the volume of capillary pores, probably also due to an air entraining effect of liquid siloxane. This feature can justify the improved resistance to freeze-thaw cycles, and thus, durability of $H_{Sila/Sil}$ and H_{Sil} treatments compared to H_{NST}.

Furthermore, the difference in the moisture transport properties of treated limestone and mortar can be attributed also to the higher roughness of the mortar, compared to flatter surface of the limestone, which accounts for better adhesion of both fissured and crack free SiO_2-TiO_2 films to the substrate [37].

All treatments induce an increase of the μ in limestone specimens. More specifically, specimens treated with $H_{Sila/Sil}$ and H_{Sil} show the highest μ increase, which significantly decreases after artificial aging in the case of the H_{NST} treatment. The H_{NST} treatment shows a lower μ after artificial aging compared with unaged specimens. Considering the mortar specimens, it can be concluded that μ variation is extremely low with all treatments, both before and after artificial aging. The $H_{Sila/Sil}$ is the only treatment which induced a slight increase of the μ after artificial aging.

Ultimately, it would be expectable that a substrate with higher DI would have also a higher μ (i.e., lower WVP). For treated limestone specimens this trend is confirmed; however, specimens treated with $H_{Sila/Sil}$ have a higher μ/DI ratio (even after artificial aging) compared to the other treatments. A different behavior is observed with treated mortar specimens: Aged and unaged specimens with H_{NST} treatment show the trend mentioned above, whereas unaged H_{Sil} treatment and aged $H_{Sila/Sil}$ treatment show a DI decrease and a μ increase.

5. Conclusions

In this paper, the effectiveness and durability of three commercially available hydrophobic products (a silicon and titanium dioxides-based nanostructured dispersion—H_{NST}; a silane/oligomeric siloxane—$H_{Sila/Sil}$; and a siloxane—H_{Sil}) when applied to a Moleanos limestone and on a cement-based mortar, were analyzed. The alteration of the moisture transport properties (water absorption by capillarity and under low pressure, drying kinetics and water vapor permeability) of the treated substrates, prior to and after artificial aging tests, was evaluated.

Results show that the effectiveness and durability of the water-repellent treatment is influenced both by the type of hydrophobic product and by the treated substrate.

Although the products were applied at different concentrations, following the recommendations of the producers, all treatments induce a significant decrease of the values of the capillary water absorption and water absorption under low pressure on the mortar specimens. A lower decrease was observed in the limestone specimens. This difference is attributed to the higher open porosity of the mortar specimens compared to limestone specimens, thus allowing a deeper penetration of the hydrophobic products, which increases the water-repellency of the treated mortar. After artificial aging, all hydrophobic treatments show a significant durability to the type and duration of the artificial aging cycles considered in this work, maintaining a reasonably low water absorption in both mortar and limestone specimens, when compared to untreated specimens. The H_{NST} treatment shows a slightly greater loss of efficacy in terms of water capillary absorption and water vapor permeability after artificial aging, mostly when applied on limestone. Therefore, it can be considered less durable.

The treatments induce also a variation of the drying index, which increases with all treatments on limestone specimens (even after artificial drying) and a general decrease on mortar specimens, except for the H_{NST} treatment before aging, which slightly increases the DI. On the other hand, the H_{NST} treatment shows a lower drying index after artificial aging.

$H_{Sila/Sil}$ and H_{Sil} treatments significantly reduce the water vapor permeability of limestone specimens, whereas the H_{NST} treatment induces a smaller decrease, with values similar to those of untreated specimens after artificial aging. The WVP of the treated mortar specimens was not significantly affected by the hydrophobic treatments, even after aging tests.

These observations confirm that the hydrophobic products are generally more effective and durable on mortar specimens, rather than on low-porosity limestone, as in the case of the studied Moleanos limestone.

When pondering the variation of all the moisture transport properties, the hydrophobic product based on siloxane (H_{Sil}) has the best performance on cement-based mortar; in fact, the molecular structure of siloxanes matches to the higher porosity of this substrate. On the other hand, although it has a lower durability compared to the other treatments, H_{NST} has the best performance when applied on Moleanos limestone, with a significant decrease of the capillary water absorption and a low variation of the drying index and of the WVP. This behavior can result from the combination of the low porosity and micro-sized pores of the stone with the surface deposition of a nanostructured layer. Additionally, the presence of TiO_2 confers antibacterial activity against the microorganism growth and pollutant absorption.

The $H_{Sila/Sil}$ treatment significantly decreases its water-repellent properties. However, it hinders the drying process and the breathability of the substrates, even after artificial aging. Thus, based on the results of this study, its use is not recommended in either limestone or mortar.

Further tests (e.g., optimization of the protocol; artificial aging cycles with UV light, pollutants and/or biological colonization; FTIR analysis of the hydrophobic products; morphological analysis by SEM-EDS; contact angle measurements of the treated substrates, among others) are ongoing to correlate the effect of the physical-chemical aging on the effectiveness of hydrophobic products, and ultimately, to verify the durability to more prolonged weathering action (from lab to real natural scale tests) of the hydrophobic products.

Author Contributions: Conceptualization, G.B.; Methodology, I.F.-C. and R.V.; Investigation, C.E.; Writing—Original draft preparation, G.B.; Writing—Review and editing, I.F.-C. and R.V.; Supervision, I.F.-C.; Project administration, I.F.-C. All authors have read and agreed to the published version of the manuscript.

Funding: This research was funded by Portuguese Foundation for Science and Technology (FCT), grant number PTDC/ECI-EGC/30681/2017 (WGB_Shield – Shielding building' facades on cities revitalization. Triple resistance for water, graffiti and biocolonization of external thermal insulation systems).

Acknowledgments: The authors acknowledge the companies CIN, Saint-Gobain Weber and NanoPhos for the supply of the hydrophobic products.

Conflicts of Interest: The authors declare no conflict of interest.

References

1. Ferreira Pinto, A.; Delgado Rodrigues, J. Assessment of durability of water repellents by means of exposure testes. In *9th International Congress on Deterioration and Conservation of Stone*; Fassina, V., Ed.; Elsevier: Amsterdam, The Netherlands, 2000; pp. 273–285.

2. Roos, M.; Konig, F.; Stadtmuller, S.; Weyershausen, B. Evolution of silicone based water repellents for modern building protection. In *5th International Conference on Water Repellent Treatment of Building Mate-Rials—Hydrophobe V*; Aedificatio Publishers: Brussels, Belgium, 2008; pp. 3–16.

3. Medeiros, M. Contribuição ao Estudo da Durabilidade de Concretos com Protecção Superficial Frente à ação de íons Cloretos. Ph.D. Thesis, Polytechnic School of the University of São Paulo, São Paulo, Brasil, 2008.

4. Jerman, M.; Černý, R. Effect of moisture content on heat and moisture transport and storage properties of thermal insulation materials. *Energ. Build.* **2012**, *53*, 39–46. [CrossRef]

5. Pavlík, Z.; Žumár, J.; Medved, I.; Černý, R. Water vapor adsorption in porous building materials: Experimental measurement and theoretical analysis. *Transp. Porous Med.* **2012**, *91*, 939–954. [CrossRef]

6. Lepech, M.; Li, V. Water permeability of engineered cementitious composites. *Cem. Concr. Comp.* **2009**, *31*, 744–753. [CrossRef]

7. Silva, L.; Flores-Colen, I.; Vieira, N.; Timmons, A.B.; Sequeira, P. Durability of ETICS and Premixed One-Coat Renders in Natural Exposure Conditions. In *New Approaches to Building Pathology and Durability*; Building Pathology and Rehabilitation; Delgado, J.M.P.Q., Ed.; Springer: Singapore, 2016; Volume 6, pp. 131–158.

8. De Vries, J.; Polder, R. Hydrophobic treatment of concrete. *Constr. Build. Mater.* **1997**, *11*, 259–265. [CrossRef]

9. Maranhão, F.; Loh, K. *The Use of Hydrophobic Products in Porous Building Materials*; Téchne Edition: São Paulo, Brazil, 2010; p. 155.

10. Ahmad, D.; van den Boogaert, I.; Miller, J.; Presswell, R.; Jouha, H. Hydrophilic and hydrophobic materials and their applications. *Energy Sources Part A* **2018**, *40*, 2686–2725. [CrossRef]

11. Edwards, P. Preservation for the Future: The need for waterproofing structures above and below ground. *J. Protect. Coat. Linings* **2002**, *7*, 42–46.

12. Finzel, W.; Vincent, H. *Silicones in Coatings. Federation Series on Coatings Technology*; Federation of Societies for Coatings Technology: Philadelphia, PA, USA, 1996.

13. Selley, D. Chemical Considerations for Making Low-VOC Silicon-Based Water Repellents. *J. Coat. Technol. Res.* **2010**, *7*, 26–35.

14. Charola, A. Water repellents and other "protective" treatments: A critical review. In *3rd International Conference on Surface Technology with Water Repellents Agents—Hydrophone III*; Aedificatio Publishers: Hannover, Germany, 2001; pp. 3–20.

15. Raupach, M.; Buttner, T. Hydrophobic treatments on concrete. Evaluation of the durability and non-destructive testing. In *2nd International Conference on Concrete Repair, Rehabilitation and Retrofitting II*; Alexander, M.G., Beushausen, H.D., Dehn, F., Moyo, P., Eds.; CRC Press: Cape Town, South Africa, 2008; pp. 907–913.

16. Justo, J.; Castro, J.; Cicero, S. Notch effect and fracture load predictions of rock beams at different temperatures using the Theory of Critical Distances. *Int. J. Rock Mech. Min.* **2020**, *125*, 104–161. [CrossRef]

17. Moura, A.; Flores-Colen, I.; de Brito, J. Study of the effect of three anti-graffiti products on the physical properties of different substrates. *Constr. Build. Mater.* **2016**, *107*, 157–164. [CrossRef]

18. CEN—European Committee for Standardization. *Methods of Test for Mortar for Masonry—Part 2: Bulk Sampling of Mortars and Preparation of Test Mortars*; EN 1015-2; CEN: Brussels, Belgium, 2006.

19. Esteves, C.; Ahmed, H.; Flores-Colen, I. The influence of hydrophobic protection on building exterior claddings. *J. Coat. Technol. Res.* **2019**, *16*, 1379–1388. [CrossRef]

20. Rutter, T.; Hutton-Prager, B. Investigation of hydrophobic coatings on cellulose-fiber substrates with in-situ polymerization of silane/siloxane mixtures. *Int. J. Adhes. Adhes.* **2018**, *86*, 13–21. [CrossRef]

21. Qiao, B.; Liang, Y.; Wang, T.J.; Jiang, Y. Surface modification to produce hydrophobic nano-silica particles using sodium dodecyl sulfate as a modifier. *Appl. Surf. Sci.* **2016**, *364*, 103–109. [CrossRef]

22. Yu, K.; Liang, Y.; Ma, G.; Yang, L.; Wang, T.Y. Coupling of synthesis and modification to produce hydrophobic or functionalized nano-silica particles. *Colloid Surf. A* **2019**, *574*, 122–130. [CrossRef]

23. Kapridaki, C.; Maravelaki-Kalaitzaki, P. TiO_2–SiO_2–PDMS nano-composite hydrophobic coating with self-cleaning properties for marble protection. *Prog. Org. Coat.* **2013**, *76*, 400–410. [CrossRef]

24. Alfieri, I.; Lorenzi, A.; Ranzenigo, A.; Lazzarini, L.; Predieri, G.; Lottici, P.P. Synthesis and characterization of photocatalytic hydrophobic hybrid TiO_2-SiO_2 coatings for building applications. *Build. Environ.* **2017**, *111*, 72–79. [CrossRef]

25. Truppi, A.; Luna, M.; Petronella, F.; Falcicchio, A.; Giannini, C.; Comparelli, R.; Mosquera, M.J. Photocatalytic activity of TiO_2/AuNRs–SiO_2 nanocomposites applied to building materials. *Coatings* **2018**, *8*, 296. [CrossRef]

26. CEN—European Committee for Standardization. *Methods of Test for Mortar for Masonry—Part 18: Determination of Water Absorption Coefficient due to Capillary Action of Hardened Mortar*; EN 1015-18; CEN: Brussels, Belgium, 2002.

27. RILEM. Water Absorption under Low Pressure, Pipe Method, Test No 11.4. In *Recommandations Provisoires*; RILEM TC 25-PEM; RILEM Publications SARL: Paris, France, 1980; pp. 201–202.

28. RILEM. *Evaporation Curve, Test No 11.5*; Recommandations Provisoires; RILEM TC 25-PEM; RILEM Publications SARL: Paris, France, 1980; pp. 205–207.

29. *NORMAL 29/88—Misure dell'Indice di Asciugamento (Drying Index), Alterazioni dei Materiali Lapidei e Trattamenti Conservative—Proposte per l'unificazione dei Metodi Sperimentali di Studio e Controllo*; ICR-CNR: Rome, Italy, 1988.

30. *CEN—European Committee for Standardization, Methods of Test for Mortar for Masonry—Part 19: Determination of Water Vapour Permeability of Hardened Rendering and Plastering Mortars*; EN 1015-19; CEN: Brussels, Belgium, 2008.

31. Velosa, A.; Veiga, M.R. Development of Artificial Ageing Tests for Renders—Application to Conservation Mortars. In *7th International Masonry Conference—7IMC*; Thompson, G., Ed.; International Masonry Society: London, UK, 2006.

32. Veiga, M.R.; Santos Silva, A. *Long-term Performance and Durability of Masonry Structures Degradation Mechanisms, Health Monitoring and Service Life Design, Chapter 6—Mortars*; Woodhead Publishing Series in Civil and Structural, Engineering; Ghiassi, B., Lourenço, P., Eds.; Elsevier: Amsterdam, The Netherlands, 2019; pp. 169–208.

33. *CEN—European Committee for Standardization, Methods of test for Mortar for Masonry—Part 21: Determination of the Compatibility of One-Coat Rendering Mortars with Substrates*; EN 1015-21; CEN: Brussels, Belgium, 2002.

34. Pascoal, P.; Borsoi, G.; Veiga, R.; Faria, P.; Santos Silva, A. Consolidation and chromatic reintegration of historical renders with lime-based pozzolanic products. *Stud. Conser.* **2015**, *60*, 321–332.

35. Maia, J.; Ramos, N.M.M.; Veiga, R. Assessment of test methods for the durability of thermal mortars exposure to freezing. *Mater. Struct.* **2019**, *52*, 112. [CrossRef]

36. Thomson, M.L.; Lindqvist, J.E.; Elsen, J.; Groot, C.J.W.P. Porosity of historic mortars. In *Characterisation of Old Mortars with Respect to Their Repair–Final Report of RILEM TC 167-COM*; Groot, C., Ashall, G., Hughes, J., Eds.; RILEM Publications SARL: Paris, France, 2004; pp. 77–106.

37. Li, Z.; Liu, L.; Yan, S.; Zhang, M.; Xia, J.; Xie, Y. Effect of freeze-thaw cycles on mechanical and porosity properties of recycled construction waste mixtures. *Constr. Build. Mater.* **2019**, *210*, 347–363. [CrossRef]

38. Calia, A.; Lettieri, M.; Masieri, M. Durability assessment of nanostructured TiO_2 coatings applied on limestones to enhance building surface with self-cleaning ability. *Build. Environ.* **2016**, *110*, 1–10. [CrossRef]

39. Nogueira, R.; Ferreira Pinto, A.P.; Gomes, A. Artificial ageing by salt crystallization: Test protocol and salt distribution patterns in lime-based rendering mortars. *J. Cult. Herit.* **2020**. [CrossRef]

40. Hunt, A.; Ewing, R. *Percolation Theory for Flow in Porous Media*, 2nd ed.; Lect. Notes in Phys.; Springer: Heidelberg, Germany, 2008; p. 771.

41. Santos, A.R.; Veiga, M.R.; Santos Silva, A.; de Brito, J. Tensile bond strength of lime-based mortars: The role of the microstructure on their performance assessed by a new non-standard test method. *J. Build. Eng.* **2020**, *29*, 101136. [CrossRef]

42. Hall, C.; Hoff, W.D. *Water Transport in Brick, Stone and Concrete*; Spon Press, Taylor & Francis Group: New York, NY, USA, 2012.

43. Tsakalofa, A.; Manoudisb, P.; Karapanagiotisc, I.; Chryssoulakisc, I.; Panayiotoub, C. Assessment of synthetic polymeric coatings for the protection andpreservation of stone monuments. *J. Cult. Herit.* **2007**, *8*, 69–72. [CrossRef]

44. Lucquiaud, V.; Courard, L.; Gérard, O.; Michel, F.; Handy, M. Evaluation of the durability of hydrophobic treatments on concrete architectural heritage. *Restor. Build. Monum.* **2015**, *20*, 395–404.

45. Guo, T.; Weng, X. Evaluation of the freeze-thaw durability of surface-treated airportpavement concrete under adverse conditions. *Constr. Build. Mater.* **2019**, *206*, 519–530. [CrossRef]

46. Falchi, L.; Zendri, E.; Müller, U.; Fontana, P. The influence of water-repellent admixtures on the behaviour and the effectiveness of Portland limestone cement mortars. *Cem. Concr. Comp.* **2015**, *59*, 107–118. [CrossRef]

 *coatings*

Article

Evaluation of the Antibacterial Activity of a Geopolymer Mortar Based on Metakaolin Supplemented with TiO$_2$ and CuO Particles Using Glass Waste as Fine Aggregate

Ruby Mejía-de Gutiérrez [1,*], Mónica Villaquirán-Caicedo [1], Sandra Ramírez-Benavides [1], Myriam Astudillo [2] and Daniel Mejía [2]

[1] Composites Materials Research Group (CENM), Universidad de Valle, calle 13 #100-00 Cali, Colombia; monica.villaquiran@correounivalle.edu.co (M.V.-C.); sandra.ramirez@correounivalle.edu.co (S.R.-B.)

[2] Biotechnology and Bacterial Infections Research Group, Universidad de Valle, calle 4b #36b-37, Cali, Colombia; myriam.astudillo@correounivalle.edu.co (M.A.); daniel.mejia@correounivalle.edu.co (D.M.)

* Correspondence: ruby.mejia@correounivalle.edu.co

Received: 29 December 2019; Accepted: 6 February 2020; Published: 9 February 2020

Abstract: Metakaolin-based geopolymer cements were produced by alkaline activation with a potassium hydroxide and potassium silicate solution. To produce the geopolymer composites, 10 wt.% titanium oxide (TiO$_2$) and 5 wt.% copper oxide (CuO) nanoparticles were used. The geopolymer mortar was prepared using glass waste as fine aggregate. The raw materials and materials produced were characterized by X-ray diffraction, electron microscopy, and Fourier-transform infrared spectroscopy techniques. Likewise, the geopolymer samples were characterized to determine their physical properties, including their density, porosity, and absorption. The photocatalytic activity of the materials was evaluated by activating the nanoparticles in a chamber with UV–Vis light during 24 h; then, different tests were performed to determine the growth inhibition of *Staphylococcus aureus*, *Escherichia coli*, and *Pseudomonas aeruginosa* bacteria in nutrient agar for times of up to 24 h. The study results showed that a geopolymer mortar containing glass waste as fine aggregate (GP-G) exhibited a water absorption 56.73% lower than that of the reference geopolymer paste without glass (GP). Likewise, glass particles allowed the material to have a smoother and more homogeneous surface. The pore volume and density of the GP-G were 37.97% lower and 40.36% higher, respectively, than those of the GP. The study with bacteria showed that, after 24 h in the culture media, the GP-G mortars exhibited a high inhibition capacity for the growth of *P. aeruginosa* from solutions of 10^{-4} mL and in solutions of 10^{-6} mL for *E. coli* and *S. aureus*. These results indicate the possibility of generating antibacterial surfaces by applying geopolymer composite.

Keywords: geopolymer; glass waste; titanium oxide; copper oxide; antibacterial surfaces

1. Introduction

The problematic relationships between environmental pollution and health preservation and prolongation are a subject of current interest and are, therefore, widely studied. Most health problems are associated with the presence of microorganisms such as bacteria, fungi, protozoa, and viruses that commonly infect humans in the living environment, leading to chronic infections and even leading to mortality [1–3]. The inanimate surfaces of different materials are often described as the source of hospital outbreaks because their contact with different microorganisms allows microbial permanence for a period of time favoring bacterial proliferation, and they are, therefore, reservoirs of bacteria that cause diseases transmitted through surfaces or fomites. This continues to be a cause of great concern for the medical domain, with a significant economic burden. One of the main causes of morbidity is infection with methicillin-resistant *Staphylococcus aureus* (MRSA) [4,5]. These microorganisms also contribute to severely deteriorating surfaces, greatly reducing the durability and increasing the repair costs of the material [6]. Considering this dynamic, a material with antibacterial surfaces would be ideal to prevent microbial proliferation and, likewise, contamination.

Therefore, surface materials and antibacterial systems are of the utmost importance not only in hospitals and sanitary environments but also for domestic, industrial, and marine applications, among others [7]. Various studies were carried out with the goal of improving the antibacterial capacity of construction materials as glazes on ceramic tiles and pastes based on alkali-activated slag [2,8,9]. Among these construction materials, the use of geopolymers is a great option due to their high alkalinity and ease of functionalization by incorporating semiconductor materials (ZnO, TiO_2, CuO, and Fe_2O_3), which, when exposed to UV and UV–Vis radiation, exhibit a functional activity, i.e., photocatalytic properties [10,11]. One of the most studied semiconductors is titanium dioxide (TiO_2), a harmless material that is highly resistant to photocorrosion, stable in aqueous solutions, inexpensive, and abundant in nature, which also exhibits a desired high photocatalytic activity under ultraviolet irradiation [9,12–14]. Semiconductor particles as TiO_2, Fe_3O_4, ZnO, and CuO can cause inactivation of bacteria (i.e., elimination of bacteria) and viruses in different types of environments [2,15–17]. These metal oxides were incorporated into the surface of different materials using advanced deposition techniques, such as chemical vapor deposition, ion implantation, sputtering, and electrochemical deposition of a solution. However, these technologies are expensive and difficult to apply to large-volume particles or complex shapes [18].

In general, the antimicrobial activity of CuO and TiO_2 metallic nanoparticles was studied with Gram-positive and Gram-negative bacteria [19–21]. Although the nanoparticles tended to inhibit the bacteria in all cases, it was observed that the effectiveness depends on the morphology and particle size. Hasmaliza et al. [9] evaluated the antibacterial properties of ceramic tiles coated with enamel mixed with anatase TiO_2 by exposing the tiles for different times (0, 2, 4, and 8 h) to the bacterium *Escherichia coli*; the TiO_2 content (5, 10, and 15 wt.%) and particle size of the oxide were also varied, and, at a longer exposure time, the number of colony-forming units (CFUs) decreased. At the same time, the nanometric particle sizes favored the antibacterial properties due to the greater surface area available to contact the bacteria; in contrast, in the presence of a greater amount of TiO_2, the antibacterial yield was lower [9]. Kumar et al. [20] evaluated the antibacterial activity of a polymer nanocomposite containing TiO_2 and CuO nanoparticles with satisfactory results. Haider et al. [19] evaluated the photocatalytic and antibacterial activity of TiO_2 nanoparticles synthesized via the sol–gel method and calcined at different temperatures (400, 600, 800, and 1000 °C). TiO_2 was exposed to two types of bacteria, *Pseudomonas aeruginosa* and *S. aureus*, and it was 100% effective in eliminating these bacteria under solar irradiation.

Geopolymers based on metakaolin, halloysite clay, and fly ash with TiO_2 nanoparticles reported photocatalytic properties. These materials allow functional ceramics to be produced for self-cleaning ability, removing dyes as B-rhodamine and methylene blue, and leading to nitric oxide degradation [11,22–28]. Nanoparticles are also used in dental applications [10,29]. However, the information on the antibacterial effect of nanoparticles incorporated in geopolymeric pastes or mortars is limited [30]. TiO_2 microparticles (20 and 50 wt.%) and ZnO nanoparticles were incorporated in geopolymer pastes based on metakaolin (MK) [31,32], where the authors reported a bacteriostatic effect in the presence of contaminated water. Similar results were obtained using MK-based geopolymer mortar doped with copper [6]. Geopolymers based on fly ash and calcined baluko shells incorporating Ag nanoparticles were also used to evaluate their antibacterial capacity [33,34]. These studies showed that the nanoparticles can be used to prepare geopolymers with satisfactory inhibition capacity for the growth of bacteria. Additionally, it was reported that Portland cement mortars with added glass containing 2 wt.% TiO_2 and the incorporation of nanosilica exhibited the ability to inhibit *E. coli* growth in just 30 min [35].

The objective of this preliminary study was to determine the effects of incorporating nanoparticles of metallic oxides, such as TiO_2 and CuO, on the antimicrobial potential of a geopolymeric binder based on metakaolin (MK), and its corresponding geopolymer mortar using glass waste as a fine aggregate. Geopolymer mortars were fabricated for their use as coatings of construction elements for environments susceptible to bacterial growth. Among the microorganisms potentially pathogenic to humans were selected Gram-negative bacteria such as *E. coli* and *Pseudomonas aeruginosa* and Gram-positive bacteria such as *S. aureus*. Infections with these bacteria are recurrent and transmissible to people close to each other [4,5,36,37]. In addition, this preliminary study proposes the direct incorporation of such nanoparticles by mechanically mixing the components of the material in order to directly apply the geopolymer mortar on the substrate surfaces instead of using conventional deposition processes to produce oxide coatings.

2. Materials and Methods

2.1. Materials

To obtain the geopolymer cement paste, the precursor used was metakaolin (MK MetaMax, BASF, Florham Park, NJ, USA). A mixture of a commercial potassium silicate (SiO_2 = 26.38%, K_2O = 13.06%, H_2O = 60.56%), analytical grade potassium hydroxide (KOH), and water was used as an alkaline activator. The cementitious material was formulated using the following molar ratios of oxides, as determined by previous studies: SiO_2/Al_2O_3 = 2.5 and K_2O/SiO_2 = 0.28 [38,39]. To produce the geopolymer composite with photocatalytic and antibacterial properties, titanium oxide (TiO_2) and copper oxide (CuO) particles were added. The TiO_2 used was high-purity and analytical grade (Merck, reference 1008081000). The CuO nanoparticles were synthesized from copper acetate (($CH_3COO)_2Cu·H_2O$) using the modified Pechini method [40] (Figure 1). To evaluate the cementitious properties, the pastes and mortars were prepared. The mortar was prepared using glass waste (G) as a fine aggregate.

To synthesize the CuO nanoparticles, citric acid and ethylene glycol were initially mixed in 100 mL of distilled water, and the mixture was heated to 70 °C. Then, copper acetate was added. These precursors were mixed with a molar ratio of 1:1:2, and the solution was constantly stirred using a magnetic stirrer. Simultaneously, an NH_4OH solution was added until a neutral pH (pH = 7) was obtained. The resulting polymeric resin was subjected to an initial thermal treatment at 350 °C; the thermally treated powder was pulverized using a ceramic mortar. The resulting fine powder was thermally treated at 450 °C. The heating rate used in the two heat treatments was adjusted to 10 °C/min.

2.2. Preparation of the Geopolymers

To prepare the geopolymers, the components were firstly mixed in a solid state; then, the activating solution was added. The components were mixed for 15 min using a Hobart mixer. Next, the mixture was poured into silicone molds, and discs with 2.5 cm diameters were prepared. The liquid/solid ratio was 0.35 for the pastes and 0.40 for the mortar. The samples were wrapped with a plastic film to prevent moisture from evaporating and kept for 24 h in a chamber with a relative humidity (RH) >90% and a temperature of 25 °C. Subsequently, the samples were demolded and stored for 28 days. Figure 2 shows the experimental methodology followed in the study in order to prepare the geopolymer composites. Table 1 shows the compositions and codes of the mixtures evaluated. Two geopolymer pastes (GP, mGP) were prepared using 10 wt.% titanium oxide (TiO_2) and 5 wt.% copper oxide (CuO) particles. The percentage of TiO_2 incorporated was selected based on previous studies [11], where a high photocatalytic capacity for degrading B-rhodamine was achieved using 10 wt.% TiO_2. Geopolymer mortars (GP-G) were manufactured with a binder/fine aggregate ratio of 1:2 by weight using glass waste (G) as fine aggregate. The photocatalytic capacity of MK-based geopolymer mortars (GP-G) increased up to 72% compared to that obtained with mortars using natural sand [41]. Geopolymer mortars with CuO particles were not prepared because the results obtained with the proportion of the oxide used in mGP were negative, as discussed later.

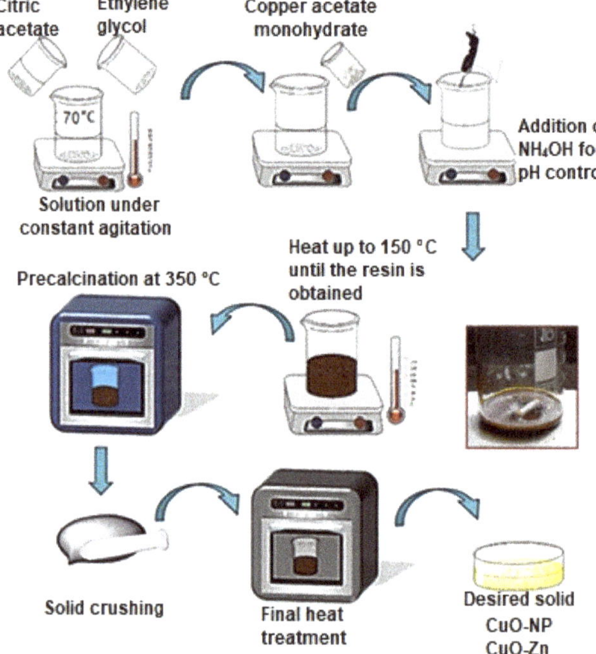

Figure 1. Diagram of the methodology used to synthesize the CuO nanoparticles via the Pechini method.

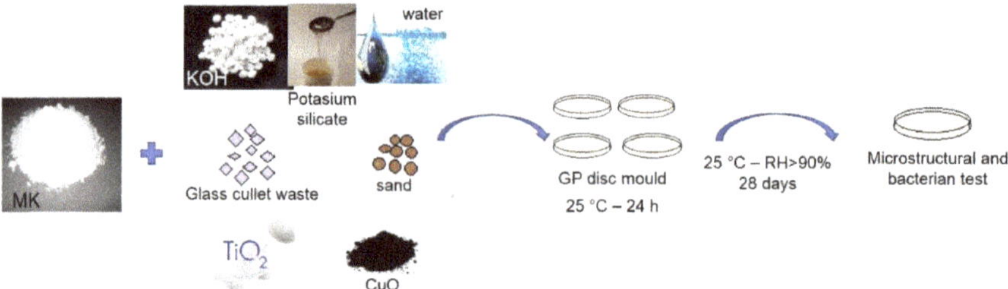

Figure 2. Diagram of the methodology used to prepare the geopolymer composites.

Table 1. Compositions of the fabricated materials. ID—identifier.

ID	Geopolymer (GP) Type	GP (wt.%)	TiO$_2$ (wt.%)	CuO (wt.%)
GP	Paste	90	10	0
mGP	Paste	95	0	5
GP-G	Mortar with glass waste (G) as fine aggregate	90	10	0

2.3. Physicochemical and Microstructural Characterization Techniques

The following instrumental techniques were used to characterize the materials:

- The chemical composition of the starting materials was determined by X-ray fluorescence (XRF) using a Magix Pro PW-2440 X-ray fluorescence spectrometer (PANalytical, Tollerton, Nottingham, UK) equipped with a rhodium tube with a maximum output of 4.0 kW and 0.02% sensitivity.
- The X-ray diffraction (XRD) patterns were obtained with a RINT2000 wide-angle goniometer using Cu Kα radiation at 45 kV and 40 mA, and data were collected with a step size of 0.02° over the range of 5°–60° 2θ using a scan speed of 5°/min.
- Fourier-transform infrared spectroscopy (FTIR) was performed with a PerkinElmer Spectrum 100 spectrometer operated in transmittance mode, and data were collected for a wavelength range of 450 cm^{-1} to 4000 cm^{-1}. The samples were prepared using the compressed KBr method.
- The morphological observations and particle sizes of TiO2 and CuO were evaluated by scanning electron microscopy (SEM) using a JEOL JSM-6490 LV SEM system operated with an accelerating voltage of 20 kV. The specimens were coated with gold, and the observations were made in vacuum mode.
- The density, pore volume, and water absorption of the geopolymers were determined following the procedures detailed in the ASTM C642-13 standard, except that the temperature used to dry the sample in the oven was 60 °C for 48 h.

2.4. Evaluation of the Antibacterial Activity of the Geopolymeric Materials

The geopolymer samples were exposed to the bacteria under study by activating the materials with UV-A light for 48 h. The UV-A radiation was provided by two mercury lamps (Electrolux T8 20 W BLB) located inside a black acrylic dome. The lamps emitted light at an intensity of 10.3 W·m^{-2}, which was measured with a Delta Ohm HD 2102.2 Photoradiometer using the filter for UV-A light range (k = 360 nm) at 5 mm [14]. The growth inhibition capacity of the bacteria in the materials was studied using *E. coli*, *S. aureus*, and *P. aeruginosa* bacteria. The tests were performed in two stages. The first stage followed the standard to evaluate the bactericidal capacity with the bacteria to be tested in the GP (with 10 wt.% TiO$_2$) and mGP (with 5 wt.% CuO) samples. This initial stage was performed by following the bacterial growth inhibition zone method (disc diffusion agar test). This method consisted of placing the sample, i.e., the GP and mGP disc, in a Petri dish and spreading nutrient agar on the sample. The Petri dishes were then covered for 24 h at 25 °C, and the bacterial growth was verified by the formation of an "inhibition halo" around the disc and under the disc. If the material is favorable to bacterial proliferation, more bacteria would be found at the site where the material rests (Figure 3).

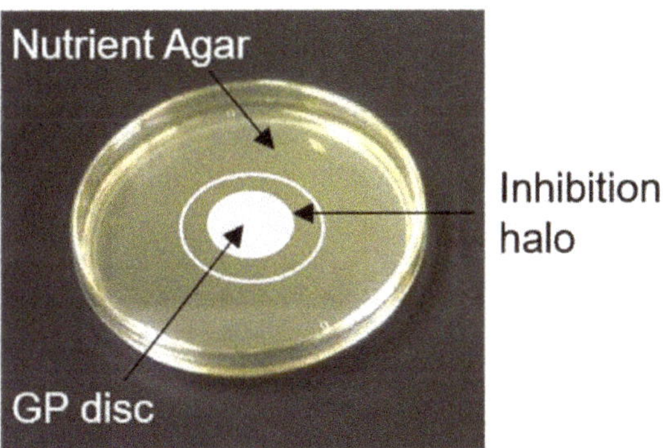

Figure 3. Assembly of the nutrient agar method.

By considering the results of the initial stage, the ability of the geopolymer paste (GP) and the mortar with glass (GP-G) to inhibit bacterial growth was evaluated in the second stage. During this stage, the methodology used was based on the Hasmaliza study [9], with a few modifications. The bacteria were cultured in tryptic soya broth (Merck, Darmstadt, Germany) for 24 h. At the beginning of the test, a count was performed in each of the broths by performing six serial dilutions in plate count agar (PCA, Scharlau 01-161). Subsequently, 100 μL of each of the selected bacteria were placed on discs, which were exposed for 2 to 24 h at room temperature (25 °C) under conditions isolated from contamination. After each of the exposure times, the discs were washed with 5 mL of a 1% peptone-buffered saline solution (Difco, Waltham, MA, USA), which was performed with vigorous stirring for 5 min; finally, the six dilutions were performed in a type-2 laminar flow cabinet (C4, Colombia) to count the number of colony-forming bacteria using the pour plate procedure with PCA. Finally, each agar with its respective inoculum was left in the incubator (WTC Binder, Germany) for 24 h at 37 °C to subsequently count the bacteria grown for each material in the different dilutions.

3. Results and Analysis

3.1. Physicochemical and Microstructural Characterization of the Starting Materials and the Produced Geopolymers

The chemical composition of MK and G is shown in Table 2. MK was basically composed of silica and alumina oxides (approximately 97 wt.%) and contains TiO_2 impurities (1.73 wt.%). This oxide was in the form of anatase, as seen in the XRD results (Figure 4). The results shown in Table 2 indicate that the glass waste used in G was from the SiO_2–Na_2O–CaO system (calcium sodium glass) and was highly amorphous, as corroborated by the XRD results based on the elevation of the baseline for 2θ angles between 20° and 30° (Figure 4).

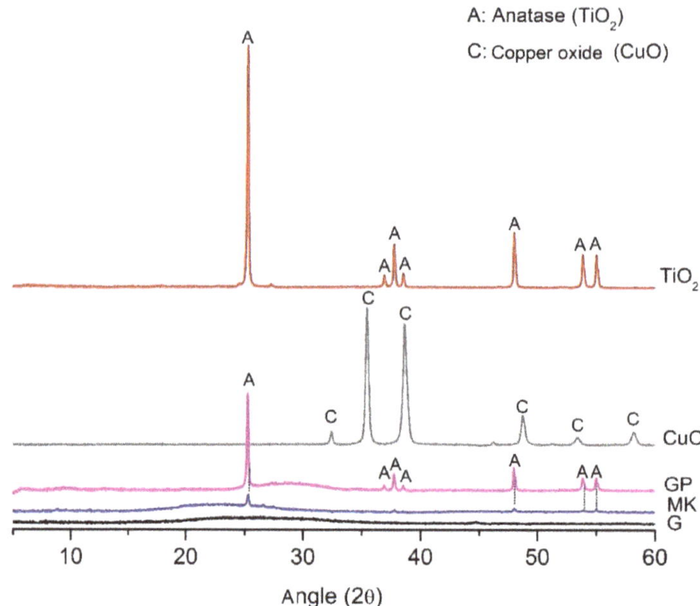

Figure 4. X-ray diffraction (XRD) patterns of the raw materials and GP.

Table 2. Chemical composition of metakaolin (MK) and glass waste (wt.%).

Oxide/Material	SiO$_2$	Al$_2$O$_3$	TiO$_2$	Fe$_2$O$_3$	Na$_2$O	MgO	K$_2$O	P$_2$O$_5$	CaO	CeO$_2$	V$_2$O$_5$	SO$_3$
MK	52.02	44.95	1.73	0.47	0.30	0.19	0.16	0.02	0.02	0.04	0.05	0.05
G	72.27	1.49	–	0.62	13.37	–	–	–	11.15	–	–	–

Titanium oxide existed in the anatase phase (TiO$_2$, Inorganic Crystal Structure Database ICSD # 154604), as identified by the peaks observed in the diffractogram (Figure 4) at approximately 2θ = 25.3°, 37.8°, 48.1°, 53.9°, and 55.1°, which also indicated the high photocatalytic potential of TiO$_2$. The XRD pattern of the copper oxide nanoparticles synthesized from copper acetate by the Pechini method showed peaks at 2θ = 34°, 36°, 38°, 49°, 53°, and 58°, which corresponded to pure monoclinic CuO (ICSD code 67850), a phase considered to have bactericidal properties [42]. Similar results were found by Román et al. [15]. The XRD pattern obtained for the GP indicated that the material was amorphous (see the lifted baseline for 2θ between 25° and 35°), with the only crystalline phase being the anatase phase, which was contributed to a high degree by the TiO$_2$ particles and to a lesser extent by MK.

SEM was used to determine the particle size of the TiO$_2$ and CuO particles (Figure 5); the average TiO$_2$ and CuO particle sizes were 0.615 µm and 0.198 µm, respectively. Figure 5a shows the elongated and irregular shape of the TiO$_2$ particles, while Figure 5b shows that the CuO particles has a rounded and more homogeneous shape.

Figure 5. SEM images of the (**a**) TiO$_2$ and (**b**) CuO particles.

The FTIR results obtained for MK, GP, and GP-G are shown in Figure 6. In the MK spectrum, a large band corresponding to asymmetric vibrations, specifically, vibrations of the Si–O–Si and O–Si–O groups, was centered at 1089 cm^{-1} [43]. The band located at 814 cm^{-1} was associated with the vibrational mode of the Al–O bond of the Al^{IV} present in MK [43,44].

When MK was activated under alkaline conditions to form a geopolymer, the main bands at wavenumbers of 1007 and 1015 cm^{-1} shifted in both the GP and GP-G mortar spectra; this band corresponded to the asymmetric vibration of the Si–OT bonds in the GP gel potassium aluminum silicate hydrates (KASH) (where T corresponds to Si or Al tetrahedrons) [45]. According to Tchakouté et al. [46], the infrared absorption bands at wavenumbers close to 1100 cm^{-1} and 950 cm^{-1} could indicate that the geopolymers contained more SiQ^3 and SiQ^2 species, respectively. Additionally, the band observed in the GP-G mortar spectrum was more intense than that observed in the GP spectrum, indicating the presence of more undissolved silica in the alkaline system [43]. This result was expected, since the mortar was designed with particles of G. The band located at approximately 703 cm^{-1} came from symmetric vibrations of the presence of Ti–O–Ti photoactive species [10].

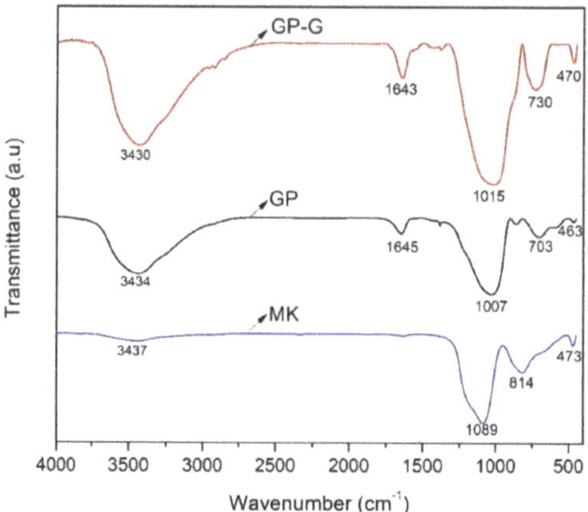

Figure 6. Fourier-transform infrared (FTIR) spectra obtained for MK, GP, and GP-G.

The peaks at approximately 3434 cm^{-1} and 1645 cm^{-1} in the MK, GP, and GP-G spectra corresponded to stress vibrations of the H–OH group, indicating that water molecules were associated with free water [38,47–49]. Finally, the peaks corresponding to the bending vibration of the Si–O–Si group [47,49] were identifiable at 473, 463, and 470 cm^{-1} for the MK, GP, and GP-G samples, respectively.

The SEM images of GP and GP-G are shown in Figure 7. Figure 7a shows that the GP had a smooth and homogeneous surface, indicating the presence of large amounts of the KASH gel. Unreacted MK that was not dissolved by the alkaline activation conditions was also identified [11,44,47]. GP showed a homogeneous distribution of the TiO$_2$ nanoparticles inside the matrix. In a previous study [11], the effect of the incorporation of TiO$_2$ in the geopolymeric matrix was investigated, and it was noteworthy that there was no significant difference between the morphologies of the systems with and without TiO$_2$, and that the formation of the KASH gel was not impeded by the addition of TiO$_2$, as these coexisted in the structure of the material. Comparing the physical properties (Table 3) of the geopolymer matrix [11], with and without TiO$_2$ addition, we found that the water absorption decreased by just 3.9%, and the density and volume of the permeable pores increased by 5.73% and 1.52%, respectively.

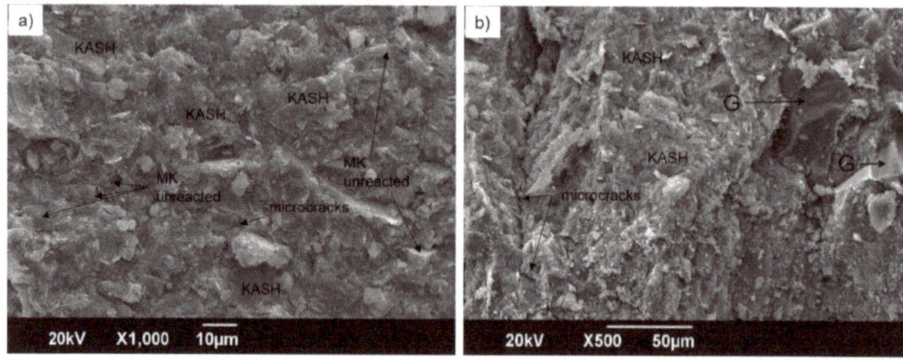

Figure 7. SEM images of the samples with 10 wt.% TiO$_2$: (**a**) GP and (**b**) GP-G mortar.

The glass particles of the GP-G mortar (Figure 7b) were identifiable by their large sizes and angular smooth surfaces [41,47], which were due to the milling processes that these particles underwent. The presence of micro-cracks in the sample was also observed with the incorporation of glass (Figure 7b); similar results were reported by Lin et al. [47] and Mejía de Gutiérrez et al. [41]. The GP-G had a more homogeneous and denser surface similar to that reported by Lin et al. [47] and a higher density (see Table 3); the material was more compact, which was reflected in its lower water absorption capacity and lower volume of permeable pores.

Table 3. Physical properties of the GP and GP-G samples.

Sample	Paste without TiO$_2$ [11]	GP (Paste) (10 wt.% TiO$_2$)	GP-G (Mortar) (10 wt.%TiO$_2$ + G)
Absorption after immersion (%)	27.25 ± 0.15	26.18 ± 0.42	11.54 ± 0.26
Dry density (kg/m^3)	1291 ± 0.25	1365 ± 0.02	1916 ± 140
Volume of permeable pores (%)	36.18 ± 0.34	36.73 ± 0.90	22.78 ± 1.97

3.2. Assays with Bacteria

Figure 8 shows the results of the first experimentation stage, corresponding to the standardization method using Petri dishes and exposure in nutrient agar with the GP and mGP. As shown in Figure 8a, for the mGP, no inhibition halo was observed for *E. coli*. Although authors such as Román et al. [15] showed the capacity of CuO nanoparticles to eliminate *Ochrobactrum anthropi*, in the present study and under the composition used (5 wt.% CuO), CuO geopolymeric compounds with antibacterial capacity for *E. coli*, *S. aureus*, and *P. aeruginosa* were not obtained.

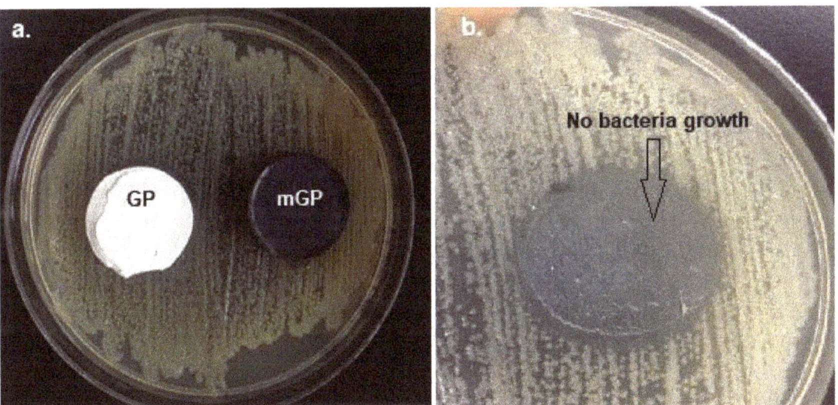

Figure 8. Standardization test for *Escherichia coli*: (**a**) GP (10 wt.% TiO$_2$) and mGP (5 wt.% CuO); (**b**) inhibition of growth of bacteria under GP sample.

Previous studies demonstrated that the quantity of nanoparticles is an important parameter. The entrapment and inhibition of bacteria on surfaces was evaluated using CuO and Ag nanoparticles [2,22,33,34,42], and a good compatibility was reported in orthodontic adhesives using CuO in high quantities [17]. The amount of CuO nanoparticles used in this preliminary study was probably low and, for this reason, it did not allow for bacterial elimination. This is according to Guo et al. [14], who evaluated different percentages of TiO$_2$ in coatings fabricated by immersion and reported that an incorporation of 5 wt.% TiO$_2$ particles failed to eliminate *E. coli*, which was attributed to the processes of initial photocatalytic inactivation step being slow with this proportion of titanium oxide.

These nanoparticles were incorporated into materials because of their high surface/volume ratio, which allowed interactions with bacterial cell membranes, preventing these microorganisms from attaching to the material surface (bacteriostatic effect); when this occurred, the material was able to eliminate these microorganisms (bactericidal effect) [13,22–24,34]. Table 4 shows the results of the initial standardization assay for GP and mGP with the three bacteria tested.

Table 4. Results of the first bactericidal test phase: standardization method.

Sample	Bacteria/Formation of Halo/Halo Size		
	Staphylococcus aureus	*Escherichia coli*	*Pseudomonas aeruginosa*
GP	No—0	Yes—0.5 mm	No—0
mGP	No—0	No—0	No—0

In the case of the geopolymer with TiO_2 nanoparticles (GP), the standardization results were satisfactory for *E. coli*, in which a bactericidal effect was in fact observed for the sample GP. The inhibition was satisfactory, and a halo did form around the sample (Figure 8a); additionally, when sample GP was removed from the agar nutrient, no bacteria were found under the sample, as shown in Figure 8b. On the contrary, for the mGP, the halo formation was not evidenced; when the mGP sample was removed from the agar nutrient, an appreciable number of colonies grew below it for the three bacteria tested.

Based on the preliminary results of the first phase of standardization, the study was continued exclusively with GP samples; additionally, a GP-G mortar was fabricated and tested. The decision to use a geopolymer mortar with glass to measure the bactericide capacity of the geopolymers was based on the results reported by Mejía de Gutiérrez et al. [41], who observed a higher photocatalytic efficiency for B-rhodamine degradation in compounds functionalized with TiO_2 when glass waste was used to prepare the geopolymer mortars. Likewise, studies by Sikora et al. [35] showed that the use of glass enhanced the photocatalytic activity for *E. coli* colony degradation, and elimination occurred in 30 min.

In the second phase, the number of CFUs was analyzed in different solutions for short times (4–5 h) and after one day in the agar with the *E. coli*, *P. aeruginosa*, and *S. aureus* bacteria.

Analyzing the GP-G mortar exposed to the different solutions for short times (Table 5) showed that, after 5 h, CFUs were not present the in the 10^{-7} mL solutions for the three bacteria evaluated. Evaluating *S. aureus* exposed to GP (Figure 9) for 4 h in all solutions showed CFU growth, while, for the GP-G in the 10^{-7} solutions, zero CFUs were evident. However, the GP exhibited significant antimicrobial activity for only the 10^{-7} solutions containing *P. aeruginosa* (Figure 9). The results reported in this study corresponding to GP and GP-G are satisfactory. The toxicity of nanoparticles for inhibiting bacterial growth was previously demonstrated for *S. aureus* using ZnO nanoparticles by Sikora [35], and for *E. coli* using CuO by Kumar [20] and TiO_2 by Sunada et al. [50]. Sunada et al. [50] explained that the photokilling reaction of *E. coli* cells using TiO_2 is initiated by a partial decomposition of the outer membrane, followed by an attack of the cytoplasmic membrane, resulting in cell death. In general, the adhesion of nanoparticles onto the cell wall and the membrane of the microorganisms produces morphological changes characterized by shrinkage of the cytoplasm and membrane detachment, finally leading to rupture of cell wall [51]. Additionally, the photocatalysis generates reactive oxygen species (ROS) and free radical species, contributing to an increase in the antibacterial potential of the nanoparticles [34,51,52].

For the agar solutions exposed to GP for longer times (Table 6), the greatest inhibition occurred in the 10^{-6} solutions for *Pseudomonas aeruginosa* (Figure 10) and *E. coli* (Figure 11). The highest efficacy for the inhibition of CFU bacterial growth was evidenced by the GP-G mortar in the 10^{-6} solutions for the three bacteria evaluated. This result confirms that the incorporation of glass wastes in geopolymeric compounds increased the photocatalytic capacity and the bactericidal effect of the MK-based geopolymer supplemented with TiO_2 and, therefore, improved the bacterial growth inhibition capacity on surfaces.

Finally, according to Tables 5 and 6, samples GP and GP-G showed a bacteriostatic effect depending on the bacteria and time.

Although TiO_2 and other nanoparticles previously attracted a lot of interest due to their antibacterial properties in different applications, according to the results, in general, samples containing glass wastes showed a higher bacteriostatic effect. The use of ground glass particles maximized the capacity for inhibiting CFU growth after 5 h for the GP-G mortars in solutions of 10^{-7} mL containing the three bacteria evaluated (*S. aureus*, *P. aeruginosa*, and *E. coli*). Furthermore, GP-G seems to be particularly effective against *P. aeruginosa*.

Table 5. Bacterial growth inhibition capacity of GP geopolymers (4 h) and GP-G (5 h) in bacteria-rich agar solutions.

Sample, Time in Agar	Dissolutions (mL)	CFU *		
		S. aureus	*P. aeruginosa*	*E. coli*
GP, 4 h	10^{-3}	++	++	++
	10^{-4}	++	++	++
	10^{-5}	++	+	+
	10^{-6}	+	+	+
	10^{-7}	+	−	+
	10^{-8}	+	−	+
GP-G, 5 h	10^{-3}	+	+	+
	10^{-4}	+	+	+
	10^{-5}	+	+	+
	10^{-6}	+	+	−
	10^{-7}	−	−	−
	10^{-8}	−	−	−

* (+), bacterial growth determined by CFUs; (++), bacterial growth determined by the formation of large colonies in all areas; (−), inhibition of bacterial growth.

Table 6. Bacterial growth inhibition capacity of GP (24 h) and GP-G (25 h) in bacteria-rich agar solutions.

Sample, Time in Agar	Dissolutions (mL)	Bacteria		
		S. aureus	*P. aeruginosa*	*E. coli*
GP, 24 h	10^{-3}	++	++	++
	10^{-4}	++	++	+
	10^{-5}	++	+	+
	10^{-6}	+	−	−
	10^{-7}	+	−	−
	10^{-8}	−	−	−
GP-G, 25 h	10^{-3}	+	+	+
	10^{-4}	+	−	+
	10^{-5}	+	−	+
	10^{-6}	−	−	−
	10^{-7}	−	−	−
	10^{-8}	−	−	−

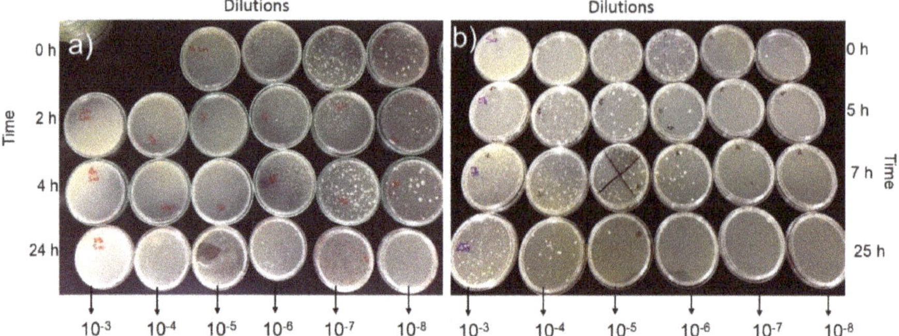

Figure 9. Reduction in the number of viable bacteria over time in the *S. aureus* agar for (**a**) GP and (**b**) GP-G.

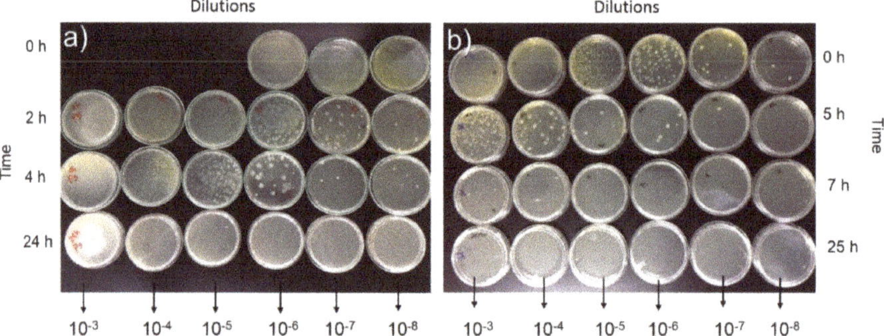

Figure 10. Reduction in the number of viable bacteria over time in the *P. aeruginosa* agar for (**a**) GP and (**b**) GP-G.

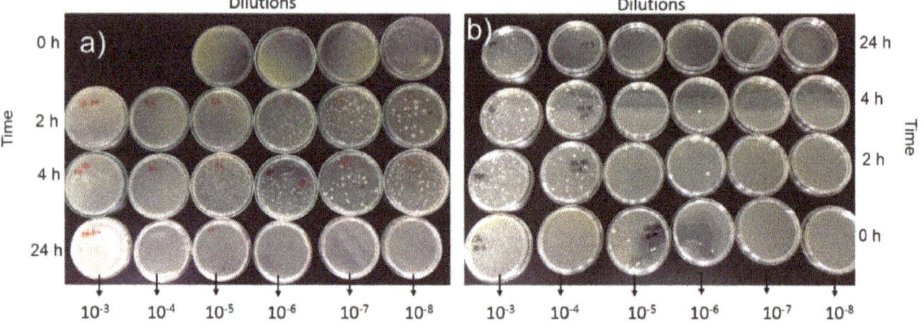

Figure 11. Reduction in the number of viable bacteria over time in the *E. coli* agar for (**a**) GP and (**b**) GP-G.

4. Conclusions

Geopolymer composites based on the alkaline activation of MK and nanoparticles of TiO_2 (10 wt.%) and CuO (5 wt.%) were produced. This study evaluated the physical and microstructural properties of the geopolymers and the functionality of these geopolymers for inhibiting the growth of the bacteria *S. aureus*, *P. aeruginosa*, and *E. coli*. Based on the results obtained, the following conclusions can be drawn:

- Here, 10 wt.% TiO_2 used for the manufacture of geopolymeric pastes and mortars showed satisfactory elimination of bacterial growth. In contrast, 5 wt.% CuO did not inhibit bacterial growth.
- The incorporation of glass wastes as fine aggregate endowed the GP-G with a 56.73% lower water absorption compared to the GP sample. Likewise, the glass particles endowed the material with a smoother and more homogeneous surface, 40.36% higher density, and 37.97% lower pore volume.
- The use of ground glass particles maximized the capacity for inhibiting CFU growth after 5 h for the GP-G mortars in solutions of 10^{-7} mL containing the three bacteria evaluated (*S. aureus*, *P. aeruginosa*, and *E. coli*).
- After 24 h in the bacterial culture, the GP-G mortars exhibited a high capacity for inhibiting the growth of *P. aeruginosa* in the 10^{-4} mL solutions and *E. coli* and *S. aureus* in the 10^{-6} mL solutions.
- These results indicate the potential for using the MK-based geopolymer composite with 10 wt.% TiO_2 for applications on surfaces with antibacterial properties.

Author Contributions: Conceptualization, R.M.-d.G.; methodology, R.M.-d.G., M.V.-C. and M.A.; formal analysis, S.R.-B., M.A. and D.M.; investigation, R.M.-d.G., M.V.-C. and M.A.; resources, project administration and funding acquisition, R.M.-d.G.; writing—original draft preparation, M.V.-C. and S.R.-B.; writing—review and editing, R.M.-d.G. and M.V.-C. All authors have read and agreed to the published version of the manuscript.

Funding: This research was funded by Universidad del Valle (Cali, Colombia), grant number 139-2017.

Conflicts of Interest: The authors declare no conflict of interest.

References

1. Ang, T.; He, Y.; Xiangxin, X.; Yong, L. Antibacterial Ceramic Fabricated by the Ti-Bearing Blast Furnace Slag. In Proceedings of the 8th Pacific Rim International Congress on Advanced Materials and Processing, Hilton Waikoloa Village in Waikoloa, HI, USA, 4–9 August 2013; Marquis, F., Ed.; Springer: Cham, Switzerland, 2016; pp. 1627–1634.
2. Baheiraei, N.; Moztarzadeh, F.; Hedayati, M. Preparation and antibacterial activity of Ag/SiO_2 thin film on glazed ceramic tiles by sol-gel method. *Ceram. Int.* **2012**, *38*, 2921–2925. [CrossRef]
3. Wang, L.; Hu, C.; Shao, L. The antimicrobial activity of nanoparticles: Present situation and prospects for the future. *Int. J. Nanomed.* **2017**, *12*, 1227–1249. [CrossRef] [PubMed]
4. Salgado, C.D.; Farr, B.M.; Calfee, D.P. Community-Acquired Methicillin-Resistant Staphylococcus aureus: A Meta-Analysis of Prevalence and Risk Factors. *Clin. Infect. Dis.* **2003**, *36*, 131–139. [CrossRef]
5. Gelatti, L.C.; Bonamigo, R.R.; Matsiko-Inoue, F.M.; Do-Carmo, S.; Becker, A.P.; Da Silva-Castrucci, F.M.; Campos Pignatari, A.C.; Dazevedo, P.A. Community-acquired methicillin-resistant staphylococcus aureus carrying sccmec type IV in southern Brazil. *Rev. Soc. Bras. Med. Trop.* **2013**, *46*, 34–38. [CrossRef]
6. Drugă, B.; Ukrainczyk, N.; Weise, K.; Koenders, E.; Lackner, S. Interaction between wastewater microorganisms and geopolymer or cementitious materials: Biofilm characterization and deterioration characteristics of mortars. *Int. Biodeterior. Biodegrad.* **2018**, *134*, 58–67. [CrossRef]
7. Crocker, J. Antibacterial Materials. *Mater. Sci. Technol.* **2007**, *22*, 238–241. [CrossRef]
8. Abdel-Gawwad, H.A.; Mohamed, S.A.; Mohammed, M.S. Recycling of slag and lead-bearing sludge in the cleaner production of alkali activated cement with high performance and microbial resistivity. *J. Clean. Prod.* **2019**, *220*, 568–580. [CrossRef]
9. Hasmaliza, M.; Foo, H.S.; Mohd, K. Anatase as Antibacterial Material in Ceramic Tiles. *Procedia Chem.* **2016**, *19*, 828–834. [CrossRef]
10. Dahiya, M.S.; Tomer, V.K.; Duhan, S. Bioactive glass/glass ceramics for dental applications. In *Applications of Nanocomposite Materials in Dentistry*; Abdullah, M., Mohammad, I., Mohammad, A., Eds.; Woodhead Publishing Series in Biomaterials: Cambridge, UK, 2018; pp. 1–25.
11. Guzmán-Aponte, L.; Mejía de Gutiérrez, R.; Maury-Ramírez, A. Metakaolin-Based Geopolymer with Added TiO_2 particles: Physicomechanical Characteristics. *Coatings* **2017**, *7*, 233. [CrossRef]
12. Mohd Adnan, M.A.; Muhd Julkapli, N.; Amir, M.N.I.; Maamor, A. Effect on different TiO_2 photocatalyst supports on photodecolorization of synthetic dyes: A review. *Int. J. Environ. Sci. Technol.* **2019**, *16*, 547–566. [CrossRef]

13. Rai, M.; Yadav, A.; Gade, A. Silver nanoparticles as a new generation of antimicrobial. *Biotechnol. Adv.* **2009**, *27*, 76–83. [CrossRef] [PubMed]

14. Ming-Zhi, G.; Tung-Chai, L.; Chi-Sun, P. Nano-TiO_2-based architectural mortar for NO removal and bacteria inactivation: Influence of coating and weathering conditions. *Cem. Concr. Compos.* **2013**, *36*, 101–108.

15. Román, L.E.; Castro, F.; Maúrtua, D.; Condori, C.; Vivas, D.; Bianclii, A.E.; Paraguay-Delgado, F.; Solis, J.L.; Gómez, M.M. CuO nanoparticles and their antimicrobial activity against nosocomial strains. *Rev. Colomb. Quim.* **2017**, *46*, 28–36.

16. Sikora, P.; Augustyniak, A.; Cendrowski, K.; Nawrotek, P.; Mijowska, E. Antimicrobial activity of Al_2O_3, CuO, Fe_3O_4, and ZnO nanoparticles in scope of their further application in cement-based building materials. *Nanomaterials* **2018**, *8*, 212. [CrossRef] [PubMed]

17. Toodehzaeim, M.H.; Zandi, H.; Meshkani, H.; Hosseinzadeh Firouzabadi, A. The Effect of CuO Nanoparticles on Antimicrobial Effects and Shear Bond Strength of Orthodontic Adhesives. *J. Dent.* **2018**, *19*, 1–5.

18. Yang, S.; Zhang, Y.; Yu, J.; Zhen, Z.; Huang, T.; Tang, Q.; Chu, P.K.; Qi, L.; Lv, H. Antibacterial and mechanical properties of honeycomb ceramic materials incorporated with silver and zinc. *Mater. Des.* **2014**, *59*, 461–465. [CrossRef]

19. Haider, A.J.; Anbari, R.; Kadhim, G.R.; Salame, C.T. Exploring potential environmental applications of TiO_2 nanoparticles. *Energy Procedia* **2017**, *119*, 332–345. [CrossRef]

20. Kumar, A.M.; Khan, A.; Suleiman, R.; Qamar, M.; Saravanan, S.; Dafalla, H. Bifunctional CuO/TiO_2 nanocomposite as nanofiller for improved corrosion resistance and antibacterial protection. *Prog. Org. Coat.* **2018**, *114*, 9–18. [CrossRef]

21. Mersian, H.; Alizadeh, M.; Hadi, N. Synthesis of zirconium doped copper oxide (CuO) nanoparticles by the Pechini route and investigation of their structural and antibacterial properties. *Ceram. Int.* **2018**, *44*, 20399–20408. [CrossRef]

22. Armayani, M.; Pratama, M.A. The Properties of Nano Silver (Ag)-Geopolymer as Antibacterial Composite for Functional Surface Materials. *Matec. Web Conf.* **2017**, *97*, 01010. [CrossRef]

23. Elbourne, A.; Crawford, R.J.; Ivanova, E.P. Nano-structured antimicrobial surfaces: From nature to synthetic analogues. *J. Colloid Interface Sci.* **2017**, *508*, 603–616. [CrossRef] [PubMed]

24. Dizaj, S.M.; Lotfipour, F.; Barzegar-Jalali, M.; Zarrintan, M.H.; Adibkia, K. Antimicrobial activity of the metals and metal oxide nanoparticles. *Mater. Sci. Eng. C* **2014**, *44*, 278–284. [CrossRef] [PubMed]

25. Åsbrink, S.; Norrby, L.J. A refinement of the crystal structure of copper (II) oxide with a discussion of some exceptional. *Acta. Crystallogr. Sect. B* **1970**, *26*, 8–15. [CrossRef]

26. Mejía, J.M.; Mendoza, J.D.; Yucuma, J.; Mejía de Gutiérrez, R.; Mejía, D.E.; Astudillo, M. Mechanical, in-vitro biological and antimicrobial characterization as an evaluation protocol of a ceramic material based on alkaline activated metakaolin. *Appl. Clay Sci.* **2019**, *178*, 105141. [CrossRef]

27. Falah, M.; MacKenzie, K.J.D. Synthesis and properties of novel photoactive composites of P25 titanium dioxide and copper (I) oxide with inorganic polymers. *Ceram. Int.* **2015**, *41*, 13702–13708. [CrossRef]

28. Syamsidar, D.; Nurfadilla, S. The Properties of Nano TiO_2-Geopolymer Composite as a Material for Functional Surface Application. *MATEC Web Conf.* **2017**, *97*, 01013. [CrossRef]

29. Rodrigues Magalhaes, A.P.; Laura, B.S.; Gonzaga Lopes, L.; Rodrigues de Araujo, E.; Estrela, C.; Torres, E.M.; Figueiroa Bakuzis, A.; Carvalho Cardoso, P.; Santos Carriao, M. Nanosilver Application in Dental Cements. *ISRN Nanotechnol.* **2012**, *2012*, 6. [CrossRef]

30. Wu, Y.; Lu, B.; Bai, T.; Wang, H.; Du, F.; Zhang, Y.; Cai, L.; Jiang, C.; Wang, W. Review. Geopolymer, green alkali activated cementitious material: Synthesis, applications and challenges. *Constr. Build. Mater.* **2019**, *224*, 930–949. [CrossRef]

31. Mondragón-Figueroa, M.; Guzmán-Carrillo, H.; Rico, M.A.; Reyez-Araiza, J.L.; Pineda-Piñón, J.; López-Naranjo, E.J.; Columba-Palomares, M.C.; López-Romero, J.M.; Gasca-Tirado, J.R.; Manzano-Ramírez, A. Development of a Construction Material for Indoor and Outdoor, Metakaolinite-Based Geopolymer, with Environmental Properties. *J. Mater. Sci. Eng. A* **2019**, *9*, 131–142.

32. Nur, Q.A.; Sari, N.U.; Harianti, S. Development of Geopolymers Composite Based on Metakaolin-Nano ZnO for Antibacterial Application. *IOP Conf. Ser. Mater. Sci. Eng.* **2017**, *180*, 012289. [CrossRef]

33. Dela Cerna, K.; Janairo, J.I.; Promentilla, M.A. Development of nanosilver-coated geopolymer beads (AgGP) from fly ash and baluko shells for antimicrobial applications. *MATEC Web Conf.* **2019**, *268*, 05003. [CrossRef]

34. Dibyendu, A.; Manas, S.; Moumita, M.; Abiral, T.; Saroj, M.; Brajadulal, C.H. Anti-microbial efficiency of nano silver–silica modified geopolymer mortar for eco-friendly green construction technology. *RSC Adv.* **2015**, *5*, 64037.

35. Sikora, P.; Augustyniak, A.; Cendrowski, K.; Horszczaruk, E.; Rucinska, T.; Nawrotek, P.; Mijowska, E. Characterization of mechanical and bactericidal properties of cement mortars containing waste glass aggregate and nanomaterials. *Materials (Basel)* **2016**, *9*, 701. [CrossRef] [PubMed]

36. Centers for Disease Control & Prevention. Four Pediatric Deaths from Community-Acquired Methicillin-Resistant Staphylococcus aureus—Minnesota and North Dakota 1997–1999. *Morb. Mortal. Wkly. Rep.* **1999**, *48*, 707–710.

37. Van Elsas, J.D.; Semenov, A.V.; Costa, R.; Trevors, J.T. Survival of Escherichia coli in the environment: Fundamental and public health aspects. *ISME J.* **2011**, *5*, 173–183. [CrossRef]

38. Villaquirán-Caicedo, M.A. Studying different silica sources for preparation of alternative waterglass used in preparation of binary geopolymer binders from metakaolin/boiler slag. *Constr. Build. Mater.* **2019**, *227*, 116621. [CrossRef]

39. Villaquirán-Caicedo, M.A.; Mejía-de Gutiérrez, R. Mechanical and microstructural analysis of geopolymer composites based on metakaolin and recycled silica. *J. Am. Ceram. Soc.* **2019**, *102*, 3653–3662. [CrossRef]

40. Sunde, T.O.L.; Grande, T.; Einarsrud, M.A. Modified Pechini Synthesis of Oxide Powders and Thin Films. In *Handbook of Sol-Gel: Science and Technology*; Klein, L., Aparicio, M., Jitianu, A., Eds.; Springer: Cham, Switzerland, 2016; pp. 1–30.

41. Mejía de Gutiérrez, R.; Villaquirán-Caicedo, M.A.; Guzmán-Aponte, L. Alkali-activated metakaolin mortars using glass waste as fine aggregate: Mechanical and photocatalytic properties. *Constr. Build. Mater.* **2020**, *235*, 117510. [CrossRef]

42. Akl, A.; Nida, M. Antibacterial Activity of synthesized Copper Oxide Nanoparticles using Malva sylvestris Leaf Extract. *SMU Med. J.* **2015**, *2*, 91–101.

43. Bernal, S.A.; Rodríguez, E.D.; Mejía de Gutierrez, R.; Provis, J.L.; Delvasto, S. Activation of Metakaolin/Slag Blends Using Alkaline Solutions Based on Chemically Modified Silica Fume and Rice Husk Ash. *Waste Biomass Valoriz.* **2011**, *3*, 99–108. [CrossRef]

44. Villaquiran-Caicedo, M.A.; Mejía de Gutiérrez, R. Synthesis of ternary geopolymers based on metakaolin, boiler slag and rice husk ash. *Dyna* **2015**, *82*, 104–110. [CrossRef]

45. Arellano-Aguilar, R.; Burciaga-Díaz, O.; Gorokhovsky, A.; Escalante-García, J.I. Geopolymer mortars based on a low grade metakaolin: Effects of the chemical composition, temperature and aggregate:binder ratio. *Constr. Build. Mater.* **2014**, *50*, 642–648. [CrossRef]

46. Tchakouté, H.K.; Rüscher, C.H.; Kong, S.; Kamseu, E.; Leonelli, C. Geopolymer binders from metakaolin using sodium waterglass from waste glass and rice husk ash as alternative activators: A comparative study. *Constr. Build. Mater.* **2016**, *114*, 276–289. [CrossRef]

47. Lin, K.L.; Shiu, H.S.; Shie, J.L.; Cheng, T.W.; Hwang, C.L. Effect of composition on characteristics of thin film transistor liquid crystal display (TFT-LCD) waste glass-metakaolin-based geopolymers. *Constr. Build. Mater.* **2012**, *36*, 501–507. [CrossRef]

48. Elimbi, A.; Tchakouté, H.K.; Njopwouo, D. Effects of calcination temperature of kaolinite clays on the properties of geopolymer cements. *Build. Mater.* **2011**, *25*, 2805–2812. [CrossRef]

49. Covarrubias, C.; García, R.; Arriagada, R.; Yánez, J.; Garland, M.T. Cr (III) exchange on zeolites obtained from kaolin and natural mordenite. *Microporous Mesoporous Mater.* **2006**, *88*, 220–231. [CrossRef]

50. Sunada, K.; Watanabe, T.; Hashimoto, K. Studies on photokilling of bacteria on TiO_2 thin film. *J. Photochem. Photobiol. A: Chem.* **2003**, *156*, 227–233. [CrossRef]

51. Dakal, T.C.; Kumar, A.; Majumdar, R.S.; Yadav, V. Mechanistic basis of antimicrobial actions of silver nanoparticles. *Front. Microbiol.* **2016**, *7*, 1–17. [CrossRef]

52. Marugan, J.; Grieken, R.V.; Pablos, C.; Sordo, C. Analogies and differences between photocatalytic oxidation of chemicals and photocatalytic inactivation of microorganisms. *Water Res.* **2010**, *44*, 789–796. [CrossRef]

Article

Fire Protection Performance and Thermal Behavior of Thin Film Intumescent Coating

Jing Han Beh [1], Ming Chian Yew [2,*], Ming Kun Yew [3] and Lip Huat Saw [2]

[1] Department of Architecture and Sustainable Design, Lee Kong Chian Faculty of Engineering Science, University of Tunku Abdul Rahman, Cheras, 43000 Kajang, Malaysia

[2] Department of Mechanical and Material Engineering, Lee Kong Chian Faculty of Engineering and Science, University of Tunku Abdul Rahman, Cheras, 43000 Kajang, Malaysia

[3] Department of Civil Engineering, Lee Kong Chian Faculty of Engineering and Science, University of Tunku Abdul Rahman, Cheras, 43000 Kajang, Malaysia

* Correspondence: yewmc@utar.edu.my or yewmingchian@gmail.com; Tel.: +603-9086-0288

Received: 27 June 2019; Accepted: 18 July 2019; Published: 31 July 2019

Abstract: This paper presents the heat release characteristics, char formation and fire protection performance of thin-film intumescent coatings that integrate eggshell (ES) as an innovative and renewable flame-retardant bio-filler. A cone calorimeter was used to determine the thermal behavior of the samples in the condensed phase in line with the ISO 5660-1 standard. The fire resistance of the coatings was evaluated using a Bunsen burner test to examine the equilibrium temperature and formation of the char layer. The fire propagation test was also conducted according to BS 476: Part 6. On exposure, the samples X, Y, and Z were qualified to be Class 0 materials due to the indexes of fire propagation being below 12. Samples Y and Z reinforced with 3.50 wt.% and 2.50 wt.% of ES bio-filler, respectively, showed a significant improvement in reducing the heat release rate, providing a more uniform and thicker char layer. As a result, the addition of bio-filler content has proven to be efficient in stopping the fire propagation as well as reducing the total heat released and equilibrium temperature of the intumescent coatings.

Keywords: acrylic resin; bio-filler; cone calorimeter; heat release rate; intumescent coating; steel

---

1. Introduction

According to world fire statistics, there are more than 322,000 people per year suffering in fire accidents due to the ineffectiveness of fire protection systems. Hence it is important to implement an effective fire protection system in every building to protect the occupants and building whenever there is a fire outbreak. Intumescent fire protective materials perform as a passive fire protection system and play a crucial role in fulfilling the fire safety building regulations to effectively stop the advance of a fire. Steel structure starts to lose its mechanical properties (temperature >500 °C) and tends to buckle, leading to failure of the building structures. Indeed, the fire safety rules and regulations in buildings are of paramount importance to ensure the evacuation time and safety for occupants [1].

Intumescent coatings are mainly designed to reduce the heat and fire propagation on the substrate. Interior decorative materials in buildings are mostly combustible products that are a serious hazard in a fire. In general, most of the flame-retardant materials were developed due to the smoke and toxic gas produced from thermal decomposition of the commonly deployed brominated flame-retardant products [2]. Fires are unavoidably used to produce a lot of energy and heat, which might lead to serious injury or death [3–5].

The applications of fire protective coatings are one of the most effective ways to protect different substrates toward a fire. The expansion process of intumescent fire protective coating is due to the physical and chemical interactions of three main flame-retardant additives: (1) Ammonium polyphosphate (APP) acts as an acid source, (2) pentaerythritol (PER) acts as a carbon source and (3) melamine (MEL) acts as an expanding agent. The use of flame-retardant ingredients may prevent a small flame towards a major catastrophe. The polymer binder turns out to be important due to two extensive properties: It contributes to the char layer growth and controls the development of even char foam structure [6,7]. Several advantages of using intumescent fire protective coatings over other approaches of structural fire protection are the artistically attractive appearance it gives to the substrates, fast application, easy to cover complex details and maintaining the intrinsic properties of steel structures [8,9].

This research highlights a renewable chicken eggshell (ES) flame-retardant bio-filler and its important role in industrial coatings. ES waste is an aviculture by-product, which causes a serious conservation risk due to its disposal constitutes. ES waste comprises about 5% organic materials and 95% calcium carbonate in calcite form [10,11]. ES waste can create new value by being converted into profitable products. Its biochemical composition and accessibility make ES a latent source for renewable flame-retardant bio-filler, which improved mechanical and thermal properties of coatings and bio-polymer composites [12–21]. ES also offers benefits for various industrial applications, as it is lightweight, inexpensive, environmentally friendly, has high thermal stability and is available in bulk quantities [22–31]. In this research work, the performance of a steel plate coated with an intumescent coating was tested by using a Bunsen burner. In addition, the fire propagation and fire behavior of intumescent coatings were tested according to BS 476: Part 6 and ISO 5660-1 [32,33], respectively. These thin-film intumescent coatings were evaluated with respect to fire behavior analysis with thermal characteristics in a cone calorimeter and the fire-resistive performance.

2. Experimental

2.1. Materials

In this experimental work, the three main components of materials used for the preparation of intumescent coatings were (1) the halogen-free flame retardant additives: Ammonium polyphosphate (particle size <15 μm), melamine (particle size <40 μm), pentaerythritol (particle size <40 μm), (2) the flame retardant fillers: Aluminum hydroxide with a specific surface area in a range between 0.5 to 50 m^2/g, magnesium hydroxide (specific surface area <8 m^2/g), titanium dioxide (particle size <40 μm, specific surface area = 150 m^2/g) and eggshell bio-filler (mean particle size = 22.99 μm and specific surface area = 148.41 m^2/g) [11] and (3) the polymer binder: Acrylic resin, which has slow-burning or even self-extinguishing behavior when exposed to fire. Moreover, it does not generate harmful smoke or gases. The eggshell powder preparation is shown in Figure 1.

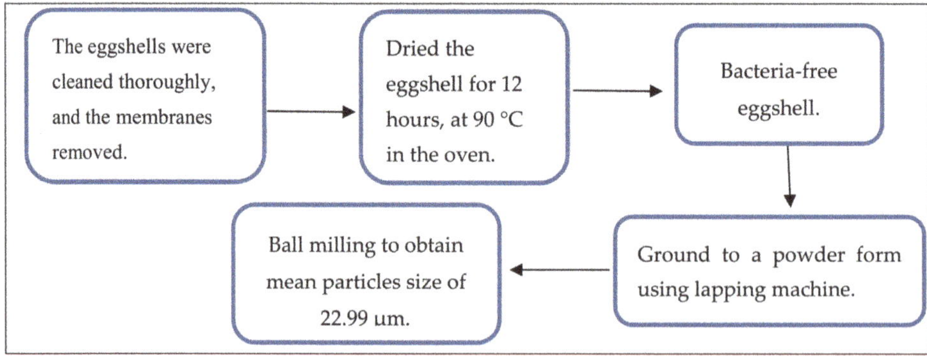

Figure 1. Flow chart of chicken eggshell (ES) powder preparation.

Four kinds of thin-film intumescent coatings W, X, Y and Z were prepared and tested to evaluate the fire protection properties and combustion performances of the coatings. The sample details are tabulated in Table 1.

Table 1. Specifications of experimental materials.

Ingredients	Parts by Weight for Formulations			
	W	X	Y	Z
APP	20	20	20	20
MEL	10	10	10	10
PER	10	10	10	10
TiO_2	5.00	3.40	3.20	3.00
$Al(OH)_3$	–	2.80	3.30	2.75
$Mg(OH)_2$	5.00	3.80	–	1.75
ES	–	–	3.50	2.50
Polymer Binder	**W**	**X**	**Y**	**Z**
Acrylic resin	50	50	50	50
Weight * (g)	26.70	24.80	25.80	25.90
Thickness (mm)	1.50	1.50	1.50	1.50
Density (g/cm^3)	1.780	1.653	1.720	1.727

* Sample sized for cone calorimeter test = 10 cm (*l*) × 10 cm (*w*) × 1.5 mm (*t*)

2.2. Fire Protective Test

The fire protective test allowed the observation of the development of the char layer and the evolution of temperature when exposed to fire to determine the performance of the intumescent fire protective coatings. The intumescent formulation was applied onto a steel plate after being grit-blasted (dimensions of steel plate: 100 × 100 × 2.6-mm^3) by using a gun sprayer. This method was repeated 3–5 times until a 2.0 ± 0.2 mm coating thickness was attained. The Bunsen burner of the gas tank consumption was about 160 g/h, and the coated steel plate with the intumescent coating was mounted vertically and tested for 60 min of fire (about 1000 °C). In this fire test, 400 °C was chosen as a critical temperature for the coated steel under the small-scale fire test. The time-temperature curves of the coated steel plates were recorded and verified using a model of 307/308 hand-held mini thermometer. The temperature profile and thickness of the char layer on the backside of the steel plates were recorded and evaluated (see Figure 2).

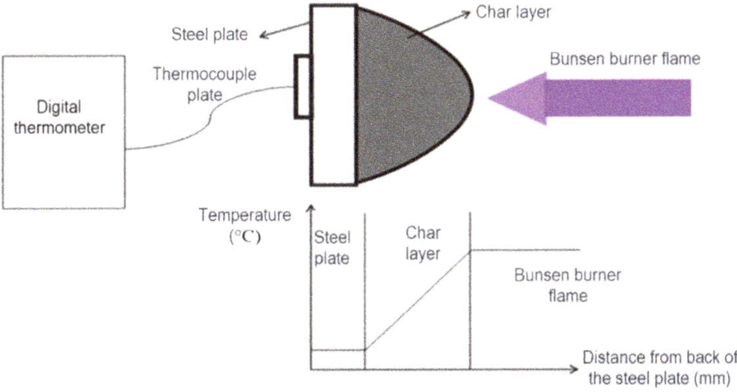

Figure 2. Schematic of the experimental setup for fire protective test [31]. Figure 2 is adapted with permission from [31]. Copyright 2018 Elsevier.

2.3. BS 476: Part 6 Fire Test

The fire propagation test according to BS 476: Part 6 was conducted and evaluated in this experimental work [32]. The experimental method consisted of exposing the coated steel with an intumescent formulation to a standard small flame for 20 min, with 2 kW of an extra irradiance from the third to the final minute of the fire propagation test. The temperature of the grown ignition intumescent coatings was recorded. This was compared to the temperatures produced from the steel plate coated with the intumescent coating. The result stated was the index of fire propagation, which offers a relative measure of the involvement to the evolution of the fire made by the flat coating surface. The coated steel plate with the intumescent fire protective coating was exposed to the fire conditions. To be Class 0 certified the fire propagation index (I) must be below 12. In this fire test, a better fire protective material was determined by a lower numerical value of the index.

In addition, the heat rate and amount of heat grown by the coating sample were measured and considered under prearranged conditions. The steel plates with a standard dimension: $225 \times 225 \times 2.3$-mm^3 were coated with a thickness of 1.5 ± 0.1 mm intumescent coating. Moreover, the index of fire propagation performance was calculated using the below equations:

$$I_1 = \sum_{t=0.5}^{t=3} \frac{\theta_m - \theta_c}{10t} \tag{1}$$

$$I_2 = \sum_{t=4}^{t=10} \frac{\theta_m - \theta_c}{10t} \tag{2}$$

$$I_3 = \sum_{t=12}^{t=20} \frac{\theta_m - \theta_c}{10t} \tag{3}$$

$$I = I_1 + I_2 + I_3 \tag{4}$$

Where:

I = index of intumescent coating performance;

t = time from the beginning at which readings were taken (min);

θ_m = temperature of the intumescent coating at time t;

θ_c = temperature of the calibration curve at time t.

2.4. Sample Preparation for the Cone Calorimeter Test

Before the cone calorimeter test, sample sizes (10 cm $\times$ 10 cm $\times$ 1.5 mm) were maintained at $50 \pm 5\%$ relative humidity (RH) and 23 ± 2 °C. The pretreated coating samples were enveloped with aluminum foil (thickness: 0.03–0.05 mm) with the shiny side of the aluminum foil facing the sample. The coating sample was wrapped without any treatment and the non-exposed surface was covered with foil, which typically forms to the cone calorimeter. One of the most acceptable and internationally recognized fire testing apparatuses is the cone calorimeter. This equipment test was conducted in accordance with the ISO 5660-1 standard. The cone calorimeter is used to examine the fire characteristics of the sample with different measurements simultaneously. The most important parameter to determine a fire's hazard level is to obtain the value of the heat release rate (HRR) of the sample. The specifications of the samples are shown in Table 1.

In this cone calorimeter test, the prepared coating sample and the holder were located on a mass measurement device. All the experiment works were evaluated by employing the coating samples in the same holder under a heater of the cone calorimeter. The fire situation comprised four stages: (1) Ignition, (2) growth, (3) fully developed, and (4) decay. The heat flux of 50 kW/m^2 was set and conducted by corresponding to the fully developed fire stage. The distance between the cone and the coating sample was 6 cm. The spark power and the igniter were removed when the ignition or temporary flame occurred and the time was recorded. If the flame went out after removing the spark power, the igniter was re-inserted within 5 s and then the spark was maintained until test completion. The coating sample and sample holder were detached after collection of all data. Each pretreated coating samples were tested three times according to the standard. The experimental data of the coating sample were calculated based on the average of three tests [33].

Flammability Test

For this flammability test, a heat flux of 50 kW/m² was irradiated to the square samples and measured in the horizontal position. The following parameters were determined during the test: The heat release rate (HRR), peak HRR, time to ignition (TTI), and total weight loss. The time to fire start on the coating surface due to heat radiation is known as the TTI. The following equations were used to calculate the HRR:

$$\dot{q}'' \, (t) = \frac{q(t)}{A_s} \tag{5}$$

$$\dot{q}(t) = \left(\frac{\Delta h_c}{r_0}\right)(1.10) \, C \, \sqrt{\frac{\Delta P}{T_e}} \, \frac{\left(X_{O_2}^o - X_{O_2}(t)\right)}{1.105 - 1.5 X_{O_2}(t)} \tag{6}$$

where $\dot{q}''$ is the rate of heat release per unit area (kW/m²), $\dot{q}$ is the HRR (kW), A_s. is the initially exposed area (m²), Δh_c is the net heat of combustion (kJ/kg), and r_0 is the stoichiometric oxygen/fuel mass ratio (-). The maximum intensity of an HRR curve is determined by peak of heat released rate (PHRR).

3. Results and Discussion

3.1. Bunsen Burner

The temperature profiles and the development of char layers of the steel plates coated with intumescent coating formulations were recorded and compared. The data of time-temperature curves of the coatings are presented in Figure 3. Samples W, X, Y, and Z showed comparable temperature profiles after the test. During the first 10 min of fire, there was no difference in the temperature of all coating samples, and the temperature increased rapidly to 181, 170, 163 and 159 °C for samples W, X, Z, and Y, respectively. After 15 min of fire, the equilibrium temperatures were reached for all coatings and remained almost unchanged until 60 min of fire. The small-scale fire test results demonstrated that the equilibrium temperatures of curves W, X, Y, and Z were 173, 168, 155 and 160 °C, respectively.

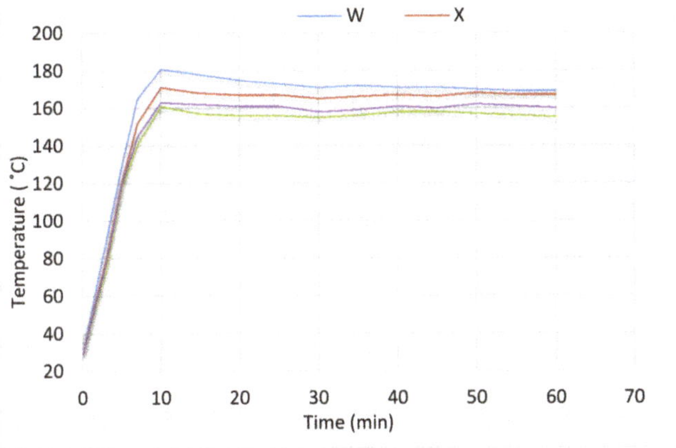

Figure 3. The time-temperature curves of the coated steel plates with coating samples.

Moreover, Figure 4 exhibited that the thicknesses and expansion rates for samples Y and Z containing ES bio-filler were 39.50 mm-0.625 mm/min and 37.50 mm-0.59 mm/min, respectively. Sample Y had the best fire protection performance in terms of its equilibrium temperature and char formation compared to samples W, X, and Z. The growth of a multicellular char layer of sample Y was mainly attributed to the decarbonation of 3.5 wt.% of ES. It formed calcium oxide by releasing non-combustible carbon dioxide gas on heating, as follows:

$$CaCO_3 \text{ (s)} \rightarrow CaO \text{ (s)} + CO_2 \text{ (g)} \tag{7}$$

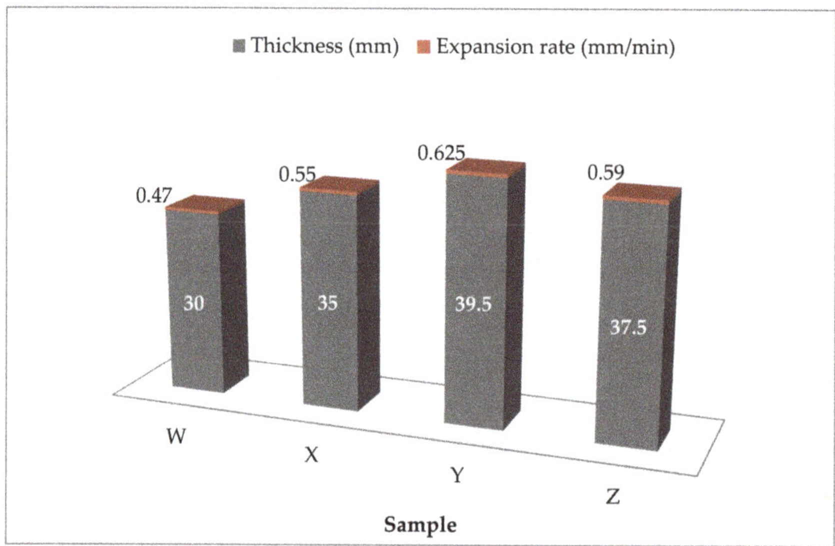

Figure 4. The expansion rate and thickness of char layer of intumescent coating samples.

In addition, the expansion of the char layer can be initiated due to physical and chemical reactions contributed by an appropriate mixture of flame-retardant materials and binder, or the development of a cohesive structure during heating. This dense char layer could trap the degradation ingredients into the residue and result in a rounded swelling. This protecting layer declines the heat transfer from the heat source to the underlying steel in maintaining the integrity of the protected substrate against fire. The outcomes demonstrated that the coating comprising phosphate, nitrogen, ES, TiO_2, $Mg(OH)_2$, $Al(OH)_3$ containing fire-retardant elements significantly contributed to a better fire protection performance, which resulted from the formation of a uniform and dense char layer. It was found that there was a correlation between the thickness of the char layer and the equilibrium temperature. This shows that the thickness of the char layer affected the fire protection performance of the coating.

3.2. BS 476: Part 6

The BS 476: Part 6 fire test found that all samples fulfilled the requirement, except the sample W ((I) = 22.3). The index and sub-index of the fire propagation test for all coating samples are presented in Table 2.

Table 2. The index and sub-index of BS 476: Part 6, fire propagation test.

Time (min)	Calibration, Temperature (°C)	Coating W (°C)	Coating X (°C)	Coating Y (°C)	Coating Z (°C)
0.5	14	18	14	12	11
1	18	21	18	16	15
1.5	23	26	23	19	19
2	27	30	27	23	22
2.5	30	34	31	26	26
3	34	38	34	30	29
4	72	122	55	54	60
5	108	212	133	129	150
6	129	274	169	181	179
7	148	321	213	202	208
8	166	364	227	219	226
9	182	378	272	234	236
10	192	405	290	244	241
12	214	417	295	249	246
14	230	418	302	253	250
16	238	416	304	258	253
18	246	403	306	260	253
20	257	385	308	263	259
Sub index 1 (I_1)		1.6	0.2	0.1	0
Sub index 2 (I_2)		15	5.1	4.6	4.4
Sub index 3 (I_3)		5.7	1.5	1.2	0.7
Index of Performance (I)		22.3	6.8	5.9	5.1

The BS 476: Part 6 test results showed that the sub-index (I_1:I_2:I_3) of coating samples W, X, Y and Z was (1.6:15:5.7), (0.2:5.1:1.5), (0.1:4.6:1.2) and (0:4.4:0.7), respectively. The index (I) results for the same coating samples were 22.3, 6.8, 5.9 and 5.1, respectively. It emphasized that the sub-index must be below 6 and the index of fire propagation must be below 12 for the coating samples to be certified as Class 0 materials. Among all coating samples, only sample W did not qualify as a Class 0 material since its index was 22.3 (I), which is out of the index of performance of this category ($I > 12$).

Evaluation of the fire propagation index for samples W, X, Y and Z revealed that samples Y ((I) = 5.9) and Z ((I) = 5.1) with 3.5 wt.% and 2.5 wt.% of ES, respectively, showed a great reduction in fire propagation index compared to sample W. It can be concluded that the incorporation of ES bio-filler into the coating formulation led to substantial inhibition of fire propagation, which could be contributed to the decarbonation of calcium carbonate at a high decomposition temperature [28].

In addition, sample X ((I) = 6.8) also exhibited a significant improvement in the reduction of fire propagation compared to sample W. This coating formulation showed an appropriate combination of TiO_2/$Al(OH)_3$/$Mg(OH)_2$ flame retardant fillers with flame retardant ingredients led to a significant improvement in stopping the fire propagation behavior. This phenomenon is due to the main phosphorus element of ammonium polyphosphate (APP), which could easily respond with different oxides during a fire to produce ceramic-like solid materials (X-O-P species, X = Ti, B, Al, Mg, etc.). This develops a more cohesive and dense char structure [1,34,35]. The properties of the char structure are associated with fire protection performance of the sample [36,37].

3.3. Cone Calorimeter Test

The results of the cone calorimeter test using 50 kW/m^2 heat fluxes are shown in Table 3. The overall burning time of all intumescent coating samples was about 700 to over 900 s, and the TTI was 8–10 s.

Table 3. Data of the cone calorimeter test of samples.

Sample	Peak of Heat Released Rate (kW/m^2)	Total Heat Released (MJ/m^2)	Thickness of Char Layer (mm)	Time to Ignition (s)	Residual Weight (wt.%)
W	106.03	22.4	21.0	9	43.85
X	111.86	21.6	30.0	8	46.12
Y	91.00	11.5	35.5	10	61.81
Z	99.98	12.0	34.0	10	58.48

According to cone measurements, the TTI values of samples W, X, Y, and Z were 9, 8, 10 and 10, respectively. The TTI of the high-density samples Y and Z, which contained ES bio-filler, had a longer time than those of the lower density sample X, demonstrating that the main factors were the density and decomposition temperature of the flame-retardant fillers [38]. In addition, the remaining mass of coating samples W, X, Y, and Z were 43.85%, 46.12%, 61.81% and 58.48%, respectively, after the test.

Therefore, samples Y and Z incorporated with ES were difficult to ignite and contribution to the TTI value and residual weight compared to samples W and X, due to its higher decomposition temperature. It is important to examine the profile of the HRR curve over time as it may reveal evidence on the varying thermal behavior of the heating process due to physical and chemical reactions of intumescent coatings.

Figure 5 displays the HRR versus time profiles of the coating samples after ignition. The burning behavior of the entire samples exhibited a single peak. Sample X showed the maximum peak of 111.86 kW/m^2 at 35 s, which was higher than other maximum peaks of samples W, Y and Z of 106.03, 91.00 and 99.98 kW/m^2, respectively. The PHRR of sample Y was the lowest among all the samples due to its positive synergistic effect in reducing the heat release rate with the addition of 3.5 wt.% ES bio-filler into flame retardant ingredients and binder. The results show that the HRR value with the incorporation of bio-filler of samples Y and Z maintained at a level below 20 kW/m^2 in the time range between 250–700 s ignition, while the HRR of samples W and X without addition of ES bio-filler decreased slowly and maintained at a level below 20 kW/m^2 after at 600 s ignition.

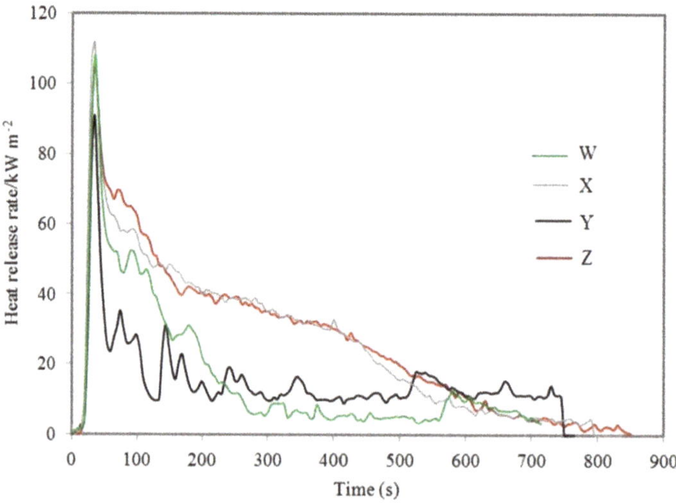

Figure 5. Heat release rate (HRR) of the samples.

Figure 6 displays the curves of total heat released (THR) against the time of the intumescent coating samples. Samples W and X showed higher values of 22.4 and 21.6 MJ/m^2, respectively, compared to samples B and D, which had lower values of 11.5 and 12.0 MJ/m^2. The THR of samples W and X rose sharply and tended to follow a smooth curve after ignition. Samples Y and Z showed a very significant improvement in the reduction of the THR with the addition of the novel ES bio-filler. This result indicated that the heat release of samples Y and Z during combustion was very small and not enough to sustain combustion without external heat flux. The excellent flame retardancy properties of samples Y and Z were probably caused by the existence of the carboxylic group and calcium ions in the calcium carbonate and the carbon source, which promoted a dehydration reaction and decarboxylation reaction to release non-burning gases, such as H_2O and CO_2.

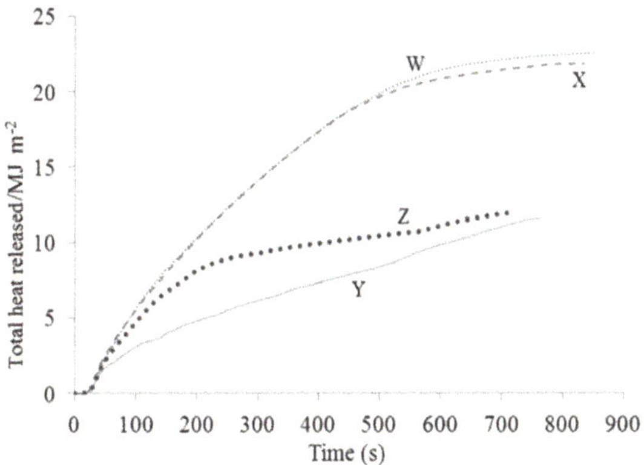

Figure 6. The curves of total heat released (THR) versus time of the samples.

The coating samples before and after the cone calorimeter test are shown in Figure 7a–d. Samples Y and Z, which comprise ES, had more effective char formation and expansion rate compared to samples W and X, due to appropriate combinations of flame retardant fillers (Y-ES/Al(OH)$_3$/TiO$_2$) and Z-ES/Al(OH)$_3$/Mg(OH)$_2$/TiO$_2$). This could be attributed to the physical and chemical integration of the flame-retardant ingredients. The decomposition of Mg(OH)$_2$ and Al(OH)$_3$ flame-retardant fillers is described in the equations below:

$$Mg(OH)_2\ (s) \rightarrow MgO\ (s) + H_2O\ (g) \tag{8}$$

$$2Al(OH)_3\ (s) \rightarrow Al_2O_3\ (s) + 3H_2O\ (g) \tag{9}$$

Figure 7. The coating samples before (W, X, Y and Z) and after (**a–d**) the cone calorimeter test.

The properties of the $Al(OH)_3$ flame-retardant filler displays strong reversibility of the dehydration reaction when exposed to heat, resulting in good fire resistance performance, since water released inside the particles recombine with the reactive surface of the freshly formed alumina [39]. However, the endothermic decomposition of the $Mg(OH)_2$ filler would attribute to a gaseous water phase, which could enclose the flame by eliminating oxygen and dilute combustible gases by reducing the total heat released [40].

Sample Y revealed the highest rate growth and thickest char layer among the coating samples, as presented in Figure 7c. The development of the multicellular layer could have been originated by the release of non-combustible CO_2 due to the decarbonation of ES bio-filler, which induces swelling by trapping the degradation products into the residue, as explained in Section 3.1.

The thermal degradation of ammonium polyphosphate can easily react with flame-retardant fillers to form a ceramic-like material, which increases the char formation by giving a dense and uniform char layer, which could insulate and protect the unprotected substrate in a fire [34,35].

4. Conclusions

The thermal characteristics of four intumescent coating formulations have been studied in accordance with the BS 476 Part 6: Fire propagation test and ISO 5660-1 cone calorimeter standard test under atmospheric conditions with a piloted ignition. The incorporation of the ES bio-filler in the intumescent formulation led to a good thermal resistance and fire protection performance. It was found that all the parameters that characterize coating thermal resistance, such as TTI, HRR, and THR, decreased when 3.50 wt.% and 2.50 wt.% ES bio-filler was added to samples Y and Z. Hence, this study revealed that the addition of ES bio-filler strongly influenced the thermal properties and formation of the char layer of intumescent coatings. The coated samples X, Y and Z showed neither fire propagation nor afterglow combustion. Appropriate combinations of $Al(OH)_3/TiO_2/ES$ in the coating formulation decreased the index value of fire propagation and HRR, whilst providing a thicker and more uniform char layer. The addition of renewable ES bio-filler showed significant enhancement in fire protection and the quality of the intumescent fire protective coatings, as well as being beneficial to the environment. In general, it can be determined that intumescent coatings display significant fire protection qualities in a practical and effective fire protective coating for steel, as shown by the findings of this study.

Author Contributions: Interpretation, J.H.B. and M.C.Y.; methodology, J.H.B.; validation, J.H.B., M.C.Y., M.K.Y. and L.H.S.; formal analysis, J.H.B.; investigation, J.H.B. and M.C.Y.; resources, J.H.B.; data collection, J.H.B.; writing—original draft preparation, J.H.B.; writing—review and editing, M.C.Y.; visualization, M.K.Y. and L.H.S.; supervision, M.C.Y.; project administration, M.C.Y.; funding acquisition, M.C.Y.

Funding: This project was funded by the University of Tunku Abdul Rahman under the UTARRF.

Acknowledgments: The authors would like to express their sincere gratitude to City University of Hong Kong for providing the laboratory services.

Conflicts of Interest: There is no conflict of interest declared by authors.

References

1. Altarawneh, M.; Saeed, A.; Al-Harahsheh, M.; Dlugogorski, B.Z. Performance-based fire safety evacuation of an auditorium facility using a theoretical calculation method. *Int. J. Build. Pathol. Adapt.* **2017**, *35*, 124–138. [CrossRef]
2. Altarawneh, M.; Saeed, A.; Al-Harahsheh, M.; Dlugogorski, B.Z. Thermal decomposition of brominated flame retardants (BFRs): Products and mechanisms. *Prog. Energy Combust. Sci.* **2019**, *70*, 212–259. [CrossRef]
3. Prabir, B. *Biomass Gasification, Pyrolysis and Torrefaction: Practical Design and Theory*, 3rd ed.; Academic Press: Cambridge, MA, USA, 2018.
4. Sigmann, S.B. Playing with fire: Chemical safety expertise required. *J. Chem. Educ.* **2018**, *95*, 1736–1746. [CrossRef]
5. Harada, T. Nuclear flash burns: A review and consideration. *Burn. Open* **2018**, *2*, 1–7. [CrossRef]

6. Duquesne, S.; Magnet, S.; Jama, C.; Delobel, R. Thermoplastic resins for thin film intumescent coatings-towards a better understanding of their effect on intumescent efficiency. *Polym. Degrad. Stab.* **2005**, *88*, 63–69. [CrossRef]

7. Yang, H.; Yu, B.; Song, P.; Maluk, C.; Wang, H. Surface-coating engineering for flame retardant flexible polyurethane foams: A critical review. *Compos. Part B Eng.* **2019**, *176*, 107185. [CrossRef]

8. Yew, M.C.; Ramli Sulong, N.H.; Yew, M.K.; Amalina, M.A.; Johan, M.R. Fire propagation performance of intumescent fire protective coatings using eggshells as a novel biofiller. *Sci. World J.* **2014**, *2014*, 805094. [CrossRef]

9. Aziz, H.; Ahmad, F. Effects from nano-titanium oxide on the thermal resistance of an intumescent fire retardant coating for structural applications. *Prog. Org. Coat.* **2016**, *101*, 431–439. [CrossRef]

10. Guru, P.S.; Dash, S. Sorption on eggshell waste—A review on ultrastructure, biomineralization and other applications. *Adv. Colloid Interface Sci.* **2014**, *209*, 49–67. [CrossRef]

11. Siriprom, W.; Sangwaranatee, N.; Hidayat, R.; Kongsriprapan, S.; Teanchai, K.; Chamchoi, N. The physicochemical characteristic of biodegradable methylcellulose film reinforced with chicken eggshells. *Mater. Today Proc.* **2018**, *5*, 14836–14839. [CrossRef]

12. Yew, M.C.; Ramli Sulong, N.H.; Yew, M.K.; Amalina, M.A.; Johan, M.R. The formulation and study of the thermal stability and mechanical properties of an acrylic coating using chicken eggshell as a novel bio-filler. *Prog. Org. Coat.* **2013**, *76*, 1549–1555. [CrossRef]

13. Hassan, S.B.; Aigbodion, V.S. Effects of eggshell on the microstructures and properties of Al–Cu–Mg/eggshell particulate composites. *J. King Saud Univ. Eng. Sci.* **2015**, *27*, 49–56. [CrossRef]

14. Arias, J.L.; Fernandez, M.S.; Dennis, J.E.; Caplan, A.I. Collagen of the chicken eggshell membranes. *Connect. Tissue. Res.* **1991**, *26*, 37–45. [CrossRef] [PubMed]

15. Hiremath, P.; Shettar, M.; Shankar, M.G.; Mohan, N.S. Investigation on effect of egg shell powder on mechanical properties of GFRP composites. *Mater. Today Proc.* **2018**, *5*, 3014–3018. [CrossRef]

16. Somdee, P.; Hasook, A. Effect of modified eggshell powder on physical properties of poly(lactic acid) and natural rubber composites. *Mater. Today Proc.* **2017**, *4*, 6502–6511. [CrossRef]

17. Nakano, T.; Ikawa, N.I.; Ozimek, L. Chemical composition of chicken eggshell and shell membranes. *Poult. Sci.* **2003**, *82*, 510–514. [CrossRef] [PubMed]

18. Shah, A.; Zhang, Y.; Xu, X.; Dayo, A.; Li, X.; Wang, S.; Liu, W. Reinforcement of stearic acid treated egg shell particles in epoxy thermosets: structural, thermal, and mechanical characterization. *Materials* **2018**, *11*, 1872. [CrossRef] [PubMed]

19. Yusuff, A.S. Preparation and characterization of composite anthill-chicken eggshell adsorbent: Optimization study on heavy metals adsorption using response surface methodology. *J. Environ. Sci. Technol.* **2017**, *10*, 120–130. [CrossRef]

20. Ashok, B.; Naresh, S.; Obi, R.K.; Madhukar, K.; Cai, J.; Zhang, L.; Varadu, R.A. Tensile and thermal properties of poly(lactic acid)/eggshell powder composite films. *Int. J. Polym. Anal. Charact.* **2014**, *19*, 245–255. [CrossRef]

21. Toro, P.; Quijada, R.; Yazdani-Pedram, M.; Arias, J.L. Eggshell: A new bio-filler for polypropylene composites. *Mater. Lett.* **2007**, *61*, 4347–4350. [CrossRef]

22. Ishikawa, S.I.; Sekine, S.; Miura, N.; Suyama, K.; Arihara, K.; Itoh, M. Removal of selenium and arsenic by animal biopolymers. *Biol. Trace. Elem. Res.* **2004**, *102*, 113–127. [CrossRef]

23. Liu, J.L.; Wu, Q.S.; Ding, Y.P.; Wang, S.Y. Biomimetic synthesis of $BaSO_4$ nanotubes using eggshell membrane as template. *J. Mater. Res.* **2004**, *19*, 2803–2806. [CrossRef]

24. Liu, J.L.; Wu, Q.S.; Ding, Y.P. Controlled synthesis of different morphologies of $BaWO_4$ crystals through biomembrane/organic-addition supramolecule Templates. *Cryst. Growth Des.* **2005**, *5*, 445. [CrossRef]

25. Tsai, W.T.; Yang, J.M.; Lai, C.W.; Cheng, Y.H.; Lin, C.C.; Yeh, C.W. Characterization and adsorption properties of eggshells and eggshell membrane. *Bioresour. Technol.* **2006**, *97*, 488–493. [CrossRef] [PubMed]

26. Xiao, D.; Choi, M.M.F. Aspartame optical biosensor with bienzyme-immobilized eggshell membrane and oxygen-sensitive optode membrane. *Anal. Chem.* **2002**, *74*, 863–870. [CrossRef] [PubMed]

27. Yi, F.; Guo, Z.X.; Zhang, L.X.; Yu, J.; Li, Q. Soluble eggshell membrane protein: Preparation, characterization and biocompatibility. *Biomaterials* **2004**, *25*, 4591–4599. [CrossRef]

28. Yew, M.C.; Ramli Sulong, N.H. Fire-resistive performance of intumescent flame-retardant coatigs for steel. *Mater. Des.* **2012**, *34*, 719–724. [CrossRef]

29. Yew, M.C.; Ramli Sulong, N.H. Effect of epoxy binder on fire protection and bonding strength of intumescent fire protective coatings for steel. *Adv. Mater. Res.* **2010**, *168–170*, 1228–1232. [CrossRef]

30. Yew, M.C.; Ramli Sulong, N.H.; Chong, W.T.; Poh, S.C.; Ang, B.C.; Tan, K.H. Integration of thermal insulation coating and moving-air-cavity in a cool roof system for attic temperature reduction. *Energy Convers. Manage.* **2013**, *75*, 241–248. [CrossRef]

31. Yew, M.C.; Yew, M.K.; Saw, L.H.; Ng, T.C.; Rajkumar, D.; Beh, J.H. Influence of nano bio-filler on the fire-resistive and mechanical properties of water-based intumescent coatings. *Prog. Org. Coat.* **2018**, *124*, 33–40. [CrossRef]

32. *BS 476 Part 6: 1989+A1: 2009 Fire Tests on Building Materials and Structures. Method of test for fire propagation for products*; British Standards Institution: London, UK, 1989.

33. *ISO 5660-1 Reaction-to-Fire Tests-Heat Release, Smoke Production and Mass Loss Rate-Part 1: Heat Release Rate (Cone Calorimeter Method)*; International Standard Organization: Geneva, Switzerland, 2002.

34. Gu, J.W.; Zhang, G.C.; Dong, S.I.; Zhang, Q.Y.; Kong, J. Study on preparation and fire-retardant mechanism analysis of intumescent flame-retardant coatings. *Surf. Coat. Technol.* **2007**, *201*, 7835–7841. [CrossRef]

35. Lim, M.; Li, B.; Li, Q.; Li, S.; Zhang, S. Synergistic effect of metal oxides on the flame retardancy and thermal degradation of novel intumescent flame-retardant thermoplastic polyurethanes. *J. Appl. Polym. Sci.* **2011**, *121*, 1951–1960. [CrossRef]

36. Zhang, L.; Wang, J.Q. A new kind of halogen-free fire retardant coatings made from expandable graphite. *J. Beijing Inst. Tech.* **2001**, *21*, 649–652.

37. Wang, Z.Y.; Han, E.H.; Ke, W. Influence of expandable graphiteon fire resistance and water resistance of flame retardant coatings. *Corros. Sci.* **2007**, *49*, 2237–2253. [CrossRef]

38. Kartal, S.N.; Green, F. Decay and termite resistance of medium density fiberboard (MDF) made from different wood species. *Int. Biodeterior. Biodegrad.* **2003**, *51*, 29–35. [CrossRef]

39. Santos, P.S.; Santos, H.S.; Toledo, S.P. Standard transition aluminas. Electron microscopy studies. *Mater. Res.* **2000**, *3*, 104–114. [CrossRef]

40. Hornsby, P.R.; Watson, C.L. A study of the mechanism of flame retardance and smoke suppression in polymers filled with magnesium hydroxide. *Polym. Degrad. Stab.* **1990**, *30*, 73–87. [CrossRef]

 *coatings*

Article

Coating of Polyetheretherketone Films with Silver Nanoparticles by a Simple Chemical Reduction Method and Their Antibacterial Activity

Andrés Felipe Cruz-Pacheco [1], Deysi Tatiana Muñoz-Castiblanco [1],
Jairo Alberto Gómez Cuaspud [1], Leonel Paredes-Madrid [2], Carlos Arturo Parra Vargas [1],
José Jobanny Martínez Zambrano [1] and Carlos Andrés Palacio Gómez [3],*

[1] Facultad de Ciencias, Universidad Pedagógica y Tecnológica de Colombia, Tunja 150003, Colombia;
andresfelipe.cruz@uptc.edu.co (A.F.C.-P.); deysi.munoz@uptc.edu.co (D.T.M.-C.);
jairo.gomez01@uptc.edu.co (J.A.G.C.); carlos.parra@uptc.edu.co (C.A.P.V.);
jose.martinez@uptc.edu.co (J.J.M.Z.)

[2] Faculty of Mechanic, Electronic and Biomedical Engineering, Universidad Antonio Nariño,
Tunja 150002, Colombia; paredes.leonel@uan.edu.co

[3] Faculty of Sciences, Universidad Antonio Nariño, Tunja 150002, Colombia

* Correspondence: carlospalacio@uan.edu.co; Tel.: +57-(8)-744-7564 or +57-(8)-744-7569 or +57-(8)-744-7566

Received: 10 December 2018; Accepted: 30 January 2019; Published: 2 February 2019

Abstract: The coating of polymeric substrate polyetheretherketone (PEEK) with silver nanoparticles (AgNPs) was carried out by a wet chemical route at room temperature. The coating process was developed from the Tollens reagent and D-glucose as reducing agent. The resulting composite exhibited antimicrobial activity. The PEEK films were coated with a single layer and two layers of silver nanoparticles in various concentrations. The crystallographic properties of the polymer and the silver nanoparticles were analyzed by X-ray diffraction (XRD). Fourier transform infrared spectra (FTIR) show the interaction between the silver nanoparticles with the polymeric substrate. Transmission electron microscope (TEM) images confirmed the obtaining of metallic nanoparticles with average sizes of 25 nm. It was possible to estimate the amount of silver deposited on PEEK with the help of thermogravimetric analysis. The morphology and shape of the AgNPs uniformly deposited on the PEEK films was ascertained by the techniques of scanning electron microscopy (SEM) and atomic force microscopy (AFM), evidencing the increase in the amount of silver by increasing the concentration of the metal precursor. Finally, the antibacterial activity of the films coated with Ag in *Escherichia coli*, *Serratia marcescens* and *Bacillus licheniformis* was evaluated, evidencing that the concentration of silver is crucial in the cellular replication of the bacteria.

Keywords: AgNPs; antibacterial coatings; polyetheretherketone films; tollens method

---

1. Introduction

Currently, bacterial growth is the main problem of air purification systems such as heaters, air-conditioning and ventilators that are used indoors [1,2]. Some microorganisms are adhered to and absorbed by the parts of the air purifying systems which return to the environment under operational conditions, spreading pathogenic bacteria not only to humans but also to plants and animals [3]. Among the bacteria that are found in the environment and are scattered by air purifying equipment are *Escherichia coli*, *Bacillus subtilis*, *Staphylococcus aureus* and others such as *Serratia marcescens* that cause diseases such as gastroenteritis, urinary tract infections, meningitis and brain abscesses, in addition to pathogenic bacteria that also affect plants and cause cell death such as *Bacillus licheniformis* [4,5]. The alternative to this problem for a long time, has been the combination of air-purifying equipment filters with nanocoatings that inhibit bacterial growth, using various polymeric materials such as polyimide, polyurethane, polyacrylonitrile and polysulfone, which has generated contamination and low resistance to exposure of the various environmental effects [6,7]. For this reason, one of the polymers that promises to be efficient as a multifunctional material in the design of filters of air purification equipments is polyetheretherketone (PEEK). This semicrystalline thermoplastic polymer formed by an aromatic chain, combines ketone and ether functional groups between the aryl rings with a mechanical strength stable at high temperatures and resistance to most inorganic and organic substances in comparison with other polymers derived from the polyaryletherketone family (PAEK) [8–10]. These characteristics make PEEK an ideal material for various applications such as components in filters for air decontamination and as complements in decontaminating systems, using effective coatings based on silver nanoparticles with high activity in low concentrations [11–14]. In this sense, several methods have been used to functionalize the PEEK surface, among the most common being the sputtering and plasma deposition techniques [9,15,16]; however, deposition by means chemical routes have attracted great interest due to its simplicity and low cost [17]. Among chemical methods, the deposition of silver nanoparticles using the Tollens reagent has been studied for its simplicity, fast deposition and control of nanoparticle size, the use of non-toxic reagents being one of the most clear benefit of this procedure [18]. This green method allows a silver surface formed by homogeneously distributed nanoparticles in the substrate to be obtained, where the reduction of the ammoniacal silver complex takes place when a monosaccharide as glucose is used as a reducing agent. This results in high antibacterial effectiveness due to the development of small silver (Ag) particle sizes [19,20].

Under the above considerations, the present study aims to obtain polyetheretherketone films coated with various concentrations of silver by a simple chemical reduction method, which contributes to the development of deposition methodologies of silver nanoparticles in polymeric substrates, without the need to use expensive equipment and avoiding the use of polluting reagents. In the same way, the synthesis method using diamonical complexes of silver and glucose as a green reducing agent has the advantage of controlling the size and distribution of the AgNPs coated in the PEEK polymer, favoring antibacterial activity against *E. coli*, *S. marcescens* and *Bacillus licheniformis* for potential biomedical applications.

2. Materials and Methods

2.1. Synthesis of Silver Nanoparticles and Coating of Polyetheretherketone (PEEK) Films

Polyetheretherketone (PEEK) films were coated with 1 and 2 layers of silver nanoparticles (AgNP), reducing ammoniacal silver complexes with glucose in aqueous medium at room temperature. $AgNO_3$ (99.9%) was purchased from Sigma-Aldrich (Saint Louis, MO, USA), NaOH (99%), Ammonia (25%) and D-Glucose (99%) was obtained from Merck (Darmstadt, Germany). The PEEK film used was Sigma-Aldrich with a thickness of 0.006 mm. The PEEK film was washed with sulfochromic solution (concentrated sulfuric acid with potassium dichromate 90%/10%), for 5 min to remove contaminants in the polymer and to facilitate the coating with silver nanoparticles. For the synthesis of silver nanoparticles, Tollens reagent was used [21,22]. Stoichiometric amounts of silver nitrate in different concentrations were available in a glass reactor previously washed with sulfochromic solution. Subsequently a NaOH dissolution (0.5 mol/L) was added to the above solutions producing silver monoxide, as shown in Equation (1) [23,24].

$$2AgNO_{3\ (aq)} + 2NaOH_{(aq)} \rightarrow Ag_2O_{(s)} + 2NaNO_{3(aq)} + H_2O_{(aq)} \tag{1}$$

The precipitated silver monoxide was completely dissolved with ammonia (2 mol/L), producing the Tollens reagent (Equation (2)) [25].

$$Ag_2O_{(s)} + 4NH_{3(aq)} + 2NaNO_3 + H_2O_{(aq)} \rightarrow 2Ag(NH_3)_2NO_{3(aq)} + 2NaOH_{(aq)} \tag{2}$$

The silver nanoparticles (Ag^0) were obtained by chemical reduction of the silver diamine complex with D-glucose (1 mol/L) as a reducing agent, forming gluconic acid [26]. Equation (3) summarizes the previous reaction. At this stage, the PEEK films were immersed in the solution containing the AgNPs in three different concentrations (0.04, 0.08 and 0.12 mol/L). The immersion was done for 5 min at room temperature and under agitation at 250 rpm; this favored the random sequential adsorption of the silver particles on the polymer surface, thus obtaining the PEEK/Ag0.04, PEEK/Ag0.08 and PEEK/Ag0.12 systems with both single and two layers [27,28]. It must be clarified that the two-layer systems were obtained by repeating the aforementioned procedure for a single-layer film.

$$2[Ag(NH_3)_2]^+_{(aq)} + RCHO_{(aq)} + 2OH^- \rightarrow 2Ag_{(s)} + RCOOH_{(aq)} + 4NH_{3(aq)} + H_2O_{(aq)} \qquad (3)$$

The PEEK films coated with one and two-layer silver nanoparticles were dried at 100 °C for 2 h to remove remaining water solvent. In this way, the synthesis route using the Tollens reagent and reducing monosaccharides such as glucose contributes to the progress in the deposition of metal nanostructures in polymeric films by chemical methods.

2.2. Characterization of PEEK Films Coated with AgNPs

X-ray diffraction (XRD) measurements of all the coated films were made on a PANalytical X'Pert PRO MPD diffractometer (Bogotá, Colombia), equipped with an Ultra-Fast X´celerator detector and a Bragg–Brentano configuration, using the Cu Kα radiation (λ = 1.5418 Å) between 20° and 90°. The measurements were developed with an acceleration voltage of 40 kV and a current of 20 mA. The average size of the crystalline domains was determined with the Debye–Scherrer equation using the highest intensity peak in the diffraction pattern of each sample [29]. The infrared (IR) spectra were obtained in the Thermo Scientific Nicolet iS50 spectrometer, by the technique of total attenuated reflection (ATR). The IR spectra were processed with a resolution of 4 cm^{-1} in the average IR (4000–400 cm^{-1}). The samples were placed directly on the cell and pressed to carry out the analysis. The spectra were collected and manipulated with the Omnic^R software (version 6.1).

Transmission electron microscopy (TEM) images were obtained from a Tecnai F20 Super Twin TMP (Medellín, Colombia) of a Field Electron and Ion (FEI) microscope, equipped with a system Ion Milling PIPS II Model 695 GATAN (Medellín, Colombia) and Ultramicrotome EM UC7 LEICA (Medellín, Colombia). The thermal analysis was developed in a Seteram Thermobalance equipment. For the analysis, 5 mg of each sample was weighed and placed in an alumina crucible. The sample was subjected to a heating rate from 25 to 700 °C, under an atmosphere of N_2 with a flow of 20 mL/min.

The morphology of the AgNPs on the PEEK films were analyzed by scanning electron microscopy (SEM) without coating in a JSM-6490 JEOL microscope (Tokyo, Japan), with an acceleration voltage of 15 KV, using secondary electron scattering under high vacuum conditions. Energy-dispersive X-ray spectroscopy (EDS) microanalysis was performed on an Inca Energy 250 EDS System LK-IE250 from Oxford, England, UK, equipped with a silicon detector for light elements and resolution of 138 eV. The surface analysis of PEEK films coated with silver was performed in an atomic force microscope (AFM) Asylum Research, model MFP-3D-BIO. The AFM images were analyzed with the Gwyddion software (version 2.49).

2.3. Antibacterial Activity

The antibacterial activity of the films of PEEK with a coating of AgNPs was studied by the disc diffusion method, for which 100 μL of bacterial suspensions in nutrient broth (this medium contains, in g/L: meat extract, 1.0; yeast extract, 2.0; peptone, 5.0 and sodium chloride, 5.0) of *Escherichia coli*, *Serratia marcescens* and *Bacillus licheniformis* with optical density of 0.01 (OD 600) were spread homogeneously on the nutrient agar plate (meat extract B, 3.0 g/L; peptone, 5.0 g/L and agar, 15.0 g/L) [30]. PEEK discs of 6 mm diameter and 0.006 mm thickness with a coating of AgNPs in different concentrations were used. The samples were placed on the agar plate and were kept in the incubator at 37 °C for 24 h [16]. Each antibacterial test was evaluated in triplicate. The ImageJ software was employed to measure the diameter of the zone of inhibition of the AgNPs deposited on the PEEK under bacterial activity.

3. Results and Discussion

3.1. X-ray Diffraction (XRD)

Figure 1a shows the X-ray diffraction patterns of the PEEK/Ag0.04, PEEK/Ag0.08 and PEEK/Ag0.12 samples with a coating layer. The semicrystalline structure of the PEEK polymer is revealed with the 2θ = 16°, 23° and 30° positions, which are assigned to the (110), (220) and (211) facets, respectively [31–33]. The diffraction peaks found at 38°, 44° and 64° corresponding to the (111), (200) and (220) crystal planes of the face-centered cubic structure of the silver (FCC), which are related with JCPDS (No. 89-3722), confirmed the successful reduction process of the Tollens reagent with D-glucose to obtain the first coating layer of the silver nanoparticles in the polyetheretherketone films [22,34,35]. Figure 1b shows the X-ray diffraction patterns of the samples coated with two layers of silver. The crystallinity of the PEEK film is preserved with the deposition of the TEM second layer of silver on the surface of the polymer since there is no decrease in the intensity of the peaks. Similarly, the intensity of the characteristic metallic silver signals is maintained and increased with the second coating process without detriment to the metallic surface.

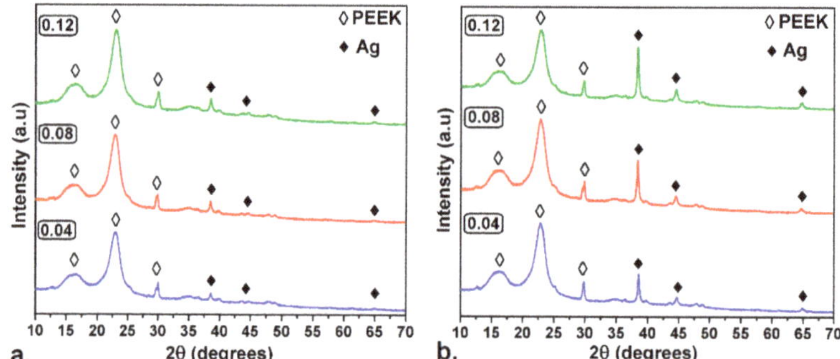

Figure 1. X-ray diffraction (XRD) patterns of polyetheretherketone (PEEK)/Ag0.04, PEEK/Ag0.08 and PEEK/Ag0.12 samples (**a**) with one coating layer and (**b**) with two coating layers.

The average sizes of the crystalline domains ($L_{(111)}$) of the silver nanoparticles deposited on the PEEK polymer surface were calculated with the Debye–Scherrer formula. The resulting values are listed in Table 1.

$$L_{(111)} = \frac{k\lambda}{\beta\cos\theta},\tag{4}$$

where k is the shape factor (0.9), λ is the X-ray wavelength, β is the full width at half maximum (FWHM) in radian, and θ is the Bragg angle in radians corresponding to the most intense (111) diffraction peak [36,37].

Table 1. Data derived from XRD and transmission electron microscopy (TEM) analyses.

Sample	Average Crystal Size by XRD (nm)	FWHM	*d*-Spacing (nm)	Average Crystal Size by TEM (nm)
PEEK/Ag0.04	23.40	0.3402	0.233	9.1
PEEK/Ag0.08	25.93	0.3066	0.234	9.7
PEEK/Ag0.12	26.48	0.3003	0.234	10.1

FWHM: Full width at half-maximum of the XRD peak.

3.2. Fourier Transform Infrared (FTIR) Spectroscopy

The Figure 2 shows the infrared spectrum of PEEK uncoated and coated with all samples obtained. Figure 2a shows the infrared spectra of PEEK polymer substrates with 1 coating layer. The Fourier transform infrared (FTIR) spectrum of the original PEEK film shows a band at 1647 cm^{-1} of the conjugated ketone stretch (C=O), the bands at 1587, 1480 and 1410 cm^{-1} corresponding to stretching vibrations of the conjugated carbons on the chains' aromatics. The signal located at 1305 cm^{-1} is characteristic of the flexion between the ketone group and the adjacent carbons. In 1275 cm^{-1} the stretching signal of the ether group is shown. The bands between 1215 and 1100 cm^{-1} are attributed to the deformation flexion in the plane of the C–H bond. The signal at 1008 cm^{-1} can be attributed to the stretched vibrations of the diphenylether bonds of the p-substituents on the aromatic ring (Ar–O–Ar). From 950 to 765 cm^{-1} corresponds to the flexure of deformation outside the plane of the C–H bond [38–40].

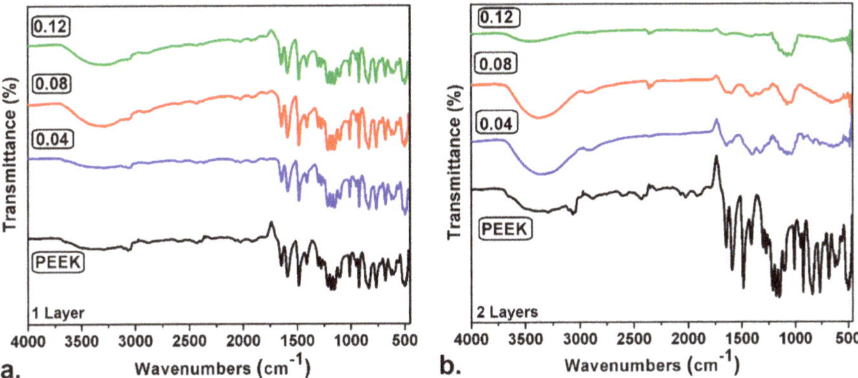

Figure 2. Fourier transform infrared (FTIR) of PEEK/Ag0.04, PEEK/Ag0.08 and PEEK/Ag0.12 samples (**a**) with one coating layer and (**b**) with two coating layers.

The main signals of the polyetheretherketone are present in all the samples obtained with a single layer, because the silver deposited in the polymer is small and does not cover the entire film. The signal in 3310 cm^{-1} corresponds to OH groups of NaOH remanent molecules, after the synthesis process of the AgNPs. The OH groups are not water molecules, since the films formed were completely dried. Likewise, in 3060 cm^{-1} the absorption band corresponding to the stretch of the C–H bond of the aromatic ring carbon is evidenced, which loses intensity in the whole spectrum of the coated films, this may be due to the interactions that are generated between the silver nanoparticles and the PEEK [41]. On the other hand, there are no signs of nitro groups (N–O), so it can be said that there are no residual nitrates after the process of synthesis and drying of the formed films.

The FTIR spectra of PEEK films with two layers of coating are shown in Figure 2b. The absorption bands below 3000 cm^{-1} of the PEEK polymer are lost, and new characteristic signals of the silver nanoparticles appear which increase the intensity with the proportion of silver deposited, confirming an optimal distribution of silver on the surface of the polymer. The absorption band at 1070 cm^{-1} is characteristic of the presence of AgNPs [42,43]. The bands at 1650 and 1400 cm^{-1} are of the ketone bond and of the conjugated carbons in the aromatic ring, respectively, which are empty spaces of the polymer where no silver was deposited. In the IR spectrum of the PEEK/Ag0.12 sample with two layers, the characteristic signals of the organic compounds are not present, and this is only the characteristic of the interaction of the silver with the polymer, demonstrating the total coating of the surface area of the polymeric substrate [44].

3.3. Transmission Electron Microscopy (TEM) Analysis

The transmission electron microscopy images of PEEK polymers coated with two layers silver nanoparticles are shown in Figure 3. The analysis of the TEM images showed a distribution of the size of the crystals by means of the ImageJ software; this is shown in Figure 4. The results derived from the statistical analysis corroborated the average sizes of the silver crystals deposited in the polymer and evidenced by the X-ray diffraction patterns using the Debye–Scherrer equation.

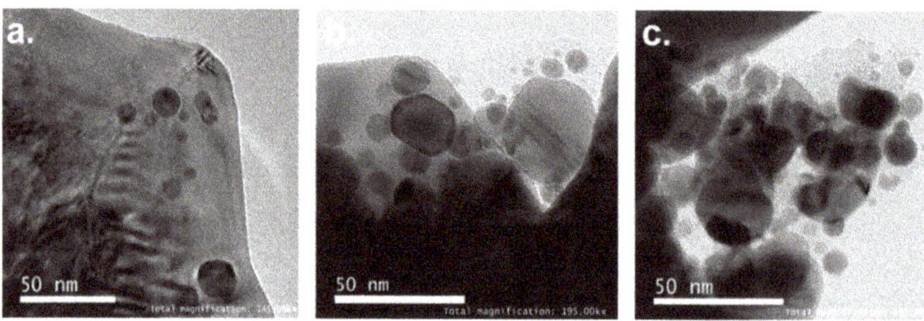

Figure 3. TEM images of (**a**) PEEK/Ag0.04, (**b**) PEEK/Ag0.08 and (**c**) PEEK/Ag0.12. The samples were coated with two layers of AgNPs.

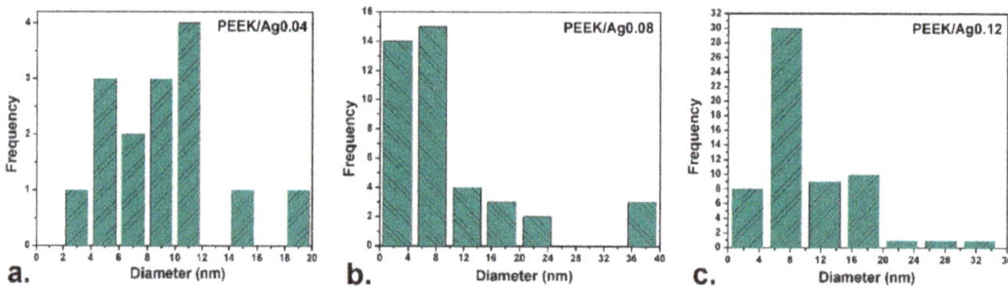

Figure 4. Distribution of the particle sizes determined from the TEM data at 50 nm of the obtained samples with 2 layers. (**a**) PEEK/Ag0.04, (**b**) PEEK/Ag0.08 and (**c**) PEEK/Ag0.12.

The comparison among particle sizes obtained by XRD and TEM are presented in Table 1. The experimental results shown that the silver nanoparticles with the polymer were coated on average by less than 30 nm. Based on previous studies [45], particle dimensions play an important role in the antibacterial activity of the coating and, therefore, this must be monitored in the final coating assembly. In the same way, the synthesis of silver nanoparticles using the Tollens reagent and an economic reducing agent, such as D-glucose, allowed control of the size and shape of the nanoparticles deposited in the PEEK [46]. In the same way, when different concentrations of $AgNO_3$ are used, the increase in the size of the synthesized nanoparticles is related to the amount of ammonia. By increasing the concentration of silver in the different samples, the amount of ammonia required for the reaction is also increased, thus favoring the stabilization of the complex $[Ag(NH_3)]^+$. This phenomenon decreases the amount of Ag^+ species, which causes the decrease of stable silver nuclei in the reduction process with glucose, and induces the formation of large particles in the growth stage [47–49].

Figure 5 shows the transmission electron microscopy images of the PEEK/Ag0.12 sample cut with ultramicrotome at different magnifications. In Figure 5a,b the obtaining of uniformly sized silver nanoparticles adhered to the PEEK polymer is corroborated by electrostatic forces generated by the high charge of the reduced silver in the synthesis process [17,50]. Likewise, the AgNPs form agglomerates of regular size, which extend throughout the PEEK substrate surface. Figure 5c shows the resistance to friction applied by the diamond tip in the cut with ultramicrotome on the polymer coated with silver nanoparticles. The cut is made in a transversal way allowing the PEEK film (thickness 6 μm) and the metallic silver adhered to the polymer by electrostatic interactions [51] to be observed.

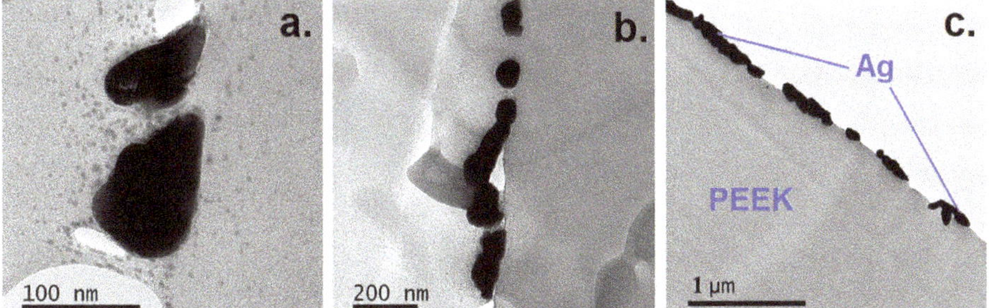

Figure 5. TEM images of the sample PEEK/Ag0.12 with two layers. (**a**) 100 nm; (**b**) 200 nm; (**c**) 1 µm.

3.4. Thermogravimetric Analysis (TGA)

Thermogravimetric analysis (TGA) of all samples is shown in Figure 6, in which it is clear that PEEK as a thermoplastic polymer exhibits a thermal decomposition above 500 °C. The amount of silver deposited is determined from the difference of pure PEEK and modified PEEK thermograms [52,53]. The temperature at which the difference was taken was 675 °C, since at this temperature the polymer is decomposed as shown in Figure 6 and as reported in the literature [54]. In this sense, for the polymers coated with a single layer of silver nanoparticles in concentrations of 0.04, 0.08 and 0.12 mol/L the amount of silver was 5.9%, 8.12% and 10.8% silver, respectively.

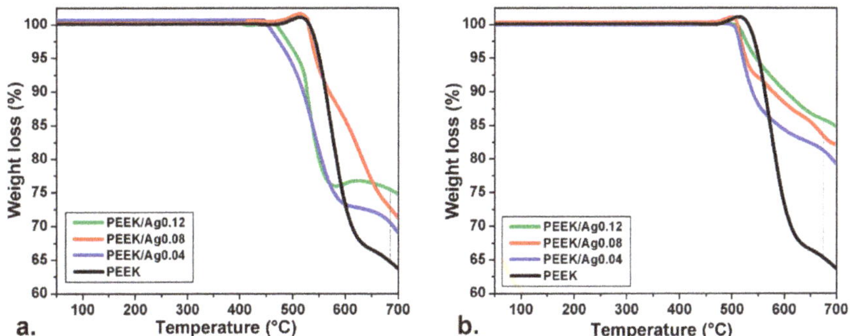

Figure 6. Thermogravimetric analysis (TGA) of uncoated PEEK, PEEK/Ag0.04, PEEK/Ag0.08 and PEEK/Ag0.12 samples (**a**) with one coating layer and (**b**) with two coating layers.

The percentages of silver deposited in the polymers coated with two layers of silver nanoparticles were 16.1, 18.5 and 20.99% Ag for the concentrations of 0.04, 0.08 and 0.12 mol/L, respectively. Due to the aromatic structure of the PEEK, a higher amount of residues are obtained (more than 60%) after heating at to 700 °C [31]. This analysis contrasts with the results of X-ray diffraction and FTIR in terms of the proportional amount of silver deposited in the PEEK, but confirm the thermal stability of materials above 450 °C.

3.5. Morphological Evolution and Particle Distribution by Scanning Electron Microscopy (SEM)

The deposition and dispersion of AgNPs in the PEEK was evaluated by scanning electron microscopy at different resolutions. Figure 7 shows the SEM images of different PEEK films with a silver layer on the surface, in which a proportional increase in the amount of silver deposited in the polymer is evidenced by the concentration of $AgNO_3$ used. Considering that the silver nanoparticles are conductive, the SEM measurements were performed without sample preparation, i.e., no gold or graphite coating was applied before sample measuring. Consequently, the black areas correspond to the polymer which is not conductive, whereas the white and gray points correspond to silver particles. The homogeneity and distribution of the AgNPs deposited on the polymer surface are quite similar to other synthesis methods; this was possible because experimental conditions allowed controlling the amount of deposited particles, and thus, the formation of agglomerates that increase the density of the solids. The SEM images are congruent with the TEM images shown in Figure 5, in which it is shown that the particles with size greater than 80 nm are composed of aggregates of smaller silver nanoparticles.

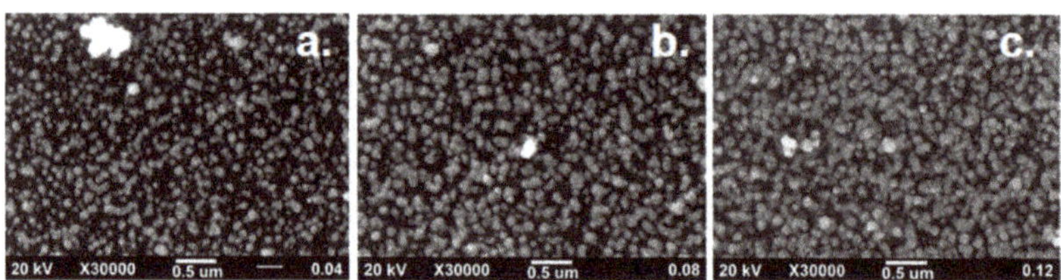

Figure 7. Scanning electron microscopy (SEM) images of the polyetheretherketone coated with one layer of silver nanoparticles in concentrations of (**a**) 0.04 mol/L, (**b**) 0.08 mol/L and (**c**) 0.12 mol/L.

Figure 8 shows SEM images of the PEEK/Ag0.04, PEEK/Ag0.08 and PEEK/Ag0.12 systems with two silver layers. The evolution of the PEEK coating is identified by increasing in the concentration of silver nanoparticles. The empty black areas shown in Figure 7 correspond to PEEK; these are homogeneously filled by silver particles, which cover most of the polymer surface as shown in Figure 8 (when two layers were applied). In the PEEK/Ag0.08 and PEEK/Ag0.12 systems, the appearance of small agglomerations is favored, which were formed by the greater amount of silver ions available in the reaction medium. In this sense, SEM micrographs correlate with the results of XRD, FTIR and TGA, in terms of increasing the proportion of silver on the surface of the polymer, showing the efficiency of the simple chemical reduction method in the deposition of metallic nanoparticles.

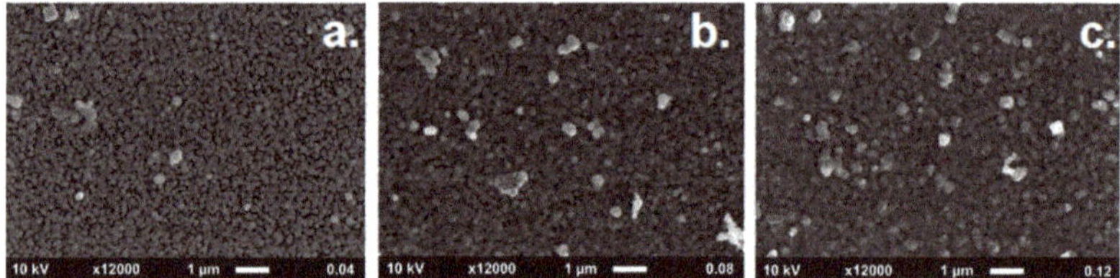

Figure 8. SEM images of the polyetheretherketone coated with two layers of silver nanoparticles in concentrations of (**a**) 0.04 mol/L, (**b**) 0.08 mol/L and (**c**) 0.12 mol/L.

3.6. Energy-Dispersive X-Ray Spectroscopy (EDS) Microanalysis

EDS microanalysis shown in Figure 9, was performed to determine the elemental composition of the PEEK/Ag0.08 system with a single and two layers of AgNPs. The analysis was performed on an area of 600 μm and 500 μm. The analysis was performed on the PEEK/Ag0.08 sample since it was the one that presented the most intense OH absorption band. According to this analysis, it is clear that the present synthesis method provides a high level of purity of the silver nanoparticles deposited in the PEEK polymer, since a residue such as NaOH is in a proportion less than 1%. In the same way, the elemental analysis does not evidence the presence of nitrogen atoms associated with remaining nitrate ions.

The weight percentage of silver shown by the EDS spectra is proportional to the result of TGA, since it is evident that the amount of silver increases with the second layer of silver nanoparticles deposited in the polymer. The difference between TGA and EDS analysis is that in the EDS, only a small and surface portion of the sample is analysed, while in the TGA a more quantitative portion is taken, being a more relevant result.

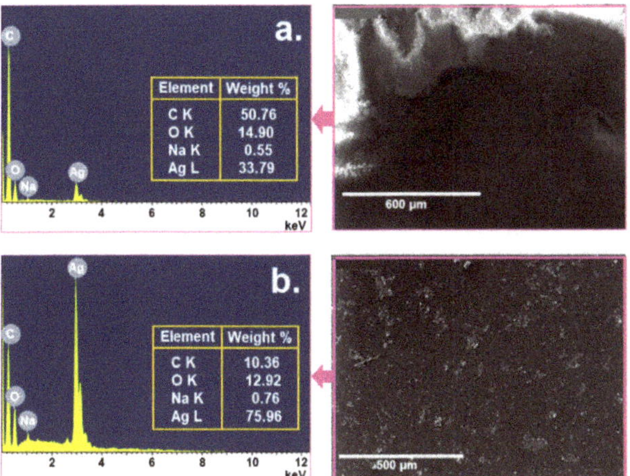

Figure 9. Energy-dispersive X-ray spectroscopy (EDS) microanalysis of the sample PEEK/Ag0.08 with a single (**a**) and with two layers (**b**).

3.7. Atomic Force Microscopy (AFM) analysis

Surface analyses of the coated Polyetheretherketone films were evaluated by AFM. Figure 10 shows AFM images of the systems with the first coating layer. The images confirm the homogeneity of the nanoparticles deposited in the PEEK substrate, just as the SEM imaging previously revealed. Likewise, the height of the images is consistent with the TEM analysis, where a coating of the nanometric order with excellent particle distribution on the surface of the polymer was obtained. Figure 10b shows a collapse caused by the piezoelectric tip of the equipment, because the PEEK film and the silver coating are very thin.

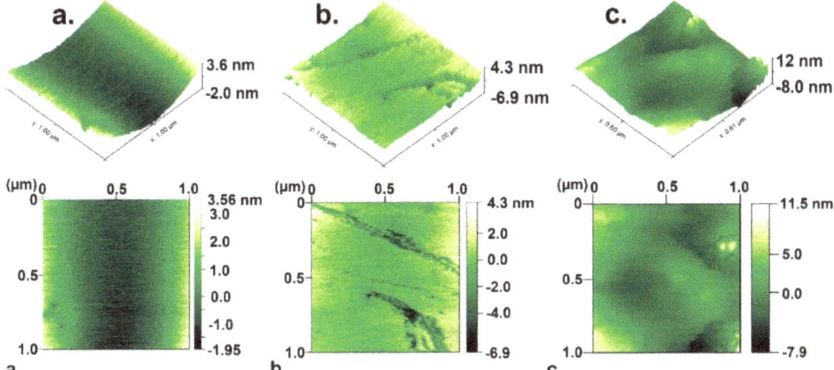

Figure 10. Atomic force microscopy (AFM) images of the (**a**) PEEK/Ag0.04, (**b**) PEEK/Ag0.08 and (**c**) PEEK/Ag0.12 systems with one layer.

The AFM images of the PEEK/Ag systems obtained with two layers of silver are shown in Figure 11. The images represent the remarkable increase in the thickness of each sample obtained with two layers of silver. The PEEK films with two layers of silver deposited on their surface at concentrations of 0.04 and 0.08 mol/L shown in Figure 11a,b exhibit a small amount of aggregate particles. Figure 11c shows the excellent homogenization in the second stage of coating of the polymer, generating small agglomerations because of the high concentration of silver.

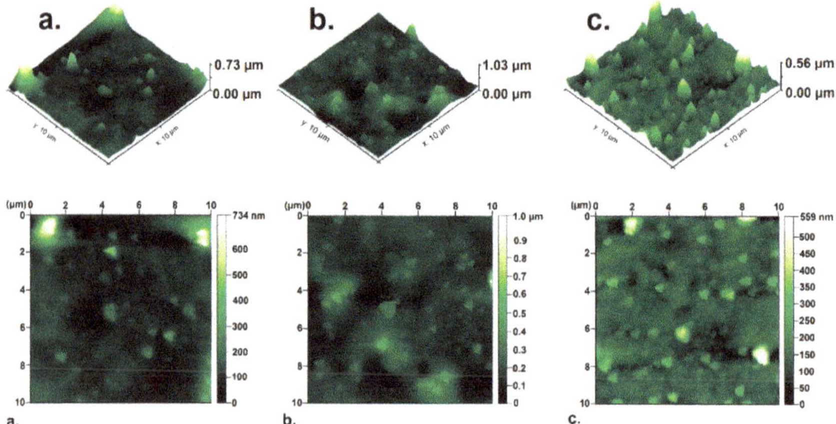

Figure 11. AFM images of the (**a**) PEEK/Ag0.04, (**b**) PEEK/Ag0.08 and (**c**) PEEK/Ag0.12 systems with two layers.

The average thickness of the three samples after the second coating process is in the range between 200 and 300 nm, and increased proportionally with the increase in Ag concentration. In general, the AFM analyses confirm the results by SEM regarding the homogeneous distribution of the silver nanoparticles deposited in the PEEK substrate.

The analysis of the roughness of PEEK materials coated with silver was evaluated with the root mean square (RMS) parameter shown in Table 2. The RMS value is related to the amount of silver deposited in the polymer. When increasing the amount of silver, the average roughness of the material increases forming several nanometric-sized valleys. The PEEK samples coated with a single layer of silver have small surface roughness values, due to the thin layer of silver deposited in the chemical synthesis method. In the same way, samples coated with two layers of silver have higher RMS values according to the quantity of silver nanoparticles deposited on the surface of the polymer [55]. The amount of nanoparticles deposited on both sides of the polymer has an inference in the antibacterial properties. This occurs because more Ag^+ species are generated which are responsible for inhibiting bacterial growth as explained in the following section and as reported by Liu et al. [55]. The surface roughness value also provides information on the adhesion of the silver nanoparticles on the polymer, corroborating the TEM analysis, since as a nanometric-sized coating, electrostatic interactions between the metallic silver and the polymer result [56]. In addition, the surface contact of the nanometer probe with the silver deposited in a single layer does not drag but leaves a groove upon movement, showing the effective adhesion with the polymer as indicated by Wenfei Li et al. [57]. On the other hand, in the samples with two layers the adhesion improves since the entire surface of the polymer is coated with AgNPs increasing the electrostatic interactions.

Table 2. Surface roughness values of the different PEEK samples coated with silver.

Sample	Surface Roughness (nm) Root Mean Square (RMS)
PEEK/Ag0.04 – 1 Layer	0.96 ± 0.3
PEEK/Ag0.08 – 1 Layer	1.54 ± 0.2
PEEK/Ag0.12 – 1 Layer	2.36 ± 0.7
PEEK/Ag0.04 – 2 Layer	56.38 ± 3.7
PEEK/Ag0.08 – 2 Layer	67.44 ± 3.0
PEEK/Ag0.12 – 2 Layer	92.63 ± 1.8

3.8. Antibacterial Test

The antibacterial activity of the PEEK films coated with various concentrations of silver nanoparticles by a green method using the Tollens reagent and a monosaccharide were tested against two Gram-negative bacteria: *Escherichia coli* and *Serratia marcescens* and one Gram-positive bacterium: *Bacillus licheniformis*, evaluating the zone of inhibition by a contact method direct with medium agar as shown in Figures 12–14. The results of the zone of inhibition measured with the ImageJ software are shown in Table 3.

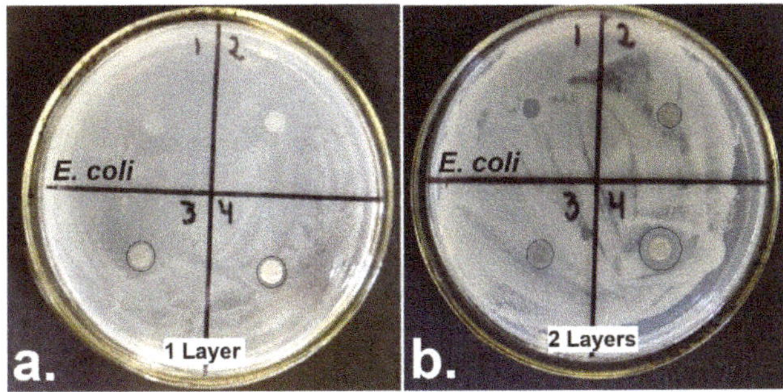

Figure 12. Inhibition zone of the PEEK (1), PEEK/Ag0.04 (2), PEEK/Ag0.08 (3) and PEEK/Ag0.12 (4) coated with one layer (**a**) and two layers (**b**) of silver with *Escherichia coli*.

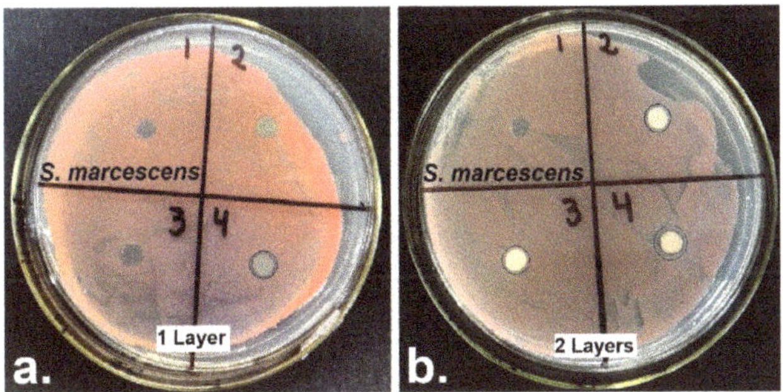

Figure 13. Inhibition zone of the PEEK (1), PEEK/Ag0.04 (2), PEEK/Ag0.08 (3) and PEEK/Ag0.12 (4) coated with one layer (**a**) and two layers (**b**) of silver with *Serratia marcescens*.

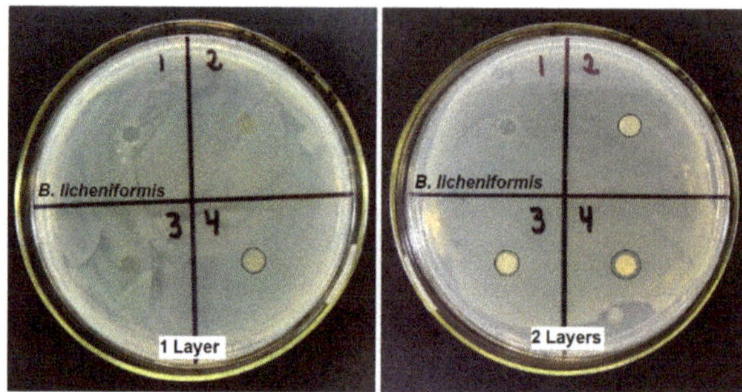

Figure 14. Inhibition zone of the PEEK (1), PEEK/Ag0.04 (2), PEEK/Ag0.08 (3) and PEEK/Ag0.12 (4) coated with one layer (**a**) and two layers (**b**) of silver with *Bacillus licheniformis*.

Table 3. Antibacterial activity of silver nanoparticles deposited in PEEK films.

Sample	Diameter of Inhibition Zone (mm) $\pm$ SD		
	E. coli	*S. marcescens*	*B. licheniformis*
PEEK	0	0	0
PEEK/Ag0.04 – 1 Layer	0	0	0
PEEK/Ag0.08 – 1 Layer	1.1 $\pm$ 0.2	0	0
PEEK/Ag0.12 – 1 Layer	1.2 $\pm$ 0.2	0.6 $\pm$ 0.2	0.5 $\pm$ 0.1
PEEK/Ag0.04 – 2 Layer	1.2 $\pm$ 0.2	0.8 $\pm$ 0.2	0.5 $\pm$ 0.1
PEEK/Ag0.08 – 2 Layer	1.4 $\pm$ 0.1	0.9 $\pm$ 0.2	0.7 $\pm$ 0.1
PEEK/Ag0.12 – 2 Layer	2.7 $\pm$ 0.3	1.2 $\pm$ 0.3	1.0 $\pm$ 0.2

SD: standard deviation.

For the antimicrobial test, an uncoated PEEK film was used as a control in the six plates, which did not present antimicrobial activity. The amount of silver deposited on the PEEK/Ag0.04 sample with a single layer was not enough to prevent the proliferation of the Gram-negative bacteria shown in Figures 12 and 13, stimulating bacterial growth in the silver-free sites that are shown in the SEM images, similar to that reported by Seuss et al. [30]. While the polymer coated with a layer AgNPs in a concentration of 0.08 mol/L presented inhibition against *E. coli* but not against *S. marcescens*. The PEEK/Ag0.12 system with a single layer and all the samples coated with two layers of AgNPs had antibacterial properties, which increased the zone of inhibition with the amount of Ag$^+$ ions deposited in the substrate, favoring the bactericidal effect.

The antibacterial efficiency of PEEK films coated with AgNPs shown in Table 3 was higher for *E. coli* compared to *S. marcescens*, and this was due to the presence of an envelope of two membranes which have different proteins and phospholipids that prevent the passage of silver nanoparticles inside the cells. [16,30,58]. Although the *S. marcescens* bacterium is resistant to traditional antibiotics, the AgNPs synthesized by a green method and deposited on a polymeric PEEK substrate had antibacterial efficiency against this microorganism, because the nanoparticles easily crossed the cytoplasmic membrane due to its small size, causing damage to the organelles of the cell and leading to the death of the microorganism, similar to that described by Baghayeri et al. [59] and Mathew et al. [4].

Figure 14 shows that the uncoated Polyetheretherketone films had no antimicrobial effect on the Gram-positive bacterium *Bacillus licheniformis* similar to a Gram-negative bacterium. Figure 14a,b show that the polyetheretherketone films coated with a single layer of silver nanoparticles in concentrations of 0.04 and 0.08 mol/L did not exhibit antibacterial activity in *B. licheniformis*, while the PEEK/Ag0.12 system and all polymeric films coated with two layers of metallic silver counteracted the growth of Gram-negative bacteria, by releasing silver ions in humid conditions. In addition, the amount of nanoparticles in the culture medium increased the zone of inhibition of bacterium B. licheniformis as shown in Table 3 [60,61]. The growth of the Gram-positive bacterium was higher in comparison with the Gram-negative bacteria, because the bacterium *Bacillus licheniformis* has a thicker peptidoglycan layer in its membrane, which regulates and prevents the path of AgNPs in low concentrations to the cell as indicated by Mathew T. et al. [4]. Similar results with a difference in antibacterial activity were observed by Sikder et al. [62,63] when exploring antibacterial surfaces on PEEK and Ti6Al4V, and such studies were performed in the case of Gram-negative (*E. coli*) and gram-positive (*S. aureus*) bacteria. Apart from the difference in the diameter of inhibition zone, their research also presented SEM images which prove the variance in interactions of Ag^+ ions with negative and positive strains of bacteria. And therefore, these results show a similar trend to the present work. Likewise, Mosselhy D.A. et al. and Ur Rehman et al. [16,64] described a similar effect to the one reported in this study, where the antibacterial properties increased with the increase of the silver nanoparticle ratio and the humid environment in which the samples were installed, such studies demonstrated the effectiveness of the AgNPs coatings in solid state as an antibacterial system.

The antibacterial mechanism of the silver nanoparticle coatings is possible because of the Ag^+ ions generated by the conversion of metallic silver into the physiological environment where the antimicrobial evaluation occurs [16]. The silver nanoparticles in cationic form penetrate the cell, deforming the cell membrane, and interacting with some proteins; the silver nanoparticles also interact with the sulfur and phosphorus bases contained in the DNA, causing an interruption in DNA replication and subsequent cell death [36,37]. Likewise, Ag^+ ions form free radicals that attack respiratory enzymes which are essential for cell replication [45,59].

The zone of inhibition of the three bacteria analyzed was similar in the PEEK/Ag0.12 samples with a single layer and PEEK/Ag0.04 with two layers. This was possible from the homogeneously distributed silver nanoparticles deposited once on the PEEK substrate when $AgNO_3$ was used in a concentration of 0.12 mol/L. Such particles had a high surface area with particle sizes below 25 nm according to the SEM and AFM analyzes. Finally, the sample with the maximum zone of inhibition against the growth of *E. coli*, *S. marcescens* and *B. licheniformis* was PEEK/Ag0.12 with two coating layers of AgNPs, corroborating the efficacy of the synthesis method by chemical reduction of ammoniacal silver complexes with glucose in obtaining coatings of metallic silver nanoparticles with high surface area [65]. In addition, the increase in the proportion of nanoparticles in the polymeric substrate favors the antibacterial effectiveness by the high liberation of silver ions in the wet conditions of the culture medium as reported by Logeswari et al and Gao et al. [66,67].

4. Conclusions

A method of coating by a chemical route with ammoniacal silver complexes was used to impregnate polyetheretherketone films with silver nanoparticles to inhibit bacterial growth. The characteristic diffraction peaks of PEEK were kept constant by completely coating the surface of the polymer. The intensity of the silver signals in the diffractograms increased with the amount of silver nanoparticles deposited. The average size of the crystalline domains of AgNPs synthesized by a simple chemical reduction method was less than 30 nm using the Debye–Scherrer formula, corroborating the results by statistical analysis with transmission electron microscopy images. The electrostatic interactions between the polymer and the deposited AgNPs were evidenced by FTIR and TEM. The thermograms showed the proportion of silver adhered to the polymeric substrate. The proportional deposition of silver on the surface of the PEEK was evaluated by scanning electron microscopy and atomic force microscopy, revealing the excellent distribution of the particles when they are synthesized with glucose. In addition, it is evident that the second deposition process leaves the polymer with a thickness around 300 nm for all samples. The action mechanism of AgNPs against the bacterial growth of pathogenic microorganisms is influenced by the conversion of metallic silver to Ag^+. The sample that shows the best antibacterial activity in Gram-positive and Gram-negative bacteria for a possible application in the design of air purification equipment is PEEK/Ag0.12 with two layers, since the excellent distribution of the coating improves contact with bacteria, inhibits the replication process, and favors cell death

Author Contributions: Conceptualization, A.F.C.-P., J.A.G.C. and J.J.M.Z.; Data Curation, A.F.C.-P., D.T.M.-C., J.A.G.C. and J.J.M.Z.; Formal Analysis, A.F.C.-P., D.T.M.-C. and J.A.G.C.; Funding Acquisition, L.P.-M., C.A.P.V., J.J.M.Z. and C.A.P.G.; Investigation, A.F.C.-P., D.T.M.-C., J.A.G.C. and L.P.-M.; Methodology, J.A.G.C.; Project Administration, C.A.P.G.; Resources, C.A.P.V., J.J.M.Z. and C.A.P.G.; Supervision, J.A.G.C., L.P.-M., C.A.P.V., J.J.M.Z. and C.A.P.G.; Validation, L.P.-M. and C.A.P.G.; Writing—Original Draft, A.F.C.-P.; Writing—Review & Editing, A.F.C.-P., J.A.G.C., L.P.-M. and C.A.P.G.

Funding: This work was supported by Universidad Antonio Nariño under the project number 2016216–PI/UAN-2018-635GFM and by Vicerrectoría de Investigación y Extensión de la Universidad Pedagógica y Tecnológica de Colombia.

Conflicts of Interest: The authors declare no conflict of interest.

References

1. Kakinuma, H.; Ishii, K.; Ishihama, H.; Honda, M.; Toyama, Y.; Matsumoto, M.; Aizawa, M. Antibacterial polyetheretherketone implants immobilized with silver ions based on chelate-bonding ability of inositol phosphate: Processing, material characterization, cytotoxicity, and antibacterial properties. *J. Biomed. Mater. Res. Part A* **2015**, *103*, 57–64. [CrossRef] [PubMed]

2. Gutarowska, B.; Stawski, D.; Skóra, J.; Herczyńska, L.; Pielech-Przybylska, K.; Połowiński, S.; Krucińska, I. PLA nonwovens modified with poly(dimethylaminoethyl methacrylate) as antimicrobial filter materials for workplaces. *Text. Res. J.* **2015**, *85*, 1083–1094. [CrossRef]

3. Ki, Y.Y.; Jeong, H.B.; Chul, W.P.; Hwang, J. Antimicrobial effect of silver particles on bacterial contamination of activated carbon fibers. *Environ. Sci. Technol.* **2008**, *42*, 1251–1255. [CrossRef]

4. Mathew, T.V.; Kuriakose, S. Studies on the antimicrobial properties of colloidal silver nanoparticles stabilized by bovine serum albumin. *Colloids Surf. B Biointerfaces* **2013**, *101*, 14–18. [CrossRef] [PubMed]

5. Burman, S.; Bhattacharya, K.; Mukherjee, D.; Chandra, G. Antibacterial efficacy of leaf extracts of Combretum album Pers. against some pathogenic bacteria. *BMC Complement. Altern. Med.* **2018**, *18*, 213. [CrossRef] [PubMed]

6. Zhu, M.; Xiong, R.; Huang, C. Bio-based and photocrosslinked electrospun antibacterial nanofibrous membranes for air filtration. *Carbohydr. Polym.* **2019**, *205*, 55–62. [CrossRef] [PubMed]

7. Le, T.S.; Dao, T.H.; Nguyen, D.C.; Nguyen, H.C.; Balikhin, I.L. Air purification equipment combining a filter coated by silver nanoparticles with a nano-TiO$_2$ photocatalyst for use in hospitals. *Adv. Nat. Sci. Nanosci. Nanotechnol.* **2015**, *6*, 015016. [CrossRef]

8. Montero, J.F.D.; Barbosa, L.C.A.; Pereira, U.A.; Barra, G.M.; Fredel, M.C.; Benfatti, C.A.M.; Magini, R.S.; Pimenta, A.L.; Souza, J.C.M. Chemical, microscopic, and microbiological analysis of a functionalized poly-ether-ether-ketone-embedding antibiofilm compounds. *J. Biomed. Mater. Res. Part A* **2016**, *104*, 3015–3020. [CrossRef] [PubMed]

9. Wiacek, A.E.; Terpiłowski, K.; Jurak, M.; Worzakowska, M. Effect of low-temperature plasma on chitosan-coated PEEK polymer characteristics. *Eur. Polym. J.* **2016**, *78*, 1–13. [CrossRef]

10. Wu, J.; Li, L.; Fu, C.; Yang, F.; Jiao, Z.; Shi, X.; Ito, Y.; Wang, Z.; Liu, Q.; Zhang, P. Micro-porous polyetheretherketone implants decorated with BMP-2 via phosphorylated gelatin coating for enhancing cell adhesion and osteogenic differentiation. *Colloids Surf. B Biointerfaces* **2018**, *169*, 233–241. [CrossRef]

11. Joe, Y.H.; Ju, W.; Park, J.H.; Yoon, Y.H.; Hwang, J. Correlation between the antibacterial ability of silver nanoparticle coated air filters and the dust loading. *Aerosol Air Qual. Res.* **2013**, *13*, 1009–1018. [CrossRef]

12. Bodden, L.; Lümkemann, N.; Köhler, V.; Eichberger, M.; Stawarczyk, B. Impact of the heating/quenching process on the mechanical, optical and thermodynamic properties of polyetheretherketone (PEEK) films. *Dent. Mater.* **2017**, *33*, 1436–1444. [CrossRef] [PubMed]

13. Schroeder, R.; Torres, F.W.; Binder, C.; Klein, A.N.; De Mello, J.D.B. Failure mode in sliding wear of PEEK based composites. *Wear* **2013**, *301*, 717–726. [CrossRef]

14. Theiler, G.; Gradt, T. Environmental effects on the sliding behaviour of PEEK composites. *Wear* **2016**, *368–369*, 278–286. [CrossRef]

15. Kvítek, O.; Fajstavr, D.; Řezníčková, A.; Kolská, Z.; Slepička, P.; Švorčík, V. Annealing of gold nanolayers sputtered on polyimide and polyetheretherketone. *Thin Solid Films* **2016**, *616*, 188–196. [CrossRef]

16. Ur Rehman, M.A.; Ferraris, S.; Goldmann, W.H.; Perero, S.; Bastan, F.E.; Nawaz, Q.; Confiengo, G.G.D.; Ferraris, M.; Boccaccini, A.R. Antibacterial and Bioactive Coatings Based on Radio Frequency Co-Sputtering of Silver Nanocluster-Silica Coatings on PEEK/Bioactive Glass Layers Obtained by Electrophoretic Deposition. *ACS Appl. Mater. Interfaces* **2017**, *9*, 32489–32497. [CrossRef]
17. Yameen, B.; Álvarez, M.; Azzaroni, O.; Jonas, U.; Knoll, W. Tailoring of poly(ether ether ketone) surface properties via surface-initiated atom transfer radical polymerization. *Langmuir* **2009**, *25*, 6214–6220. [CrossRef]
18. Khan, Z.; Hussain, J.I.; Kumar, S.; Hashmi, A.A. Silver nanoplates and nanowires by a simple chemical reduction method. *Colloids Surf. B Biointerfaces* **2011**, *86*, 87–92. [CrossRef]
19. Le, A.T.; Huy, P.T.; Tam, P.D.; Huy, T.Q.; Cam, P.D.; Kudrinskiy, A.A.; Krutyakov, Y.A. Green synthesis of finely-dispersed highly bactericidal silver nanoparticles via modified Tollens technique. *Curr. Appl. Phys.* **2010**, *10*, 910–916. [CrossRef]
20. Kvitek, L.; Panacek, A.; Prucek, R.; Soukupova, J.; Vanickova, M.; Kolar, M.; Zboril, R. Antibacterial activity and toxicity of silver—Nanosilver versus ionic silver. *J. Phys. Conf. Ser.* **2011**, *304*, 012029. [CrossRef]
21. Zienkiewicz-Strzałka, M.; Pasieczna-Patkowska, S.; Kozak, M.; Pikus, S. Silver nanoparticles incorporated onto ordered mesoporous silica from Tollen's reagent. *Appl. Surf. Sci.* **2013**, *266*, 337–343. [CrossRef]
22. Luo, Y.; Shen, S.; Luo, J.; Wang, X.; Sun, R. Green synthesis of silver nanoparticles in xylan solution via Tollens reaction and their detection for Hg^{2+}. *Nanoscale* **2015**, *7*, 690–700. [CrossRef] [PubMed]
23. Michalcová, A.; Machado, L.; Marek, I.; Martinec, M.; Sluková, M.; Vojtěch, D. Properties of Ag nanoparticles prepared by modified Tollens' process with the use of different saccharide types. *J. Phys. Chem. Solids* **2018**, *113*, 125–133. [CrossRef]
24. Montazer, M.; Allahyarzadeh, V. Electroless plating of silver nanoparticles/nanolayer on polyester fabric using $AgNO_3$/NaOH and ammonia. *Ind. Eng. Chem. Res.* **2013**, *52*, 8436–8444. [CrossRef]
25. Kyriakidou, E.A.; Alexeev, O.S.; Wong, A.P.; Papadimitriou, C.; Amiridis, M.D.; Regalbuto, J.R. Synthesis of Ag nanoparticles on oxide and carbon supports from Ag diammine precursor. *J. Catal.* **2016**, *344*, 749–756. [CrossRef]
26. Le, A.T.; Tam, L.T.; Tam, P.D.; Huy, P.T.; Huy, T.Q.; Van Hieu, N.; Kudrinskiy, A.A.; Krutyakov, Y.A. Synthesis of oleic acid-stabilized silver nanoparticles and analysis of their antibacterial activity. *Mater. Sci. Eng. C* **2010**, *30*, 910–916. [CrossRef]
27. Kubiak, K.; Adamczyk, Z.; Oćwieja, M. Kinetics of Silver Nanoparticle Deposition at PAH Monolayers: Reference QCM Results. *Langmuir* **2015**, *31*, 2988–2996. [CrossRef]
28. Oćwieja, M.; Adamczyk, Z. Controlled Release of Silver Nanoparticles from Monolayers Deposited on PAH Covered Mica. *Langmuir* **2013**, *29*, 3546–3555. [CrossRef]
29. Cruz Pacheco, A.F.; Gómez Cuaspud, J.A.; Parra Vargas, C.A.; Carda Castello, J.B. Combustion synthesis, structural and magnetic characterization of $Ce_{1-x}Pr_xO_2$ system. *J. Mater. Sci. Mater. Electron.* **2017**, *28*, 16358–16365. [CrossRef]
30. Seuss, S.; Heinloth, M.; Boccaccini, A.R. Development of bioactive composite coatings based on combination of PEEK, bioactive glass and Ag nanoparticles with antibacterial properties. *Surf. Coat. Technol.* **2016**, *301*, 100–105. [CrossRef]
31. Hasegawa, S.; Sato, K.; Narita, T.; Suzuki, Y.; Takahashi, S.; Morishita, N.; Maekawa, Y. Radiation-induced graft polymerization of styrene into a poly(ether ether ketone) film for preparation of polymer electrolyte membranes. *J. Membr. Sci.* **2009**, *345*, 74–80. [CrossRef]
32. Gupta, B.; Gautam, D.; Ikram, S. Preparation of proton exchange membranes by radiation-induced grafting of alpha methyl styrene-butyl acrylate mixture onto polyetheretherketone (PEEK) films. *Polym. Bull.* **2013**, *70*, 2691–2708. [CrossRef]
33. Gümüş, S.; Polat, Ş.; Waldhauser, W.; Lackner, J.M. Biodegradation of anti-microbial titanium-magnesium-silver coatings on polyetheretherketone for bone-contact applications. *Surf. Coat. Technol.* **2017**, *320*, 503–511. [CrossRef]
34. Vaiano, V.; Matarangolo, M.; Murcia, J.J.; Rojas, H.; Navío, J.A.; Hidalgo, M.C. Enhanced photocatalytic removal of phenol from aqueous solutions using ZnO modified with Ag. *Appl. Catal. B Environ.* **2018**, *225*, 197–206. [CrossRef]

35. Chook, S.W.; Chia, C.H.; Zakaria, S.; Ayob, M.K.; Huang, N.M.; Neoh, H.M.; Jamal, R. Antibacterial hybrid cellulose-graphene oxide nanocomposite immobilized with silver nanoparticles. *RSC Adv.* **2015**, *5*, 26263–26268. [CrossRef]

36. Pal, S.; Nisi, R.; Stoppa, M.; Licciulli, A. Silver-Functionalized Bacterial Cellulose as Antibacterial Membrane for Wound-Healing Applications. *ACS Omega* **2017**, *2*, 3632–3639. [CrossRef] [PubMed]

37. El-Naggar, N.E.-A.; Hussein, M.H.; El-Sawah, A.A. Phycobiliprotein-mediated synthesis of biogenic silver nanoparticles, characterization, in vitro and in vivo assessment of anticancer activities. *Sci. Rep.* **2018**, *8*, 8925. [CrossRef]

38. Hwang, M.L.; Song, J.M.; Ko, B.S.; Sohn, J.Y.; Nho, Y.C.; Shin, J. Radiation-induced grafting of vinylbenzyl chloride onto a poly(ether ether ketone) film. *Nucl. Instrum. Methods Phys. Res. Sect. B Beam Interact. Mater. Atoms* **2012**, *281*, 45–50. [CrossRef]

39. Kim, A.R.; Vinothkannan, M.; Yoo, D.J. Sulfonated-fluorinated copolymer blending membranes containing SPEEK for use as the electrolyte in polymer electrolyte fuel cells (PEFC). *Int. J. Hydrogen Energy* **2017**, *42*, 4349–4365. [CrossRef]

40. Jean-Fulcrand, A.; Masen, M.A.; Bremner, T.; Wong, J.S.S. Effect of temperature on tribological performance of polyetheretherketone-polybenzimidazole blend. *Tribol. Int.* **2019**, *129*, 5–15. [CrossRef]

41. Girard, J.; Joset, N.; Crochet, A.; Tan, M.; Holzheu, A.; Brunetto, P.S.; Fromm, K.M. Synthesis of new polyether ether ketone derivatives with silver binding site and coordination compounds of their monomers with different silver salts. *Polymers* **2016**, *8*, 208. [CrossRef]

42. Anupama, N.; Madhumitha, G. Green synthesis and catalytic application of silver nanoparticles using Carissa carandas fruits. *Inorg. Nano-Met. Chem.* **2017**, *47*, 116–120. [CrossRef]

43. Ismail, M.; Gul, S.; Khan, M.I.; Khan, M.A.; Asiri, A.M.; Khan, S.B. Medicago polymorpha-mediated antibacterial silver nanoparticles in the reduction of methyl orange. *Green Process. Synth.* **2018**. [CrossRef]

44. Hamciuc, C.; Hamciuc, E.; Popovici, D.; Danaila, A.I.; Butnaru, M.; Rimbu, C.; Carp-Carare, C.; Kalvachev, Y. Biocompatible poly(ether-ether-ketone)/Ag-zeolite L composite films with antimicrobial properties. *Mater. Lett.* **2018**, *212*, 339–342. [CrossRef]

45. Sana, S.S.; Dogiparthi, L.K. Green synthesis of silver nanoparticles using Givotia moluccana leaf extract and evaluation of their antimicrobial activity. *Mater. Lett.* **2018**, *226*, 47–51. [CrossRef]

46. Yu, D.; Yam, V.W.W. Hydrothermal-induced assembly of colloidal silver spheres into various nanoparticles on the basis of HTAB-modified silver mirror reaction. *J. Phys. Chem. B* **2005**, *109*, 5497–5503. [CrossRef] [PubMed]

47. Sharma, V.K.; Yngard, R.A.; Lin, Y. Silver nanoparticles: Green synthesis and their antimicrobial activities. *Adv. Colloid Interface Sci.* **2009**, *145*, 83–96. [CrossRef] [PubMed]

48. Kvítek, L.; Prucek, R.; Panáček, A.; Novotný, R.; Hrbáč, J.; Zbořil, R. The influence of complexing agent concentration on particle size in the process of SERS active silver colloid synthesis. *J. Mater. Chem.* **2005**, *15*, 1099–1105. [CrossRef]

49. Panáček, A.; Kvítek, L.; Prucek, R.; Kolář, M.; Večeřová, R.; Pizúrová, N.; Sharma, V.K.; Nevěčná, T.; Zbořil, R. Silver Colloid Nanoparticles: Synthesis, Characterization, and Their Antibacterial Activity. *J. Phys. Chem. B* **2006**, *110*, 16248–16253. [CrossRef]

50. Sun, M.; Feng, J.; Bu, Y.; Luo, C. Nanostructured-silver-coated polyetheretherketone tube for online in-tube solid-phase microextraction coupled with high-performance liquid chromatography. *J. Sep. Sci.* **2015**, *38*, 3119–3304. [CrossRef]

51. Corni, I.; Neumann, N.; König, K.; Veronesi, P.; Chen, Q.; Ryan, M.P.; Boccaccini, A.R. Electrophoretic deposition of PEEK-nano alumina composite coatings on stainless steel. *Surf. Coat. Technol.* **2009**, *203*, 1349–1359. [CrossRef]

52. Rolim, W.R.; Pelegrino, M.T.; de Araújo Lima, B.; Ferraz, L.S.; Costa, F.N.; Bernardes, J.S.; Rodigues, T.; Brocchi, M.; Seabra, A.B. Green tea extract mediated biogenic synthesis of silver nanoparticles: Characterization, cytotoxicity evaluation and antibacterial activity. *Appl. Surf. Sci.* **2019**, *463*, 66–74. [CrossRef]

53. El-Faham, A.; Atta, A.M.; Osman, S.M.; Ezzat, A.O.; El-saeed, A.M.; AL Othman, Z.A.; Al-Lohedan, H.A. Silver-embedded epoxy nanocomposites as organic coatings for steel. *Prog. Org. Coat.* **2018**, *123*, 209–222. [CrossRef]

54. Chen, J.; Li, D.; Koshikawa, H.; Zhai, M.; Asano, M.; Oku, H.; Maekawa, Y. Modification of ultrathin polyetheretherketone film for application in direct methanol fuel cells. *J. Membr. Sci.* **2009**, *344*, 266–274. [CrossRef]
55. Liu, X.; Gan, K.; Liu, H.; Song, X.; Chen, T.; Liu, C. Antibacterial properties of nano-silver coated PEEK prepared through magnetron sputtering. *Dent. Mater.* **2017**, *33*, 348–360. [CrossRef] [PubMed]
56. Felix, T.; Cassini, F.A.; Benetoli, L.O.B.; Dotto, M.E.R.; Debacher, N.A. Morphological study of polymer surfaces exposed to non-thermal plasma based on contact angle and the use of scaling laws. *Appl. Surf. Sci.* **2017**, *403*, 57–61. [CrossRef]
57. Li, W.; Chen, Y.; Wu, S.; Zhang, J.; Wang, H.; Zeng, D.; Xie, C. Preparing high-adhesion silver coating on APTMS modified polyethylene with excellent anti-bacterial performance. *Appl. Surf. Sci.* **2018**, *436*, 117–124. [CrossRef]
58. Cervantes-García, E.; García-González, R.; Salazar-Schettino, P.M. Proteínas de membrana externa de Serratia marcescens. *Revista Latinoamericana de Patología Clínica y Medicina de Laboratorio* **2014**, *61*, 224–228.
59. Baghayeri, M.; Mahdavi, B.; Hosseinpor-Mohsen Abadi, Z.; Farhadi, S. Green synthesis of silver nanoparticles using water extract of Salvia leriifolia: Antibacterial studies and applications as catalysts in the electrochemical detection of nitrite. *Appl. Organomet. Chem.* **2017**, *32*, e4057. [CrossRef]
60. Thaya, R.; Malaikozhundan, B.; Vijayakumar, S.; Sivakamavalli, J.; Jeyasekar, R.; Shanthi, S.; Vaseeharan, B.; Ramasamy, P.; Sonawane, A. Chitosan coated Ag/ZnO nanocomposite and their antibiofilm, antifungal and cytotoxic effects on murine macrophages. *Microb. Pathog.* **2016**, *100*, 124–132. [CrossRef]
61. Umoren, S.A.; Nzila, A.M.; Sankaran, S.; Solomon, M.M.; Umoren, P.S. Green synthesis, characterization and antibacterial activities of silver nanoparticles from strawberry fruit extract. *Pol. J. Chem. Technol.* **2017**, *19*, 128–136. [CrossRef]
62. Sikder, P.; Grice, C.R.; Lin, B.; Goel, V.K.; Bhaduri, S.B. Single-Phase, Antibacterial Trimagnesium Phosphate Hydrate Coatings on Polyetheretherketone (PEEK) Implants by Rapid Microwave Irradiation Technique. *ACS Biomater. Sci. Eng.* **2018**, *4*, 2767–2783. [CrossRef]
63. Sikder, P.; Koju, N.; Ren, Y.; Goel, V.K.; Phares, T.; Lin, B.; Bhaduri, S.B. Development of single-phase silver-doped antibacterial CDHA coatings on Ti6Al4V with sustained release. *Surf. Coat. Technol.* **2018**, *342*, 105–116. [CrossRef]
64. Mosselhy, D.; Granbohm, H.; Hynönen, U.; Ge, Y.; Palva, A.; Nordström, K.; Hannula, S.-P. Nanosilver–Silica Composite: Prolonged Antibacterial Effects and Bacterial Interaction Mechanisms for Wound Dressings. *Nanomaterials* **2017**, *7*, 261. [CrossRef] [PubMed]
65. Yue, X.; Zhang, T.; Yang, D.; Qiu, F.; Li, Z.; Wei, G.; Qiao, Y. Ag nanoparticles coated cellulose membrane with high infrared reflection, breathability and antibacterial property for human thermal insulation. *J. Colloid Interface Sci.* **2018**, *535*, 363–370. [CrossRef] [PubMed]
66. Logeswari, P.; Silambarasan, S.; Abraham, J. Synthesis of silver nanoparticles using plants extract and analysis of their antimicrobial property. *J. Saudi Chem. Soc.* **2015**, *19*, 311–317. [CrossRef]
67. Gao, L.; Gan, W.; Xiao, S.; Zhan, X.; Li, J. A robust superhydrophobic antibacterial Ag-TiO$_2$ composite film immobilized on wood substrate for photodegradation of phenol under visible-light illumination. *Ceram. Int.* **2016**, *42*, 2170–2179. [CrossRef]

 *coatings*

Article

Biofilm Formation of a Polymer Brush Coating with Ionic Liquids Compared to a Polymer Brush Coating with a Non-Ionic Liquid

Hideyuki Kanematsu [1,*], Atsuya Oizumi [1], Takaya Sato [2], Toshio Kamijo [2], Saika Honma [2], Dana M. Barry [3,4], Nobumitsu Hirai [1], Akiko Ogawa [1], Takeshi Kogo [1], Daisuke Kuroda [1], Katsuhiko Sano [5], Katsuhiko Tsunashima [6], Seung-Hyo Lee [7] and Myeong-Hoon Lee [7]

[1] National Institute of Technology, Suzuka College, Suzuka, Mie 510-0294, Japan;
    h25s11@ed.cc.suzuka-ct.ac.jp (A.O.); hirai@chem.suzuka-ct.ac.jp (N.H.); ogawa@chem.suzuka-ct.ac.jp (A.O.);
    kougo@mse.suzuka-ct.ac.jp (T.K.); daisuke@mse.suzuka-ct.ac.jp (D.K.)
[2] National Institute of Technology, Tsuruoka College, Tsuruoka 997-8511, Japan;
    takayasa@tsuruoka-nct.ac.jp (T.S.); kamijo@tsuruoka-nct.ac.jp (T.K.); saika@tsuruoka-nct.ac.jp (S.H.)
[3] Department of Electrical & Computer Engineering, Clarkson University, Potsdam, NY 13699, USA;
    dmbarry@clarkson.edu
[4] Science/Math Tutoring Center, SUNY Canton, Canton, NY 13617, USA
[5] D&D Corporation, 7870, Sakura-cho, Yokkaichi, Mie 512-1211, Japan; sano@ddcorp.co.jp
[6] National Institute of Technology, Wakayama College, Gobo, Wakayama 644-0023, Japan;
    tsunashima@wakayama-nct.ac.jp
[7] Division of Marine Engineering, Korea Maritime University, Busan 49112, Korea;
    seunghyo0121@gmail.com (S.-H.L.); leemh@kmou.ac.kr (M.-H.L.)
* Correspondence: kanemats@mse.suzuka-ct.ac.jp; Tel.:+81-59-368-1848

Received: 28 August 2018; Accepted: 27 October 2018; Published: 13 November 2018

Abstract: N,N-diethyl-N-(2-methancryloylethy)-N-methylammonium bis(trifluoromethylsulfonyl) imide polymer (DEMM-TFSI) brush coated specimens (substrate: glasses) and a liquid ion type of polymer brush coating were investigated for their antifouling effect on biofilms. Biofilms were produced by two kinds of bacteria, *E. coli* and *S. epidermidis*. They were formed on specimens immersed into wells (of 12-well plates) that were filled with culture liquids and bacteria. The biofilm formation was observed. Also, brush coated specimens and glass substrates were investigated in the same way. DEMM polymer brush coated specimens formed more biofilm than PMMA (polymethyl methacrylate) polymer brush coated specimens and glass substrates. A greater amount of polarized components of biofilms was also observed for DEMM polymer brush coated specimens. The polar characteristics could be attributed to the attraction capability of bacteria and biofilms on DEMM polymer brush coated specimens. When considering the ease of removing biofilms by washing it with water, the ionic liquid type polymer brush (coated specimens) could be used for antifouling applications. If an initial antifouling application is needed, then the polar characteristics could be adjusted (design of the components and concentrations of ionic liquids, etc.) to solve the problem.

Keywords: biofilms; antifouling; DEMM; DEME; PMMA; polymer brush coating; polysaccharide; protein; lipid; nucleic acid; Raman spectroscopy

---

1. Introduction

For antifouling properties of materials, the surface plays an important role [1–10]. This is because fouling phenomena occur at materials' surfaces as a result of interactions between various materials and environments. The phenomena should be controlled from both the environmental surroundings and materials' surfaces. The causes of fouling can be grouped into two main categories of animate and inanimate causes [11]. As for the latter, one can mention many types of contamination from organic to inorganic matter. Usually, they are just called contaminants. The former is usually called biofouling [12,13]. Biofouling is a process where organisms generally attach to materials' surfaces and cause their function and/or characteristics to change and deteriorate in many cases. It can be further classified into microfouling (where microbes such as bacteria and microalgae attach to materials' surfaces) and macrofouling like the attachments of barnacles, oysters, and other bigger organisms in marine environments. For the biofouling in marine environments, microfouling is usually followed by macrofouling. Therefore, the microfouling should be controlled to suppress macrofouling [14,15].

In the case of microfouling, microbial attachments are generally followed by biofilm formations [16–19]. Biofilms are matters formed by bacterial activities. Since microfouling is a process involving bacteria and other micro-organisms, biofilms play an important role. This formation process is well known, and it is one of biofouling. Biofilms are formed on materials' surfaces by multiple steps. Firstly, bacteria attach to materials' surfaces to get nutrition (carbon compounds) which generally exists on materials' surfaces as conditioning film. The repeated process of attachment and detachment occurs. If the attachment phenomenon exceeds the detachment process, then the number of bacteria on materials' surfaces begins to increase. When this number reaches a threshold value, a signal transduction phenomenon called quorum sensing occurs. At this time, the attached bacteria simultaneously excrete polysaccharides. As a result, materials' surfaces would be covered with sticky water films. These sticky water films are called biofilms and are actually the product of microfouling. Then, the biofilm with high water content, becomes a very complex matrix that contains microorganisms (extracellular polymeric substances, EPS). In most cases, they are proteins, nucleic acids, ions (organic, inorganic) and molecules, accumulated from the aqueous environment. Since the biofilm formation makes materials' surfaces sticky, various organic and inorganic contaminants could be attached and kept on materials' surfaces. To keep the materials' surfaces free from contaminants, biofilm formation should be controlled.

From the environmental side, chemical agents such as biocides, etc. might be effective countermeasures to control fouling. On the other hand, fouling could be controlled by using materials with appropriate coatings. The coating is important because it can make a new performance (anti-fouling in this case) expressed, while the inherent properties of the material would be kept.

We have investigated the biofilm formation behaviors for many kinds of materials using some unique evaluation methods and applied coating processes to antifouling effects [20–58]. In this experiment, we focused on a polymer brush coating for the anti-fouling effect. There are already many polymer brush coatings which have been investigated and proposed so far [59–61]. All of them have future possibilities from various viewpoints. The polymer brush we selected for this experiment was one made from an ionic liquid. By applying living radical polymerization to graft polymerization with the ionic liquid, some of the authors succeeded in producing concentrated polymer brushes [62–64]. The polymer brush has lots of attractive properties such as good adherence and water repellency. Particularly, this type of polymer brush has a very low friction coefficient, which may be useful to the automobile industry. Several of the authors have tried to apply it to the decoration of solid-state polymer electrolyte.

While such advanced functions and characteristics have been investigated, antifouling properties have not been closely studied yet. However, we are gradually learning that some researchers have proposed polymer brush coatings. A number of researchers believe that most of the fouling on materials' surfaces is a direct or indirect result of biofilms. Biofilms are formed by bacterial activities as previously described. However, they are different from bacteria. Since they are actually sticky water films on materials' surfaces, they may incorporate organic and inorganic components as they grow. This might result in serious fouling. Some researchers think that polymer brush coatings could be used to repel biofilms and related contaminants [65]. However, this strongly depends on the type of brush coating used and the results of the investigations performed. Therefore, we decided to use biofilms to test the ionic liquid brush coating that we developed. Such an additional characteristic (to make a material's surface free from contaminants due to anti-fouling effects) will improve the overall performance and broaden the field of applications.

2. Materials and Methods

In this experiment, we issued a polymer brush coating made from ionic liquid. N,N-diethyl-N-methyl-N-(2-methoxyethyl) ammonium bis(trifluoromethanesulfonyl) imide (DEME-TFSI). Some substituents were added to it and N,N-diethyl-N- (2-methancryloylethy)-N-methylammonium bis(trifluoromethylsulfonyl) imide (DEMM-TFSI) was grafted on glass specimens (10 mm × 15 mm) through surface-initiated living radical polymerization to get a densely grafting polymer brush coating [66,67]. Since the polymer coating was ionic, it was basically composed of a cationic part (DEMM) and an anionic one (TFSI). However, for this paper, it was called DEMM coating. On the other hand, a PMMA was coated as a reference [68]. In this paper, it was referred to as PMMA coating. It was made on the same size glass specimens through polymerization of methyl methacrylate by Activators Regenerated by Electron Transfer Atom Transfer Radical Polymerization (ARGET ATRP) process [69,70]. Glass has always been used for the general and start-up investigations by our research groups, since they are basically inactive and can avoid reactions between the substrates and solutions.

Our polymer coatings had to be swollen so that the surface would be brush-like. The polymer coating glasses were swollen using the following processes. DEMM coating specimens were immersed in acetonitrile solution for about 12 h. Then the specimens were immersed in 50% water-50% acetonitrile mixed solution for an hour. Finally, the specimens were immersed in water for 1 h. As for the PMMA polymer coated specimens, they were immersed in tetrahydrofuran (THF) solution for about 12 h. Then they were immersed in 50% water–50% THF solution for one hour and finally, they were immersed in water for one hour. Through these steps, surface coated polymers became brush-like coating on glass specimens. And since both polymer coatings were hydrophobic, such step-by-step substitution processes were needed. The number of specimens per each measurement was three ($N = 3$). We designed the polymer brush coating according to our previous studies [62–64,66–68]. The glass plate with the concentrated ionic liquid type polymer brush (the brush length: 500 nm in dry state and the graft density: 0.15 polymer chains/nm^2) was used in this test. The graft chain length, the graft density and the brush layer thickness were determined by the gel permeation chromatography [64], thermal decomposition analysis [64] and spectroscopic ellipsometry [66], respectively.

Two kinds of bacteria were used for biofilm formation and evaluation. One of them was *Escherichia coli* (*E. coli*, K-12, G6), a Gram-negative bacteria and the other was *Staphylococcus epidermidis* (*S. epidermidis*, ATCC 35984). Both are typical non-pathogenic bacteria as Gram-negative and Gram-positive bacteria, respectively. Therefore, they were the most suitable for the general investigation of biofilm formation. *E. coli* was cultured in LB (1% tryptone–0.5% yeast extract–1% NaCl) liquid broth in advance for 18 h (±2 h). On the other hand, *S. epidermidis* was cultured in Heart Infusion (HI) liquid broth (1% heart extract–1% peptone–0.5% NaCl) for 18 h (±2 h) in advance. Both were cultured at 37 °C in a shaking incubator.

A biofilm study is generally composed of two steps. One of them is to artificially produce biofilms. On a laboratory scale, the environment to produce biofilms is called a "laboratory biofilm reactor". In this experiment, we chose a simple screening process using the 12-well plates. It is generally called the microtiter plate method. This method might provide bacteria with too much nutrition and may be a sort of deflection from the real environment. However, it could give us information for a simple, and rapid data screening process. Each specimen was placed in the well of a sterilized 12-Well plate. Each well was filled with liquid broths containing bacteria. In the case of *E. coli*, the plates were kept at 25 °C, so that the biofilm formation would be accelerated. On the other hand, the plates for *S. epidermidis* were kept at 37 °C. After one day (24 h) passed, the specimens were removed from the wells. Then, they were evaluated by using Raman spectroscopy combined with optical microscopy (NRS-3100, JASCO, Halifax, NS, Canada). Raman spectroscopy is a useful method to detect exopolymeric substances

(EPS), which are excreted from bacteria and exist as one of the biofilm components. We decided to use the Raman method for this study because ionic interactions could take place using the staining technique with crystal violet solution.

3. Results

3.1. Optical Microscopic Images

All of the specimens were observed by a technique that used optical microscopy combined with the Raman spectroscopic analyzer. In Figure 1, optical microscopic images before and after the swelling process were shown for DEMM polymer brush specimens. All of them were observed in the perpendicular direction to specimens' surfaces. Before swelling (Figure 1a), the surface was pretty smooth. However, the roughness increased after swelling and the top of fiber-like brushes could be observed in Figure 1b. In the same way, PMMA specimens showed the top of brushes after swelling (Figure 2b), while they did not show any brushes before swelling (Figure 2a). These photos clearly show that polymer brush specimens were made on glass surfaces.

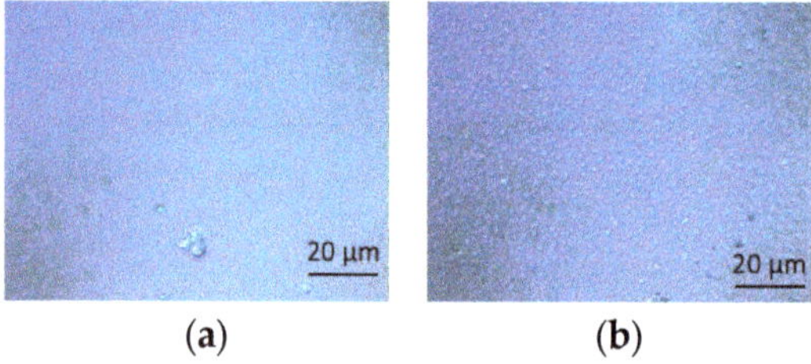

Figure 1. Optical microscopic images for N,N-diethyl-N-(2-methancryloylethy)-N-methylammonium bis(trifluoromethylsulfonyl) imide (DEMM-TFSI) specimens (**a**) before and (**b**) after selling process.

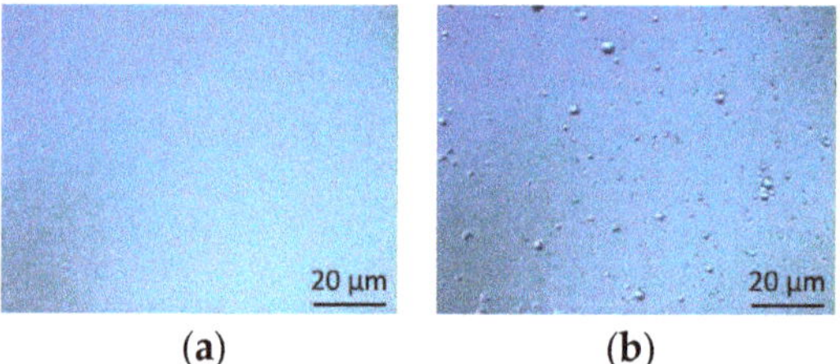

Figure 2. Optical microscopic images for Polymethyl methacryrate (PMMA) specimens (**a**) before and (**b**) after selling process.

These specimens were immersed in liquid cultures with bacteria. Figure 3 shows the surfaces of glass substrates as references after biofilm formation by *E. coli* and *S. epidermidis*, respectively. It was not easy for biofilms to form on glass specimens. However, biofilms were observed in some parts of the substrates. The scattered individual bacteria could be observed. However, some aggregates composed of bacteria were also observed. The latter should be biofilms. The light points (in the photos) are reflections of laser beams on the specimens. In both cases of bacteria (Figure 3a for the case of *E. coli* and Figure 3b for the case of *S. epidermidis*), the photos show biofilm formations on glass specimens.

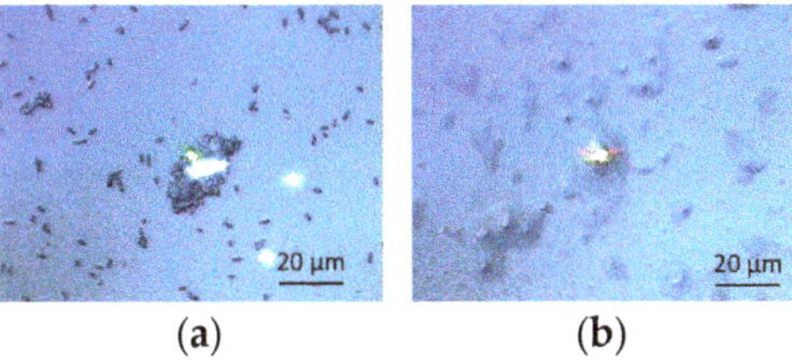

(a)   (b)

Figure 3. Optical microscopic images for glass substrates after biofilm formation: (**a**) *E. coli* and (**b**) *S. epidermidis*.

Figure 4 shows the optical microscopic images for surfaces of DEMM polymer brush specimens after biofilm formations by *E. coli* and *S. epidermidis*, respectively. Being compared with the results shown in Figures 1 and 2, surfaces of DEMM polymer brush specimens appear to be filled with contaminants. The results clearly show that biofilms were formed on the entire surfaces of the specimens.

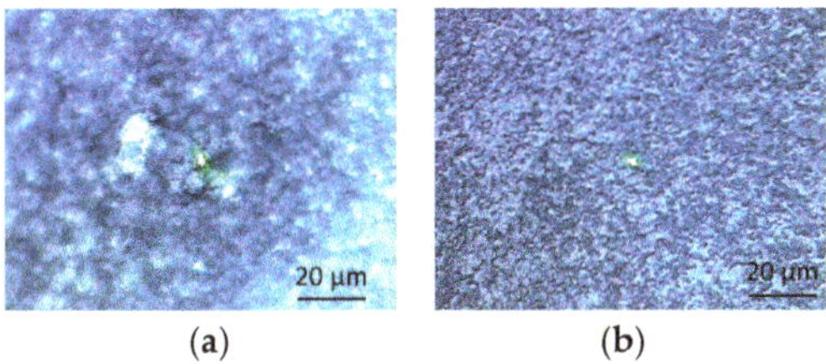

(a)   (b)

Figure 4. Optical microscopic images for DEMM polymer brush specimens after biofilm formation: (**a**) *E. coli* and (**b**) *S. epidermidis*.

Figure 5 shows the optical microscopic images for the surfaces of PMMA polymer brush specimens after biofilm formations (by these same bacteria). Figure 5a, corresponds to the result of *E. coli*, while Figure 5b refers to that of *S. epidermidis*. In both cases, biofilms can be observed in the same way. However, they seem to exist locally. There were still some places where no biofilms were observed. Even brushes could be observed in the backgrounds. The brushes in Figure 5b appear swollen through the biofilm formation process. These results show that it is more difficult for the biofilms to form on PMMA polymer brush specimens than on those of DEMM.

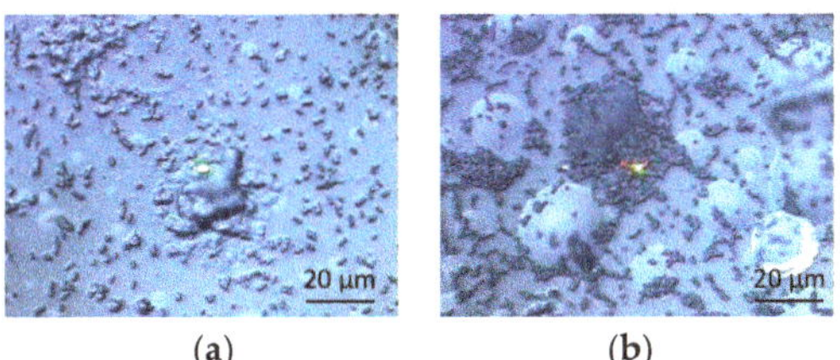

(a)   (b)

Figure 5. Optical microscopic images for PMMA polymer brush specimens after biofilm formation: (**a**) *E. coli* and (**b**) *S. epidermidis*.

3.2. Raman Scpetroscopy

All of the specimens were subjected to Raman spectroscopy and the results were analyzed to confirm the presence of biofilm by "finger print method" (Identification by comparing the data with those in the previous papers [71–81]). For each case, the same measurements were repeated three times for three specimens and the three results for each case were overlapped in the Raman shift charts basically, except for some problematic exceptional cases. Even though the measurement points on each specimen were chosen randomly, bacteria assembled and EPS was confirmed by the optical microscopy for all of those measurement points.

Before biofilm formation, the DEMM coating and PMMA coating did not show any characteristic Raman peaks. In both cases, Raman displayed small peaks of glass substrate that were observed at 1080 and 560 cm^{-1}. On the contrary, many characteristic peaks appeared for the specimens after the biofilm formation processes. Figure 6 shows the results for glass substrates. Figure 6a corresponds to the results of *E. coli*. In this case, unfortunately, the third specimen gave no robust data due to a signal failure. The first specimen showed a slightly different result from the second. From the two results, we could judge that biofilms were formed on glass substrates, since most of them could be attributed to polymers derived from biofilms or bacteria. The peak around 570 and 800 cm^{-1} could be attributed to polysaccharides or lipids [71,72]. The peak around 1100 cm^{-1} could be assigned also to that of polysaccharides or lipids [71,73]. Also, the peak around 2800 cm^{-1} could be assigned to polysaccharides [74] and lipids [75]. Figure 6b shows the results of *S. epidermidis* for glass specimens. The peak relating to nucleic acids around 2400 cm^{-1} [73] was observed in the case of *S. epidermidis*. The peak of polysaccharides and/or lipids at 2800 cm^{-1} was observed clearly. Lipid's peak was observed also at 1400 cm^{-1} for the first specimen [75]. The peak for polysaccharides/lipids at 1100 cm^{-1} [71,73] was also observed. Two peaks between 500 and 800 cm^{-1} would be related to polysaccharides/lipids [71,72].

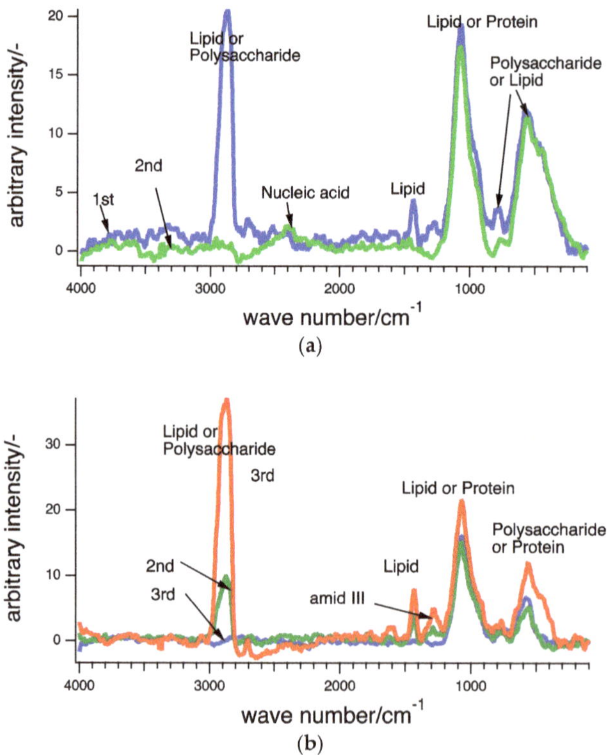

Figure 6. Raman peaks on glass substrates after biofilm formation: (**a**) *E. coli* and (**b**) *S. epidermidis*.

Figure 7 shows the results for DEMM polymer brush specimens. Figure 7a shows the results in the case of *E. coli*. All of the specimens show similar tendencies for Raman peaks, except for the broad one around 2000 cm^{-1} (peaks for stretching or vibration of triple bonds between carbon-carbon or carbon-nitrogen) [76]. However, the peak was not observed for the third specimen. For other peaks, most of them were overlapped, even though the intensities were different from each other. The peak around 2800 cm^{-1} could be attributed to those for polysaccharides and/or lipids [74,75]. The peak around 1400 cm^{-1} was considered that for lipids [74]. The small peak at 1300 cm^{-1} is highly likely that for amid III (protein) [72], while that at 1100 cm^{-1} could be assigned to that for lipids or proteins [71,73]. The broad peak around 2000 cm^{-1} corresponding of triple bonds between carbon atoms or carbon-nitrogen has been still undecidable. It suggests nitrile compounds or alkyne molecules would exist. Both might be formed through some components derived from biofilms, since the broad peaks were sometimes found in the past by us [79]. However, in this case, we used acetonitrile to swell the brush coating. It was highly likely that the chemical was remained at the upper side of coating. Figure 7b shows the results in the case of *S. epidermidis*. In this case, peaks for three specimens were almost overlapped, even though the strengths were different for each of them. The broad peak was also observed around 2000 cm^{-1} [76]. The peak around 2800 cm^{-1} could be attributed to that for lipids/polysaccharides [74,75], while lipids or proteins peaks were observed at 1100cm^{-1} [75].

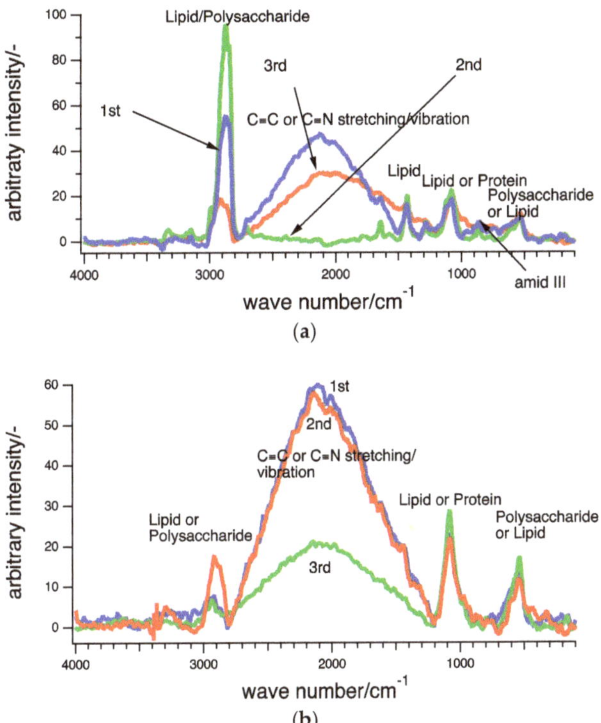

Figure 7. Raman peaks on DEMM polymer brush specimens after biofilm formation: (**a**) *E. coli* and (**b**) *S. epidermidis*.

Figure 8 shows the results for PMMA specimens. Also in these cases, most of the peaks were overlapped in both cases. In the case of *E. coli* (Figure 8a), the second specimens could not be shown due to the unintentional signal failures. In the case of *S. epidermidis*, more complicated peaks (than those in other cases) appeared, as shown in Figure 8a,b. Peaks around 2800 cm^{-1} could be attributed to either of polysaccharides [74] (at higher wave number) or lipids [75] (at lower wave number). Both peaks were almost overlapped around 2800 cm^{-1}. The peak around 1600 cm^{-1} was considered the one for polysaccharides [74]. The peak at 1400 cm^{-1} could be assigned to that for lipids [75]. Three peaks seen from 1100 to 1400 cm^{-1} belong to lipids or proteins [72,79]. We presume that the new peak just over 3000 cm^{-1} would belong to lipid [79] (However, we could not deny the possibility of nucleic acids [73]). Other new ones at 1000 cm^{-1} would be assigned to amino acid (phenylalanine) [71]. Small peaks from 550 to 800 cm^{-1} could usually be assigned to polysaccharides or lipids [71,72]. Therefore, we presume that biofilms formed on the specimen (particularly PMMA) and the sticky surface as a result of having the components of culture media incorporated into them. All of these Raman spectroscopic analyses suggest that biofilms formed on specimens' surfaces. However, they indicate that biofilms were more difficult to form on PMMA than on DEMM polymer brush coated specimens. This tendency was also supported by the results of microscopic observations. Particularly, the tendency was very remarkable for nucleic acids derived from biofilms.

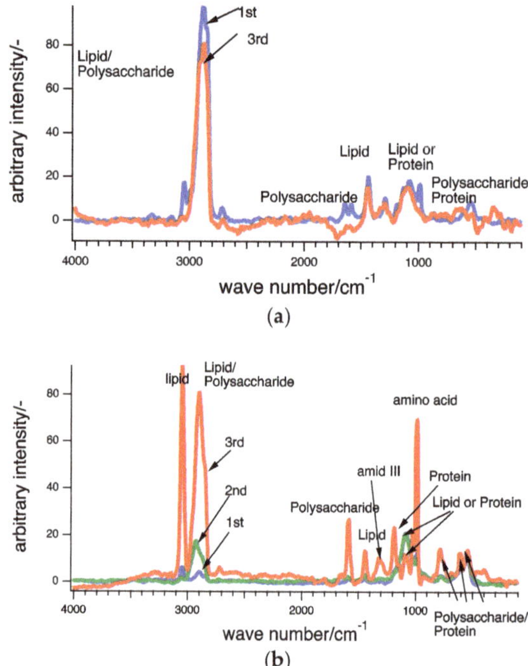

Figure 8. Raman peaks on PMMA polymer brush specimens after biofilm formation: (**a**) *E. coli* and (**b**) *S. epidermidis*.

4. Discussion

In these experiments, we vividly observed the biofilms formed on DEMM brush coated specimens. However, the contaminants derived from biofilms could be washed away easily by rinsing with water. From this viewpoint, the ionic liquid type polymer brush like DEMM would be convenient and favorable. This is because it could easily attract biofilms and the related contaminants on the materials' surfaces.

Figure 9 schematically illustrates the mechanism of antifouling for a polymer brush coating. In water and some other liquids, the polymer brush would contain water droplets between brushes (Figure 9a,b). Due to the static electrical force or mechanical one caused by geographical configurations, bacteria could get trapped and result in the formation of biofilm. However, the hydrophobic parts of the biofilms act repulsively against water drops trapped between brushes. Then they could be washed away easily as a whole (Figure 9b,c). In addition, brushes would be swollen in the aqueous solution further. The swollen effect would be added to the buoyancy and repulsion forces. Thus, biofilms could be removed. However, being compared with PMMA polymer brush coated specimens, why would DEMM polymer brush coating specimens attract bacteria-biofilms-contaminants so much?

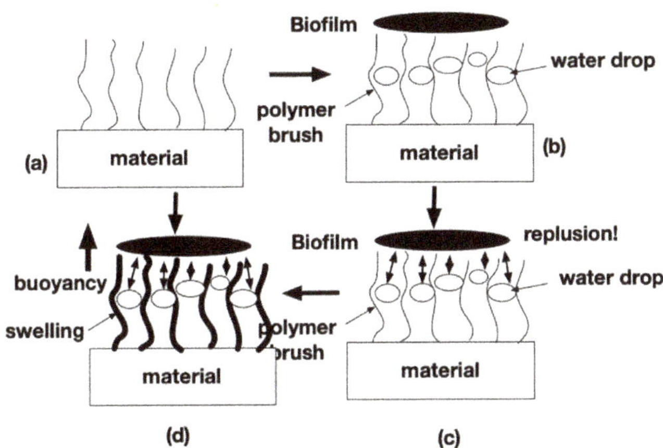

Figure 9. The mechanism of antifouling for polymer brush coating. (**a**) Polymer brush coating before immersion into a liquid solution. (**b**) Water drops are trapped in polymer brush coating, while biofilms formed by bacterial activities are attached to the top of polymer brush. (**c**) Repulsion forces are produced between water drops and hydrophobic parts of biofilms. (**d**) Water drops push biofilms up with its buoyancy and swelling action of polymer brush.

The most likely answer is due to the polar characteristics and the differences between the two types of polymer brush coated specimens. The surface of a PMMA brush coated specimen is generally neutral, while that of a DEMM brush coated specimen is polarized. Therefore, a DEMM polymer brush would tend to attract biofilms more.

From the viewpoint of merits as the ionic liquid, we could conclude that ionic liquid type polymer brush coating could attract biofilms and contaminants as a result much more in the vicinity of surfaces. Then they could be washed away by water washing [57]. The characteristic would lead to various applications such as automobiles and medical instruments. In addition, according to the practical purposes, some persons may want to remove any contaminants from the beginning. The attachment and attraction of bacteria and the following biofilm formation seem to depend on the polar characters [82]. Fortunately, the ionic liquid could arrange the polar characteristics easily. Therefore, the appropriate combination of ionic liquids and polymer brush production processes might lead to some changes of anti-biofouling effect and properties in the future, if they would be required and needed. When we come to think about those characteristics (about the application of ionic liquids to a polymer brush coating), this type of polymer brush would have a promising future as an advanced coating and material.

5. Conclusions

DEMM polymer brush coating as a liquid ionic polymer brush was investigated from the viewpoint of its biofilm formation and antifouling effect. It was also compared with the results for the PMMA polymer brush coating. Our results are demonstrated.

- Optical microscopic observations and Raman spectroscopic analyses confirmed that all specimens could form biofilms on their surfaces (more or less).
- DEMM polymer brush coating tended to form biofilms on the surfaces more than PMMA polymer brush specimens.

- Nucleic acids in biofilms could be confirmed more often for DEMM polymer brush coating specimens.
- These results suggest that the polar characteristics would play an important role in the antifouling effect related to biofilm formation.
- Ionic liquid type polymer brush coating would be beneficial to remove contaminants as a whole, due to the trapping and removing capabilities.
- Ionic liquid type polymer brush coating may change their polar characteristics leading to the antifouling effect of materials' surfaces in the future.

Author Contributions: Planning, experiments for biofilms, writing drafts, data curation, H.K. and A.O. (Atsuya Oizumi); Producing polymer brush coating specimens, discussion, advice, T.S., T.K. (Toshio Kamijo) and S.H.; Writing-review & editing, proof-reading, discussion, D.M.B.; Discussion and advice for biofilm formation and evaluation, N.H., A.O. (Akiko Ogawa), T.K. (Takeshi Kogo), D.K. and K.S., Discussion and advice for ionic liquids, K.T.; Discussion and advice from the viewpoint of application, S.-H.L. and M.-H.L.

Funding: This research was funded by JSPS KAKENHI (Grants-in-Aid for Scientific Research from Japan Society for the Promotion of Science) Grant Number 17K06826. This study was funded also by Mitsubishi Electric Corporation and the Society of Industrial Technology for Antimicrobial Articles (SIAA).

Acknowledgments: The current investigation was carried out under the project entitled Network Formation Project, 2017 by the headquarters of the National Institute of Technology (Kosen) financially (Network Formation Project, 2017). We would like to thank them for their constant encouragement. We very highly appreciate Mitsubishi Electric Corporation and its research center of advanced technology for their useful advice and financial support. Also we would like to thank the Society of Industrial Technology for Antimicrobial Articles (SIAA) as well as the Japan Food Research Laboratory (JFRL).

Conflicts of Interest: The authors declare no conflict of interest. The funders had no role in the design of the study; in the collection, analyses, or interpretation of data; in the writing of the manuscript, or in the decision to publish the results.

References

1. Fusetani, N. Biofouling and antifouling. *Nat. Prod. Rep.* **2004**, *21*, 94–104. [CrossRef] [PubMed]
2. Yebra, D.M.; Kiil, S.; Dam-Johansen, K. Antifouling technology—Past, present and future steps towards efficient and environmentally friendly antifouling coatings. *Prog. Org. Coat.* **2004**, *50*, 75–104. [CrossRef]
3. Chambers, L.D.; Stokes, K.R.; Walsh, F.C.; Wood, R.J. Modern approaches to marine antifouling coatings. *Surf. Coat. Technol.* **2006**, *201*, 3642–3652. [CrossRef]
4. Banerjee, I.; Pangule, R.C.; Kane, R.S. Antifouling coatings: Recent developments in the design of surfaces that prevent fouling by proteins, bacteria, and marine organisms. *Adv. Mater.* **2011**, *23*, 690–718. [CrossRef] [PubMed]
5. Lejars, M.; Margaillan, A.; Bressy, C. Fouling release coatings: A nontoxic alternative to biocidal antifouling coatings. *Chem. Rev.* **2012**, *112*, 4347–4390. [CrossRef] [PubMed]
6. Swain, G.W.; Schultz, M.P. The testing and evaluation of non-toxic antifouling coatings. *Biofouling* **1996**, *10*, 187–197. [CrossRef] [PubMed]
7. Brady, R.F., Jr. A fracture mechanical analysis of fouling release from nontoxic antifouling coatings. *Prog. Org. Coat.* **2001**, *43*, 188–192. [CrossRef]
8. Gudipati, C.S.; Greenlief, C.M.; Johnson, J.A.; Prayongpan, P.; Wooley, K.L. Hyperbranched fluoropolymer and linear poly(ethylene glycol) based amphiphilic crosslinked networks as efficient antifouling coatings: An insight into the surface compositions, topographies, and morphologies. *J. Polym. Sci. Part A Polym. Chem.* **2004**, *42*, 6193–6208. [CrossRef]
9. Hellio, C.; Yebra, D. *Advances in Marine Antifouling Coatings and Technologies*, 1st ed.; Woodhead Publishing: Cambridge, UK, 2009.
10. Casse, F.; Swain, G.W. The development of microfouling on four commercial antifouling coatings under static and dynamic immersion. *Int. Biodeterior. Biodegrad.* **2006**, *57*, 179–185. [CrossRef]
11. Mittal, K.L. Surface contamination: An overview. In *Surface Contamination—Genesis, Detection, and Control*, 1st ed.; Springer: Boston, MA, USA, 1979; pp. 3–45.

12. Dobretsov, S.; Thomason, J.; Williams, D.N. *Biofouling Methods*, 1st ed.; Wiley-Blackwell: Hoboken, NJ, USA, 2014.
13. Dürr, S.; Thomason, J.C. *Biofouling*, 1st ed.; Wiley-Blackwell: Hoboken, NJ, USA, 2009.
14. Corpe, W.A. Microfouling: The role of primary film forming marine bacteria. In Proceedings of the 3rd International Congress of Marine Corrosion and Fouling, Gaithersburg, MD, USA, 2–6 October 1972; Northwestern University Press: Evanston, IL, USA, 1973.
15. Claudi, R.; Mackie, G.L. *Practical Manual for the Monitoring and Control*; Lewis Publishers: Boca Raton, FL, USA, 1994.
16. Kanematsu, H.; Barry, D.M. *Biofilm and Materials Science*; Springer: New York, NY, USA, 2015.
17. Lewandowski, Z.; Beyenal, H. *Fundamentals of Biofilm Research*, 2nd ed.; CRC Press: Boca Raton, FL, USA, 2014; p. 642.
18. Kalmbach, S.; Manz, W.; Szewzyk, U. *Biofilms—Investigative Methods & Applications*; CRC Press: Boca Raton, FL, USA, 2000.
19. Flemming, H.C.; Murthy, S.P.; Venkatesan, R.; Cooksey, K. *Marine and Industrial Biofouling*; Springer: Berlin, Germany, 2009.
20. Kanematsu, H.; Ikigai, H.; Campbell, S.; Beech, I. Sn-Ag alloy plating films mitigating biofilm formation. In Proceedings of the National Association for Surface Finishing Annual Technical Conference 2009, Louisville, KY, USA, 16–17 June 2009; Curran Associates, Inc.: Red Hook, NY, USA, 2009; pp. 406–415.
21. Kanematsu, H.; Ikigai, H.; Yoshitake, M. Evaluation of various metallic coatings on steel to mitigate biofilm formation. *Int. J. Mol. Sci.* **2009**, *10*, 559–571. [CrossRef] [PubMed]
22. Kanematsu, H.; Ikegai, H.; Kuroda, D. Biofilm and metallic materials. *Rust Prev. Control Jpn.* **2011**, *55*, 369–377.
23. Kanematsu, H.; Kougo, T.; Kuroda, D.; Itoh, H.; Noorani, R. Prevention coating against biofilm formation. In Proceedings of the Sur/Fin Asia-Pacific 2012, Singapore, 4–7 December 2012; pp. 52–56.
24. Kanematsu, H.; Kuroda, D.; Itoh, H.; Ikigai, H. Biofilm formation of a closed loop system and its visualization. *CAMP-ISIJ* **2012**, *25*, 753–754.
25. Kanematsu, H.; Kuroda, D.; Itoh, H.; Kobayashi, T.; Noorani, R.; Yamamoto, M.; Tsukuda, I. Chemical surface treatment for prevention against biofilm formation. In Proceedings of the Sur/Fin Asia-Pacific 2012, Singapore, 4–7 December 2012; pp. 57–62.
26. Kanematsu, H.; Kuroda, D.; Koya, S.; Itoh, H. Development of production process on labo scale for biofilm formation by immersion into closed circulation water system. *J. Surf. Finish. Soc. Jpn.* **2012**, *63*, 459–461. [CrossRef]
27. Kanematsu, H.; Miura, Y.; Itoh, H.; Ikigai, H.; Tanaka, M. Condensation behavior of silicon into biofilm formed by germs in ambient atmosphere in a laboratory and its observation. *CAMP-ISIJ* **2012**, *25*, 810–811. (In Japanese)
28. Kanematsu, H.; Hihara, T.; Ishihara, T.; Imura, K.; Kogo, T. Evaluation for corrosion resistance of nano-cluster layer and biofilm formation. *Int. J. Eng. Sci. Res. Technol.* **2013**, *2*, 2424–2432.
29. Kanematsu, H.; Hirai, N.; Miura, Y.; Itoh, H.; Kuroda, D.; Umeki, S. Biofilm leading to corrosion on material surface and the moderation by alternative electro-magnetic field. In Proceedings of the Materials Science and Technology Conference and Exhibition 2013 (MS&T 2013), Monatreal, QC, Canada, 27–31 October 2013; pp. 2761–2767.
30. Kanematsu, H.; Hirai, N.; Miura, Y.; Itoh, H.; Masuda, T.; Kuroda, D. Evaluation technique for biofilm formed on biomaterials. In Proceedings of the 4th International Symposium on Advanced Materials Development and Integration of Novel Structured Metallic and Inorganic Materials, Nagoya, Japan, 13–15 December 2013; Nagoya University: Nagoya, Japan, 2013.
31. Kanematsu, H.; Hirai, N.; Miura, Y.; Tanaka, M.; Kogo, T.; Itoh, H. Various metals from water by biofilm from an ambient germs in a reaction container. In Proceedings of the Materials Science and Technology Conference and Exhibition 2013 (MS&T 2013), Montreal, QC, Canada, 27–31 October 2013; pp. 2154–2161.
32. Kanematsu, H.; Itoh, H.; Miura, Y.; Masuda, T.; Kuroda, D.; Hirai, N.; Barry, D.M.; McGrath, P.B. Biofilm formation on polymer materials by a laboratory acceleration reactor. In Proceedings of the 4th International Symposium on Advanced Materials Development and Integration of Novel Structured Metallic and Inorganic Materials, Nagoya, Japan, 13–15 December 2013; Nagoya University: Nagoya, Japan, 2013.

33. Kougo, T.; Yamamoto, Y.; Naiki, A.; Ogino, Y.; Kanzaki, T.; Kanematsu, H.; Ikigai, H.; Itoh, H. Proposal for Evaluation to Biofilm Formation on Metal Oxides in Atmospheric Exposure Type Circulation Water System. *Mem. Suzuka Natl. Coll. Technol.* **2013**, *46*, 81–84.

34. Kanematsu, H.; Kogo, T.; Itoh, H.; Wada, N.; Yoshitake, M. Fogged glass by biofilm formation and its evaluation. In Proceedings of the Materials Science and Technology Conference and Exhibition 2013 (MS&T 2013), Montreal, QC, Canada, 27–31 October 2013; pp. 2427–2433.

35. Kanematsu, H.; Kogo, T.; Kuroda, D.; Itoh, H.; Kirihara, S. Biofilm formation and evaluation for spray coated metal films on laboratory scale. In Proceedings of the Thermal Spray 2013—Innovative Coating Solutions for the Global Economy, Busan, Korea, 13–15 May 2013; pp. 520–525.

36. Kanematsu, H.; Kougo, T.; Kuroda, D.; Ogino, Y.; Yamamoto, Y. Biofilm formation derived from ambient air and the characteristics of apparatus. *J. Phys. Conf. Ser.* **2013**, *433*, 012031. [CrossRef]

37. Kanematsu, H.; Okura, Y.; Hirai, N.; Miura, Y.; Itoh, H.; Tanaka, M. Condensation of some metal elements in biofilm. *CAMP-ISIJ* **2013**, *26*, 417. (In Japanese)

38. Ikegai, H.; Kobayashi, M.; Ken-ichi Iimura; Hosokawa, A.; Uesugi, K.; Korda, D.; Kanematsu, H.; Toda, H. X-ray computed tomography for observing microbiologically influenced corrosion of carbon steel caused by Pseudomonas aeruginosa biofilm sig synchrotron radiation in Spring-8. *Bact. Adherence Biofilm* **2014**, *28*, 67–70. (In Japanese)

39. Kanematsu, H.; Kogo, T.; Noda, M.; Hirai, N.; Ogawa, A.; Miura, Y.; Itoh, H.; Yoshitake, M. Composite coating to control biofilm formation and MIC. In Proceedings of the 17th International Congress on Marine Corrosion and Fouling (ICMCF), National University of Singapore, Singapore, 6–10 July 2014; p. 101.

40. Kanematsu, H.; Kogo, T.; Sano, K.; Noda, M.; Wada, N.; Yoshitake, M. Nano-composite coating on glasses for biofilm control. *J. Mater. Sci. Surf. Eng.* **2014**, *1*, 58–63.

41. Kanematsu, H.; Noda, M.; Hirai, N.; Ogawa, A.; Kogo, T.; Miura, Y.; Ito, H. A trial for MIC study using a circulation-type laboratory biofilm reactor. In Proceedings of the 17th International Congress on Marine Corrosion and Fouling (ICMCF), National University of Singapore, Singapore, 6–10 July 2014; p. 75.

42. Kanematsu, H.; Tanaka, M. Biofilm analyses and their importance in materials science and engineering. *Bunseki Kagaku* **2014**, *63*, 569–580. (In Japanese) [CrossRef]

43. Ikegai, H.; Hirai, N.; Kobayashi, M.; Toda, H.; Moroboshi, T.; Ikeda, T.; Ikeda, S.; Uesugi, K.; Kuroda, D.; Kanematsu, H. Microbiologically influenced corrosion to carbon steel and biomineralization caused by pseudomonas aeruginosa biofilm. *Bact. Adherence Biofilm* **2015**, *29*, 93–96. (In Japanese)

44. Kanematsu, H.; Sasaki, S.; Miura, Y.; Kogo, T.; Sano, K.; Wada, N.; Yoshitake, M.; Tanaka, T. Composite coating to control biofilm formation and effect of alternate electro-magnetic field. *Mater. Technol.* **2015**, *30*, 21–26. [CrossRef]

45. Ogawa, A.; Noda, M.; Kanematsu, H.; Sano, K. Application of bacterial16S rRNAgene analysis to a comparison of the degree of biofilm formation on the surface of metal coated glasses. *Mater. Technol.* **2015**, *30*, 61–65. [CrossRef]

46. Kanematsu, H.; Kudara, H.; Kanesaki, S.; Kogo, T.; Ikegai, H.; Ogawa, A.; Hirai, N. Application of a loop-type laboratory biofilm reactor to the evaluation of biofilm for some metallic materials and polymers such as urinary stents and catheters. *Materials* **2016**, *9*, 824–834. [CrossRef] [PubMed]

47. Kanematsu, H.; Umeki, S.; Hirai, N.; Miura, Y.; Wada, N.; Kogo, T.; Tohji, K.; Otani, H.; Okita, K.; Ono, T. Verification of effect of alternative electromagnetic treatment on control of biofilm and scale formation by a new laboratory biofilm reactor. *Ceram. Trans.* **2016**, *259*, 199–212. [CrossRef]

48. Kanematsu, H.; Umeki, S.; Ogawa, A.; Hirai, N.; Kogo, T.; Tohji, K. The cleaning effect on metallic materials under a weak alsternating electromagnetic field and biofilm. In Proceedings of the Ninth Pacific Rim International Conference on Advanced Materials and Processing (PRICM9), Kyoto, Japan, 1 August 2016; pp. 1–4.

49. Sano, K.; Kanematsu, H.; Kogo, T.; Hirai, N.; Tanaka, T. Corrosion and biofilm for a composite coated iron observed by FTIR-ATR and Raman spectroscopy. *Trans. Inst. Mater. Finish.* **2016**, *94*, 139–145. [CrossRef]

50. Kanematsu, H. Evaluation and countermeasure for biofilms on inorganic surface. In *Biofilm Structure and Formation for Biocontrol and Countermeasure of Biofilms and Its Growth*; Matsumura, Y., Ed.; CMC Publisher: Osaka, Japan, 2017; pp. 162–189.

51. Kanematsu, H.; Barry, D.M.; Ikegai, H.; Yoshitake, M.; Mizunoe, Y. Biofilm evaluation methods outside body to inside—Problem presentations for the future. *Med. Res. Arch.* **2017**, *5*, 1469.

52. Kanematsu, H.; Saito, T.; Barry, D.M.; Hirai, N.; Kogo, T.; Ogawa, A.; Tsunashima, K. Effects of ionic liquids on biofilm formation in a loop-type laboratory biofilm reactor. *ECS Trans.* **2017**, *80*, 1147–1155. [CrossRef]

53. Kanematsu, H.; Satoh, M.; Shindo, K.; Barry, D.M.; Hirai, N.; Ogawa, A.; Kogo, T.; Utsumi, Y.; Yamaguchi, A.; Ikegai, H.; et al. Biofilm formation behaviors on graphene by *E. coli* and *S. epidermidis*. *ECS Trans.* **2017**, *80*, 1167–1175. [CrossRef]

54. Sano, K.; Masuda, T.; Kanematsu, H.; Yokoyama, S.; Hirai, N.; Ogawa, A.; Kougo, T.; Yamazaki, K.; Tanaka, T. Biofouling on mortar mixed with steel slags in a laboratory biofilm reactor. In Proceedings of the Irago Conference 2016, Irago, Aichi, Japan, 1–2 November 2016; p. 02004.

55. Kanematsu, H. A new international standard for testing antibacterial effects. *Adv. Mater. Process.* **2017**, *175*, 26–29.

56. Kanematsu, H.; Maeda, S.; Barry, D.M.; Umeki, S.; Tohji, K.; Hirai, N.; Ogawa, A.; Kogo, T.; Ikegai, H.; Mizunoe, Y. Effects of elastic waves at several frequencies on biofilm formation in circulating laboratory biofilm reactors. *Ceram. Trans.* **2018**, *265*, 43–51.

57. Kanematsu, H.; Oizumi, A.; Sato, T.; Kamijo, T.; Honma, S.; Barry, D.M.; Hirai, N.; Ogawa, A.; Kogo, T.; Kuroda, D.; et al. Polymer brush made by ionic liquids and the inhibition effects for biofilm formation. *ECS Trans.* **2018**, *85*, 1089–1095. [CrossRef]

58. Kanematsu, H.; Shindo, K.; Barry, D.M.; Hirai, N.; Ogawa, A.; Kuroda, D.; Kogo, T.; Sano, K.; Ikegai, H.; Mizunoe, Y. Electrochemical responses of graphene with biofilm formation on various metallic substrates by using laboratory biofilm reactors. *ECS Trans.* **2018**, *85*, 491–498. [CrossRef]

59. Milner, S.T.; Witten, T.A.; Cates, M.E. Theory of the grafted polymer brush. *Macromolecules* **1988**, *21*, 2610–2619. [CrossRef]

60. Milner, S.T. Polymer brushes. *Science* **1991**, *251*, 905–914. [CrossRef] [PubMed]

61. Zhou, F. *Antifouling Surfaces and Materials—From Land to Marine Environment*, 1st ed.; Springer: Berlin/Heidelberg, Germany, 2015.

62. Sato, T.; Morinaga, T.; Marukane, S.; Narutomi, T.; Igarashi, T.; Kawano, Y.; Ohno, K.; Fukuda, T.; Tsuji, Y. Novel solid-state polymer electrolyte of colloidal crystal decorated with ionic-liquid polymer brush. *Adv. Mater.* **2011**, *23*, 4868–4872. [CrossRef] [PubMed]

63. Arafune, H.; Kamijo, T.; Morinaga, T.; Honma, S.; Sato, T.; Tsuji, Y. A robust lubrication system using an ionic liquid polymer brush. *Adv. Mater. Interfaces* **2015**, *2*, 1500187. [CrossRef]

64. Morinaga, T.; Honma, S.; Ishizuka, T.; Kamijo, T.; Sato, T.; Tsuji, Y. Synthesis of monodisperse silica particles grafted with concentrated ionic liquid-type polymer brushes by surface-initiated atom transfer radical polymerization for use as a solid state polymer electrolyte. *Polymers* **2016**, *8*, 146–159. [CrossRef]

65. Hideyuiki, K.; Barry, D.M.; Ikegai, H.; Yoshitake, M.; Mizunoe, Y. Nanofibers and biofilm in materials science. In *Handbook of Nanofibers. Vol. I: Fundamental Aspects, Experimental Setup, Synthesis, Properties and Physicochemical Characterization*; Barhoum, A., Bechelany, M., HamdyMakhlouf, A.S., Eds.; Springer Nature Switzerland AG: Basel, Switzerland, 2018; pp. 1–21.

66. Yamamoto, S.; Ejaz, M.; Tsujii, Y.; Fukuda, T. Surface interaction forces of well-defined, high-density polymer brushes studies by atomic force microscopy. 2. Effect of graft density. *Macromolecules* **2000**, *33*, 5608–5612. [CrossRef]

67. Tsujii, Y.; Ohno, K.; Yamamoto, S.; Goto, A.; Fukuda, T. Structure and properties of high-density polymer brushes prepared by surface-initiated living radical polymerization. In *Surface-Initiated Polymerization*; Jordan, R., Ed.; Springer Nature: Heidelberg, Germany, 2006; pp. 1–45.

68. Nomura, A.; Ohno, K.; Fukuda, T.; Sato, T.; Tsujii, Y. Lubrication mechanisms of concentrated polymer brushes in solvents: Effect of solvent viscosity. *Polym. Chem.* **2012**, *3*, 148–153. [CrossRef]

69. Zhu, B.; Edmondson, S. ARGET ATRP: Procedure for PMMA polymer brush growth. Available online: https:www.sigmaaldrich.com/technical-documents/articles/crp-guide/arget-atrp-procedure-for-pmma-polymer-brush-growth.html (accessed on 7 November 2018).

70. Zhu, B.; Edmondson, S. Polydopamine-melanin initiators for surface-initiated ATRP. *Polymer* **2011**, *52*, 2141–2149. [CrossRef]

71. Chao, Y.; Zhang, T. Surface-enhanced Raman scattering (SERS) revealing chemical variation during biofilm formation: From initial attachment to mature biofilm. *Anal. Bioanal. Chem.* **2012**, *404*, 1465–1475. [CrossRef] [PubMed]

72. Lakin, P. General Outline and Strategies for IR and Raman Spectral Interpretation. In *Infrared and Raman Spectroscopy–Principles and Spectral Interpretation*; Elsevier: Burlington, MA, USA, 2011; pp. 117–133.

73. George, J.; Thomas, J. Raman spectroscopy of protein and nucleic acid assmeblies. *Annu. Rev. Biophys. Biomol. Struct.* **1999**, *28*, 1–27.

74. Yuen, S.N.; Choi, S.M.; Phillips, D.L.; Ma, C.Y. Raman and FTIR spectroscopic study of carboxymethylated non-starch polysaccharides. *Food Chem.* **2009**, *114*, 1091–1098. [CrossRef]

75. Czamara, K.; Majzner, K.; Pacia, M.Z.; Kochan, K.; Kaczor, A.; Baranska, M. Raman spectroscopy of lipids: A review. *J. Raman Spectrosc.* **2015**, *46*, 4–20. [CrossRef]

76. Lakin, P. IR and Raman spectra-structure correlations: Characteristic group frequencies. In *Infrared and Raman Spectroscopy–Principles and Spectral Interpretation*; Elsevier: Amsterdam, The Netherland, 2011; pp. 73–115.

77. Cao, B.; Shi, L.; Brown, N.R.; Xiong, Y.; Fredrickson, K.J.; Fomine, F.M.; Marshall, M.J.; Lipton, M.S.; Beyenal, H. Extracellular polymeric substances from Shewanella sp. HRCR-1 biofilms: Characterization by infrared spectroscopy and proteomics. *Environ. Microbiol.* **2011**, *13*, 1018–1031. [CrossRef] [PubMed]

78. Sano, K.; Kanematsu, H.; Hirai, N.; Tanaka, T. Preparation and its anti-biofouling effect observation of organic metal dispersed silane based composite coating. *J. Surf. Finish. Soc. Jpn.* **2016**, *67*, 268–273. [CrossRef]

79. Ogawa, A.; Kiyohara, T.; Kobayashi, Y.H.; Sano, K.; Kanematsu, H. Nickel, moluybdenum, and tungsten nanoparticle-dispersed alkylalkoxysilane polymer for biomaterial coating: evaluation of effects on bacterial biofilm formation and biosafety. *Biomed. Res. Clin. Pract.* **2017**, *2*, 1–7. [CrossRef]

80. Nyquist, R.A. *Interpreting Infrared, Raman, and Nuclear Magnetic Resonance Spectra*; Academic Press: San Diego, CA, USA, 2001.

81. Hamasha, K.M. Raman Spectroscopy for the Microbiological Characterization and Identification of Medically Relevant Bacteria. Ph.D. Thesis, Wayne State University, Detroit, MI, USA, 2011.

82. Rzhepishevska, O.; Hakobyan, S.; Ruhal, R.; Gautrot, J.; Barbero, D.; Ramstedt, M. The surface charge of anti-bacterial coatings alters motility and biofilm architecture. *Biomater. Sci.* **2013**, *1*, 589–602. [CrossRef]

 *coatings* 

Article

Anti-Fouling Ceramic Coating for Improving the Energy Efficiency of Steel Boiler Systems

Minh Dat Nguyen [1,2], Jung Won Bang [1], Young Hee Kim [1], An Su Bin [1], Kyu Hong Hwang [2], Vuong-Hung Pham [3] and Woo-Teck Kwon [1,*]

1   Energy Materials Centre, Korea Institute of Ceramic Engineering and Technology, Jinju 52851, Korea; mdat.hut@gmail.com (M.D.N.); bangchnocrat@kicet.re.kr (J.W.B.); yhkokim@kicet.re.kr (Y.H.K.); asb882@gmail.com (A.S.B.)
2   Department of Ceramic Engineering, Gyeongsang National University, Jinju 52828, Korea; khhwang@gnu.ac.kr
3   Advanced Institute for Science and Technology (AIST), Hanoi University of Science and Technology (HUST), No. 01, Dai Co Viet Road, Hanoi, Vietnam; vuong.phamhung@hust.edu.vn (V.-H.P.)
*   Correspondence: wtkwon@kicet.re.kr; Tel.: +82-10-8839-4690

Received: 26 July 2018; Accepted: 28 September 2018; Published: 2 October 2018

Abstract: Boilers are systems used mainly to generate steam in industries and waste-to-energy facilities. During operation, heat transfer loss occurs because a fouling layer with low thermal conductivity is deposited on the external surfaces of the boiler tube system, which contributes to the overall poor energy efficiency of waste-to-energy power plants. To overcome the fouling problem, a ceramic coating was developed and applied to carbon steel with a simple and inexpensive coating method. Anti-fouling testing, thermal conductivity measurement, and microstructure observation were performed to evaluate the performance of the coating. All evaluated properties of the coating were found to be excellent. The developed ceramic coating can be applied to boiler tubes in a real facility to protect them from the fouling problem and improve their energy efficiency.

Keywords: ceramic coating; fly ash; anti-fouling; slagging; boiler; energy efficiency

---

1. Introduction

Energy conservation and environmental protection are currently very important issues worldwide. Municipal solid waste (MSW), which is an inevitable product of modern society, is one of the most serious urban pollution sources and one of the greatest challenges for future generations. Deep concerns over global warming have led to numerous studies on energy efficiency and renewable energy [1]. MSW can be converted to an eco-friendly renewable resource that not only produces energy but also significantly reduces the greenhouse gas emissions from landfills. A waste-to-energy (WTE) incineration plant recovers energy from MSW and produces electricity and/or steam for heating [2]. For a WTE plant, the energy efficiency can be improved and CO_2 emissions be reduced by improving the heat efficiency of the boiler system because this is the main method for heat transfer. However, one of the most serious problems in WTE facilities is that unwanted mineral matter such as slagging and fouling is always deposited on the heating surfaces of the boiler. As a general definition, fouling is the deposition and accumulation of unwanted materials such as scale, algae, suspended solids, and insoluble salts on the internal or external surfaces of processing equipment, including boilers and heat exchangers [3]. In WTE plants, ash fouling is the result of several physical and chemical processes. Ash fouling, which causes serious corrosion and erosion problems, builds up on the heat transfer surfaces of boilers to act as an insulator; this greatly reduces the global heat transfer of the boiler and minimizes the yield of the plant [4]. In one study, a typical 1–1.5 mm fouling build-up on the boiler surface can increase fuel consumption by about 3%–8% [4,5]. In another study, a 0.03 in (0.8 mm) thick fouling layer caused a 9.5% reduction in the heat transfer, and a fouling layer thickness of 0.18 in (4.5 mm) caused a 69% reduction in an extreme case [6]. When this occurs, the boiler heat transfer surface should be cleaned. However, the operating costs from working to clean fouling is extremely high. Therefore, it is necessary to prevent the deposition and accumulation of unwanted matter on the boiler.

A small improvement in the boiler efficiency will clearly help save a large amount of fuel and to reduce CO_2 emissions. In a literature review, many researchers were found to have focused on improving boiler efficiency. For example, Gopal et al. [7] studied losses in boilers. Nussbaumer et al. [8] reported a measure to save boiler energy by improving the combustion efficiency. Gao et al. [9] and Karell et al. [10] investigated the role of maintenance on boiler energy conservation. Most of these researchers did not study methods to control or prevent the formation of a fouling layer on the heating surface of a boiler. To address this shortcoming of previous approaches, an anti-fouling coating to prevent ash from attaching was developed in this study to protect boilers from fouling agents and increase their thermal conductivity efficiency.

Among the various coating systems used for boiler systems, polymer-derived ceramic coatings (PDCs), glass, and glass–ceramic coatings have the advantages of chemical inertness, high temperature stability, and superior mechanical properties compared to other materials suitable for spraying [11]. Moreover, advanced thermal spraying processes such as plasma spray and high-velocity oxy-fuel (HVOF) are also typically used to deposit coatings on the surface of boilers to enhance their high-temperature oxidation resistance but they are quite expensive because of high energy consumption [12]. Until now, however, most coating techniques just focused on improving the wear, oxidation, and corrosion resistance of the boiler surface [13–17]. Few have been used to mitigate and prevent fouling problems. In this study, an anti-fouling ceramic coating was developed for application to the external surfaces of a boiler by using a low-cost slurry spray coating method.

2. Material and Methods

The experimental procedure was performed as follows: (i) prepare the starting materials and ceramic coating materials, (ii) apply the ceramic coating materials to the steel substrates, (iii) cure and heat-treat the coated substrates, and (iv) characterize and evaluate the coatings.

2.1. Starting Materials and Ceramic Coating Preparation

JIS S45C steel is the most common material used in boiler tubes and was used in this experiment. Table 1 presents the chemical composition. Steel plates with dimensions of 10 mm × 10 mm × 2 mm were cut and then cleaned in ethanol to remove all dust and oil from their surfaces prior to coating deposition.

Table 1. Chemical composition of the steel substrate (wt %).

C	Si	Mn	P	S	Cr	Ni	Cu	Fe
0.42–0.48	0.15–0.35	0.6-0.9	<0.03	<0.035	<0.2	<0.2	<0.3	96.61−97.17

Potassium silicate (water glass with concentration of 37 wt %–a product of Young Il Chemical Co., Ltd., City, Country) was used as a binder because of its strong bond with the steel substrate. To enhance the properties of the coating slurry, several kinds of active and passive fillers with particle sizes of 2–25 μm were added to the binder. Table 2 lists some of their properties. Günthner et al. [18] selected some passive fillers (BN; Si_3O_4; ZrO_2) to enable the processing of thick, dense, and crack-free composite coating systems. In our present study, passive fillers (Al_2O_3; SiO_2) were used due to their good compatibility with the alkali silicate glass and low oxygen permeability to enhance high-temperature oxidation resistance of steel. By adding these fillers, it is possible not only to generate coating with high thermal conductivity than glass coating but also to reduce the volume change of glass binder during crystallization at high temperature [18]. The dissolution of the Al_2O_3 and SiO_2 inclusions turned the binary glass (K_2O–SiO_2) into a ternary glass (K_2O–Al_2O_3–SiO_2) which may improve the chemical stability of the glass [19]. Active filler (flake Al) was added to achieve crack-free coating. In the meantime, CoO and NiO was doped to improve the adhesion between the coating and substrate and reduce the porous structure of the coating [11,20]. Various parameters such as the mass fraction of the fillers, the viscosity, the dispersing methods were varied to optimize the coating system. As a result, a mixture of binder and fillers with the chemical composition listed in Table 3 was then mixed with a mixer. The mixture was ground at 400 rpm for 240 min using a ball mill such that the fillers were dispersed homogeneously in the binder. Finally, a coating slurry with solid concentration of 43 wt % was filtered and stored.

Table 2. Some properties of chosen binders and fillers (manufacturer's data).

Binder and Fillers	Average Particle Size (μm)	SiO_2–K_2O Molar Ratio	Viscosity (MPa·s) at 20 °C	Softening Temperature
Potassium silicate (PS)	—	3.2–3.4	Low (50)	640–680 °C
Flake Al	25	—	—	—
Al_2O_3	11	—	—	—
SiO_2	12.5	—	—	—
NiO	1.7	—	—	—
CoO	3.6	—	—	—

Table 3. Chemical composition (wt %) of the slurry mixture.

PS	Al_2O_3	Al	NiO	CoO	SiO_2
66.1	6.6	0.8	3.3	3.3	19.8

The ceramic coating was sprayed on the steel plates by an air gun (W-71, Anest Iwata, Taiwan). The steel specimens were placed about 15–20 cm in front of the gun nozzle. All surfaces of the steel substrates were sequentially covered with the glass ceramic slurry by changing the surface facing the gun nozzle. To obtain a thicker glass ceramic coating, the spraying process was performed twice. Next, the specimens were cured for 2 h at 95 °C and finally sintered in a box furnace at 800 °C for 24 h. The sintered temperature was selected based on the real operating temperature of a boiler in a WTE plant.

2.2. Fly Ash and Coating Characterization

Four fly ashes from four different Korean incinerators were considered because they were very rich in alkali, chlorine, and calcium. The fly ash samples were analyzed at the Test & Standard Center of Korea Institute of Ceramic and Engineering (KICET) by scanning electron microscopy (SEM) equipped with backscattered and secondary electron detectors coupled with energy-dispersive X-ray spectroscopy (EDS) (SEM; JSM-6700F, JEOL, Tokyo, Japan), X-ray diffraction (XRD), a laser diffraction particle size analyzer (PSA) (Mastersizer S Ver. 2.15, Malvern Instruments, Malvern, UK), and inductively coupled plasma optical emission spectrometry (ICP-OES) (Perkin Elmer OPTIMA 8300, PerkinElmer, Inc., Waltham, MA, USA).

The coatings were observed with a camera (EOS 600D, Canon, Tokyo, Japan) to determine the macroscopic morphology and then further investigated by an optical microscopy. SEM-EDS and XRD were used for the microstructure characterization. To prepare cross-sectional surfaces for SEM observation, standard metallographic polishing techniques were performed; the coated specimens were cold-mounted in epoxy resin and then sequentially polished with 200, 800, 1200, and 2000 grit SiC abrasive paper.

2.3. Adhesion and Thermal Shock Testings

In the present work we used cross-cut tape test (ASTM D3359) [21] and pull-off test (ASTM D4541) [22] to obtain the adhesion between the coating and the substrate. For thermal shock testing, the coated specimen was subjected to 800 °C in a furnace for 5 min and subsequently removed using a pair of pliers and placed in a water tank (20 °C) for 5 min. The process was repeated and the total number of thermal cycles was determined up to the time when macroscopic cracks or failure appeared.

2.4. Anti-Fouling Testing

The anti-fouling performance of the developed coating against fly ash was studied. To study the anti-fouling ability of the coating, the slurry containing ash was used to observe the ash attachment on the coating surface. To do this, coated and non-coated steel plates were immersed in the fly ash solution and then dried at 50 °C for 1 h. After 24 h of sintering in the box furnace at 800 °C for 24 h, the specimens were cooled to room temperature (20–25 °C). A compressed air gun (0.5 MPa) was then used to blow the fly ash powder-deposited surface of each specimen for 1 min. The air gun nozzle was placed 20 cm away from the specimens. Subsequently, the air-blown specimens were examined with SEM-EDS. Figure 1 shows the entire experimental process. In addition, the prepared composite coatings were applied to a real-life boiler system to assess their effectiveness.

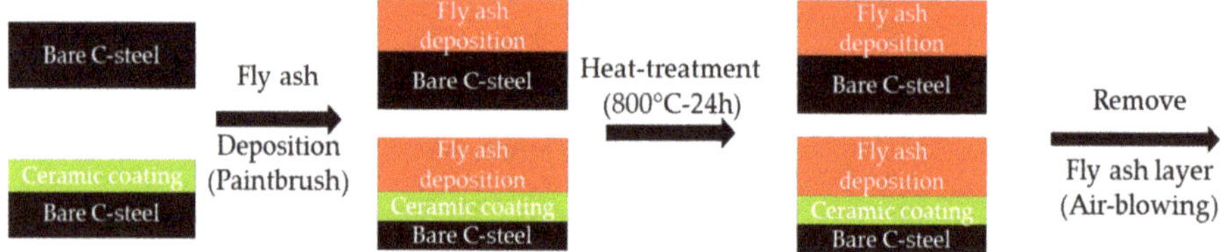

Figure 1. Anti-fouling resistance testing procedure.

2.5. Thermal Conductivity Measurement

To evaluate the energy efficiency, four different specimens were prepared comprising: (i) steel with ceramic coating; (ii) steel without ceramic coating; (iii) ceramic coating materials and (iv) corrosion specimen. Here, specimens (i) and (ii) were then subjected to anti-fouling testing. Specimen (iii) was a bulk material made from the mixture of slurry coating. Specimen (iv) was separated from the ash deposits attached to the boiler systems at JINJU INDUSTRY power plant. All the specimens were made in size of 10 mm × 10 mm × 1.2 mm and their thermal conductivity were measured with the laser-flash method (Netzsch LFA 427, Netzsch, Wittelsbacherstraße, Germany). To assess the thermal conductivity λ (W/m·K) of a specimen, the thermal diffusivity a (mm^2/s), density ρ (g/cm^3), and specific heat C_p (J/g·K) of the specimen must be known. The thermal diffusivity was measured with the heating rate 5 K/min. The data was recorded after each 100 °C from 25 to 800 °C. The specific heat C_p of the specimen was estimated by simple rule of mixture of the fillers and the coating matrix. The density of the specimen was determined using Archimedes principle.

$$\lambda(T) = C_p(T) \times a(T) \times \rho(T) \tag{1}$$

3. Results and Discussion

3.1. Properties of Fly Ashes

To date, various studies have investigated the ash deposition characteristics from different perspectives to clarify the influences of individual factors, including the chemical/mineral compositions of the ashes. Wibberley et al. [23] concluded that fine particles (<10 µm) with a high content of alkali vapors, including sodium and potassium, are the main reason for the inner layer formed in the initial stage of ash deposition. Lee et al. [24] stated that high concentrations of chlorine and sulfur in MSW affect the rate of corrosion in WTE boilers. Furthermore, accumulated ash with a high chlorine concentration on tube surfaces may lead to corrosion underneath the deposit [25,26].

The composition of MSW varies because of differences in lifestyle, so the ash content also varies. Generally, the chemical compositions of the fly ashes showed that SiO_2, Al_2O_3, Fe_2O_3, CaO, and Na_2O were the main oxides (Table 4). Lead and copper were the most common heavy metals in the ashes (Table 5). Based on previous studies, fly ash 1 (Tables 4 and 5), which has high sodium and chlorine contents, was selected as the fouling matter. The physical and chemical properties of fly ash 1 were examined in more detail. The mean size of fly ash 1 was 25 μm (Figure 2). The elemental concentrations of fly ash 1 as determined by SEM-EDS (Figure 3) were consistent with the results from ICP-OES (Tables 4 and 5).

Table 4. Oxide compositions in fly ashes 1–4 (wt %).

Fly Ash	SiO_2	Al_2O_3	Fe_2O_3	CaO	Na_2O	Cl	K_2O	MgO	SO_3	P_2O_5
1	24.9	13.2	2.56	17.1	13.1	12.8	2.36	1.82	1.29	2.96
2	85.0	2.90	0.87	7.27	1.01	0.20	0.35	0.50	0.18	0.94
3	19.7	9.06	16.6	25.3	1.91	2.88	0.89	11.2	11.1	0.15
4	20.9	5.19	12.9	23.1	1.28	0.86	0.61	7.94	24.7	0.13

Table 5. Heavy metals in fly ashes 1–4 (mg/kg).

Fly Ash	Pb	Cd	As	Hg	Cu
1	785	33	N.D	N.D	5620
2	74	N.D	N.D	N.D	2240
3	N.D	N.D	N.D	N.D	265
4	N.D	N.D	N.D	N.D	149

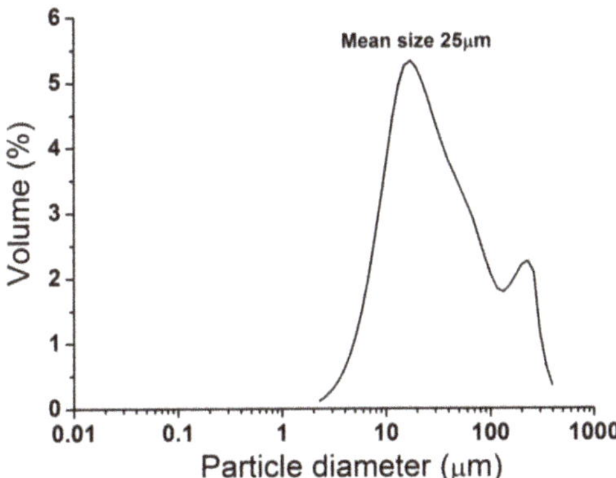

Figure 2. Particle size distribution of fly ash 1.

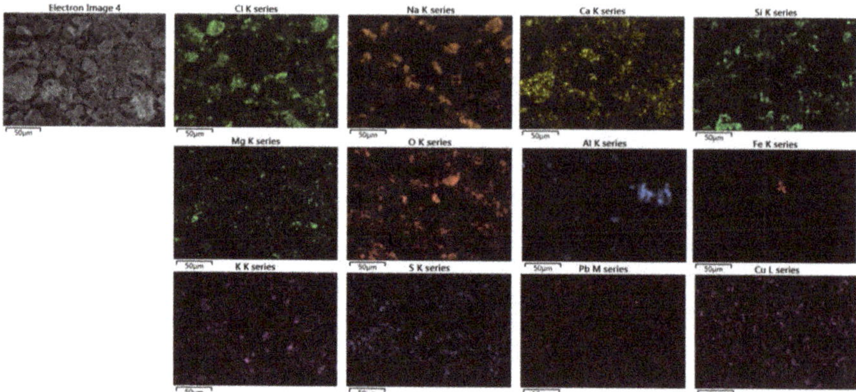

Figure 3. EDS mapping of fly ash 1.

3.2. Microstructure and Morphology of Ceramic Coating

The XRD patterns of the composite coating on steel substrate are shown in Figure 4. SiO_2, Al_2O_3, and abite $Na(Si_3Al)O_8$ peaks were detected. Abite may be the new phases formed during thermal treatment. Because the softening point of glass (640–680 °C) is lower than its heat treatment temperature (800 °C), the glass would soften, flow, and spread well on the steel surface. The fillers may be partially dissolved into the glass matrix and it is not surprising that the reactions between glass and fillers may be occurred during thermal treatment.

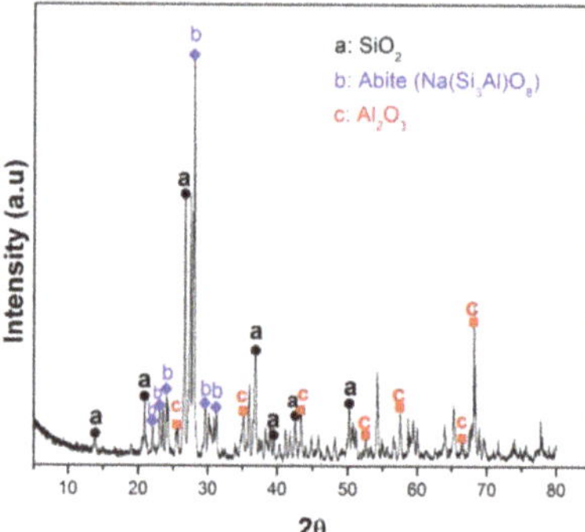

Figure 4. XRD patterns of the composite coating after thermal treatment at 800 °C.

Generally, the coating should have a relatively low surface energy and a dense structure to significantly reduce the sticking of molten ash particles to the surfaces. Moreover, it should have an appropriate thickness and have no chemical affinity with the fouling matter. Finally, it must have good adhesion with the boiler. Figure 5 shows the cross-section morphology of the heat-treated ceramic coating with a thickness of 150–160 μm. A layer enriched with silicon and aluminum in the cross-section was observed. Additionally, the coating was intact and only a few closed pores with spherical shape were seen. These certain pores can help alleviate thermal stress and thermal expansion mismatch between the coating and substrate. No cracks in the coating or interlayer at the coating–steel interface were observed, which suggests that the coating has a thermal expansion coefficient (CTE) similar to that of the steel and had a strong chemical bond with the steel [26,27]. The adhesion between the coating and steel was measured in detail as follows.

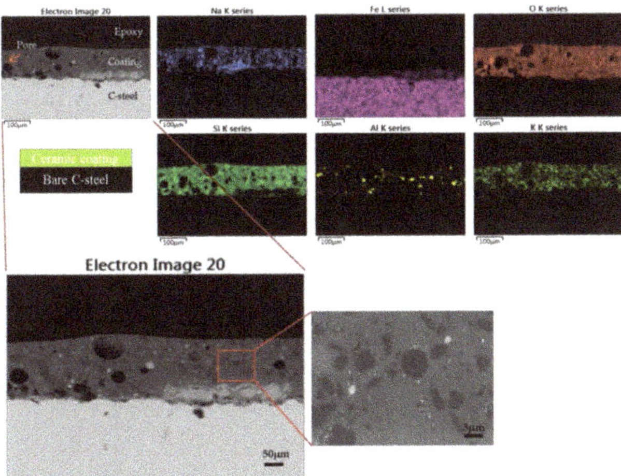

Figure 5. Morphology and EDS maps for the cross-section of the ceramic coating.

3.3. Adhesion and Thermal Shock Properties

Cross-cut tape test (ASTM D3359) [21] was first chosen for qualitative estimation, and pull-off test (ASTM D4541) [22] was then carried out to quantify the adhesion of the coatings. In cross-cut tape test, a grid of square scratches was made on the coating surface. Optical microscopy was used to evaluate about the adhesion of the coating. As shown in Figure 6a, the edges of the scratches are completely smooth and none of them is detached. According to standard ASTM D3359, the highest adhesion scale (5B) is referred to the coating. In pull-off test, the adhesive glue used in this work is Araldite 2021 with flexural strength more than 26 MPa. The glued specimens were cured at room temperature for 30 min. The specimens were mounted on testing machine and were pulled in a direction normal to the coating surface at a constant speed 0.013 mm/s until failure occurs. When failure occurs, the maximum load was recorded. As a result, the interface failure occurred in the 27–29 MPa range. Figure 6b shows cross-sectional SEM of coating after the pull-off test. A delamination appeared near the coating–steel interface, i.e., failure surface occurred within the coating. This is referred as a cohesive failure of the coating. Good adhesion of the coating is ascribed to an interdiffusion at the glass-steel interface due to the mobility of some metal ions in the glass and the metal atoms of the steel at high temperature. Metal ions from the glass will diffuse into the metal while metal atoms will diffuse into the glass.

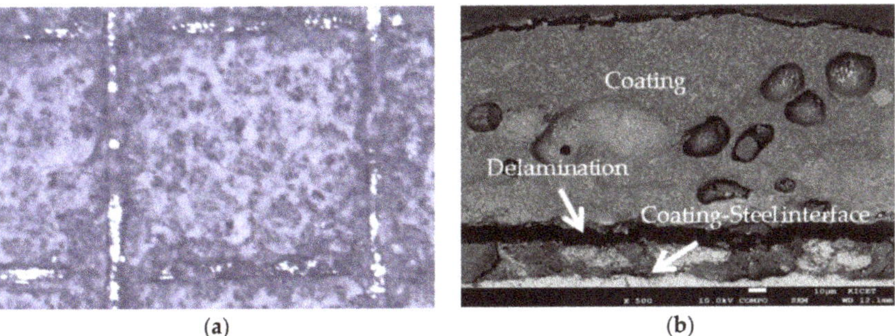

(a)   (b)

Figure 6. (**a**) Optical microscope image after cross-cut tape test; (**b**) Cross-sectional SEM image after pull-off test.

The thermal shock resistant test was examined in water quenching. The coated specimen was heated to 800 °C for 5 min and then cooled rapidly in water (20 °C) for 5 min. After ten uninterrupted thermal cycles no macroscopic cracks or failure are visible. The good thermal shock resistance of the coating is an evidence to suggest that the CTE composite coating is close to that of the steel [18]. Some small spherical pores distributed in the coating are the reason for relaxing thermal stresses and improve the thermal shock resistance of the coated samples.

3.4. Anti-Fouling

Figure 7a,b present the cross-sections of the non-coated steel plate without and with ash particulate deposited, respectively. As depicted in Figure 7a, bare C-steel experienced breakaway oxidation behavior at 800 °C with the top surface of oxide layer was very intact. On the contrary, in case of exposure with fly ashes (Figure 7b), the top surface of oxide layer was seriously damaged with many hollow blisters. The fly ash adhered strongly to the steel, which rapidly increased the fouling rate. A top view of the surfaces (Figure 8) showed that much ash was attached all over the surface of the non-coated steel (Figure 8a). This suggests that the following chemical reaction takes place between the fouling particulates and steel at high temperatures [28]:

$$3Na_2O + Fe \rightarrow Na_4FeO_3 + 2Na\uparrow \tag{2}$$

In contrast, almost no fouling ash adhered on the whole surface of the coated steel (Figure 8b); only some fine fly ash particles stuck in tiny holes were observed. Figure 9 shows that the outer area of the coating was empty of fly ash, while inner area still adhered well to the steel. This demonstrates that the developed ceramic coating provides excellent resistance against the fouling precipitation of fly ash. It could also withstand the high-pressure air jet, which means that a strong bond formed between the coating and steel.

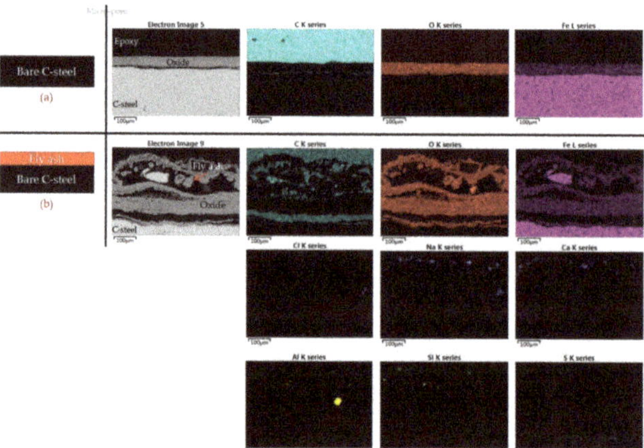

Figure 7. EDS maps for the cross-sectional surface of (**a**) the non-coated steel without fly ash deposition and (**b**) with fly ash deposition at 800 °C.

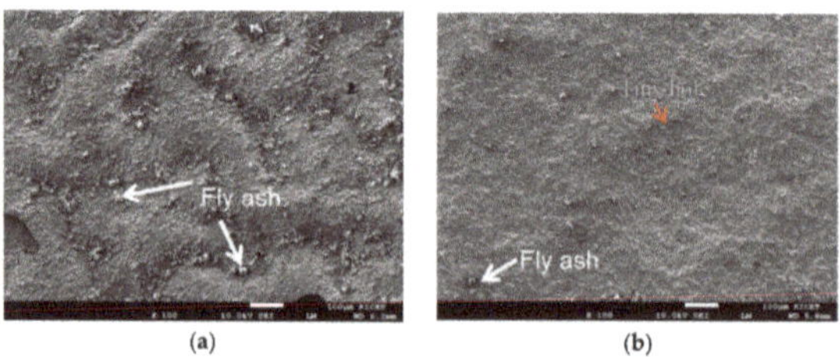

(**a**)　　　　(**b**)

Figure 8. Top-view SEM images of (**a**) non-coated steel and (**b**) coated steel after the anti-fouling testing.

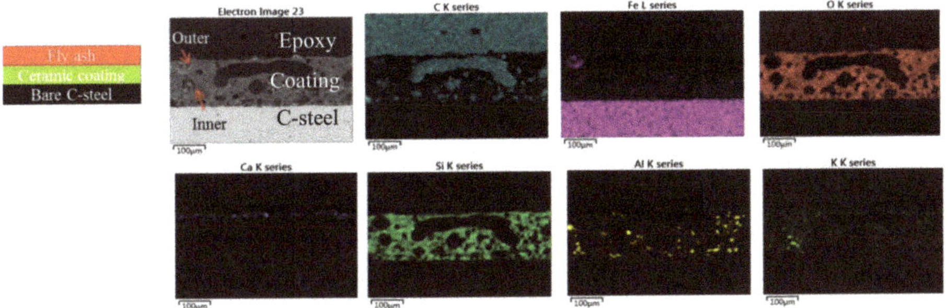

Figure 9. EDS maps of the cross-sectional surface of the coated steel after the anti-fouling testing.

3.5. Thermal Conductivity

Figure 10a describes the thermal conductivity of the coated and non-coated samples as a function of the temperature up to 800 °C. The thermal conductivity of the non-coated steel was ~5 W/mK less than that of the coated steel at 800 °C. Figure 10b indicates the thermal conductivity of the ceramic coating itself and the ash deposited specimen. At 800 °C, the thermal conductivity of the composite coating was 1.2 W/mK which is about twice as high as that of the corrosion specimen (0.6 W/mK). Those values can be explained as follows. The physical structure and microstructure of a surface is believed to affect its thermal conductivity [29,30]. As shown in Figure 7b, the surface of the non-coated steel with highly porous deposits of loose particulate matter had low thermal conductivity. The deposits that formed on the steel surface limited the absorption of the incident radiation and the transfer of this energy to tubes containing the working fluid. Meanwhile, Figure 9 shows that the surface of the coated steel had a dense interconnected structure without fouling deposits, so the thermal conductivity was high. Therefore, thanks to the ceramic coating, the energy efficiency of the boiler tubes was significantly improved.

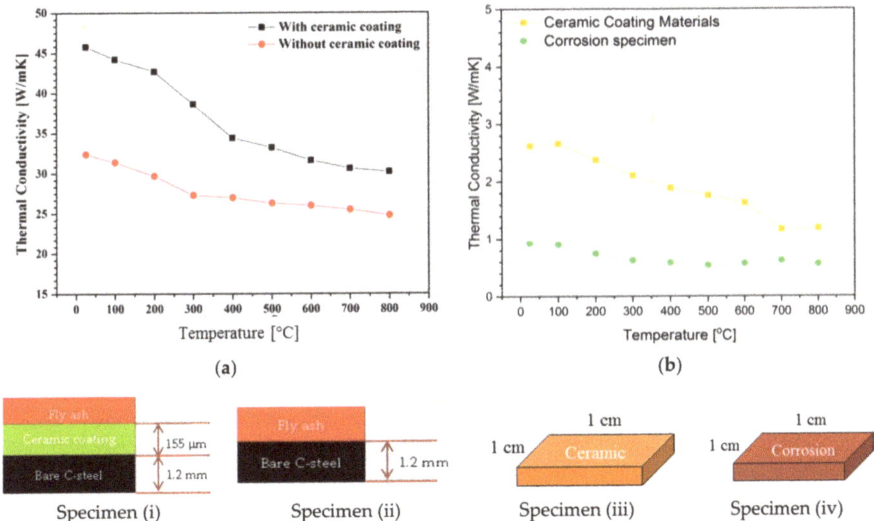

Figure 10. Thermal conductivities of all prepared samples as a function of the temperature up to 800 °C: (**a**) Specimens with and without ceramic coating; (**b**) Ceramic coating materials and corrosion specimen.

3.6. Application of the Coating

In engineering application, the ceramic coating was applied to the heating surfaces of boiler water-wall tubes of tower 2 at JINJU INDUSTRY power plant in Chungcheongbuk-do province of Korea (Figure 11). The effect of ceramic coating on fouling resistance and energy efficiency was evaluated after 3 months of operation of the boiler. As seen in Figures 11 and 12, a significant difference in the coated and uncoated areas of the boiler surfaces was observed. The coated surfaces were relatively clean with only small amount of unwanted matters attached on a small part of boiler surface. However, it should be noted that, these matters are bonded weakly with boiler surface and they can be easily removed by a low-pressure air gun. Meanwhile, fouling and slagging severely deposited on the uncoated surfaces and it is quite hard to remove them even with a high-pressure air gun.

Figure 11. Application of the ceramic coating in real boiler systems: (**a**) slurry spraying process and (**b**) anti-fouling ceramic coating after 3 months operation.

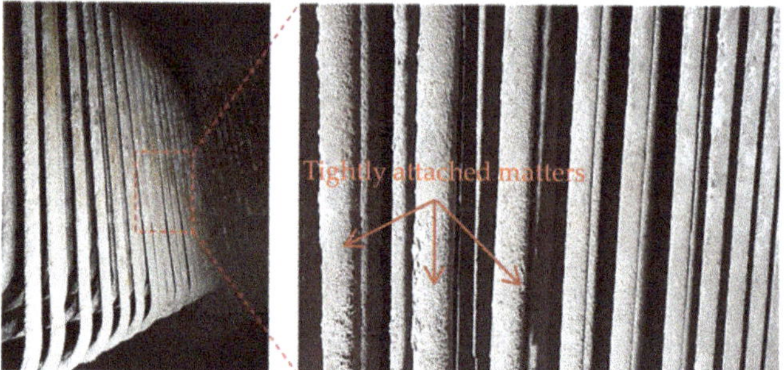

Figure 12. Photography of the uncoated areas of the boiler surfaces.

The thermal efficiency of boilers was calculated before and after applying the composite ceramic coating. To do this, it is useful to check the amount of incinerated waste (input amount or burn fuel) and the amount of produced steam (production amount). For simplicity, thermal efficiency (η) before and after using the coating is defined as the rate of total steam production over the fuel consumption. Figure 13 collected the thermal efficiency without and with using the ceramic coating in the same period, respectively. As can be seen in Figure 13, the thermal efficiency increased about 62.34% after applying the ceramic coating. Of course, there are some other operating conditions can affect the result, but it is worth concluding that the developed ceramic coating has great potential to be applied in real boiler systems to improve their overall thermal efficiency.

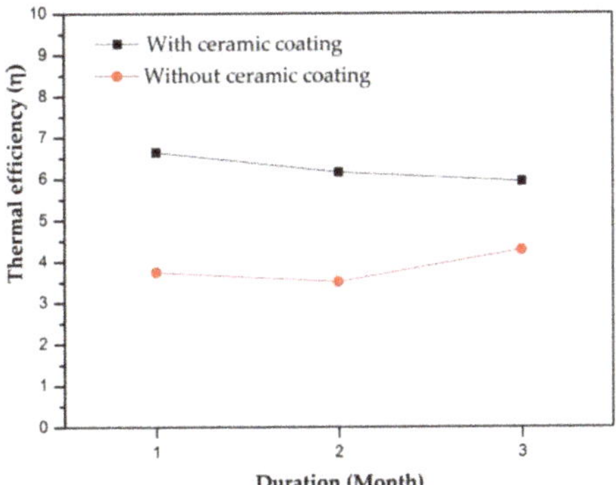

Figure 13. The thermal efficiency before and after applying the ceramic coating in boiler systems at JINJU INDUSTRY power plant in Chungcheongbuk-do (Korea).

4. Conclusions

The purpose of this study was to develop a ceramic coating to prevent the fouling deposition of fly ash. The results can be summarized as follows.

- A ceramic coating with a thickness of 150–160 μm was successfully developed and applied to carbon steel.
- Fly ash with high concentrations of sodium and chlorine was selected as the fouling matter. In the anti-fouling testing, the developed ceramic coating with a dense structure performed well at preventing fly ash fouling. In comparison, the bare steel without coating was severely fouled.
- The ceramic coating showed a significant improvement in the thermal conductivity of the boiler at high temperature (800 °C). Hence, it can help increase the overall energy efficiency for actual application to real boiler systems in WTE facilities.

Supplementary Materials: The following are available online at http://www.mdpi.com/2079-6412/8/10/353/s1. Table S1: Input and output production amount of tower 2 in period of 2017.04–2017.06 (JINJU INDUSTRY's data); Table S2: Input and output production amount of tower 2 in period of 2018.04–2018.06 (JINJU INDUSTRY's data).

Author Contributions: Conceptualization, M.D.N. and W.T.K.; Methodology, M.D.N. and J.W.B.; Software, M.D.N. and A.S.B.; Validation, M.D.N., K.H.H. and V.-H.P.; Formal Analysis, M.D.N. and A.S.B.; Investigation, M.D.N., Y.H.K. and K.H.H.; Resources, M.D.N. and J.W.B.; Data Curation, V.-H.P., Y.H.K. and K.H.H.; Writing-Original Draft Preparation, M.D.N.; Writing-Review & Editing, M.D.N. and W.T.K.; Visualization M.D.N. and J.W.B.; Supervision, W.T.K. and Y.H.K.; Project Administration, W.T.K.; Funding Acquisition, W.T.K.

Funding: This research were supported by the National Strategic Project-Carbon Upcycling of the National Research Foundation of Korea (NRF) funded by the Ministry of Science and ICT (MSIT), the Ministry of Environment (ME), and the Ministry of Trade, Industry and Energy (MOTIE) (Grant No. 2017M3D8A2086035); the Korea Institute for Advancement of Technology (KIAT) (Grant No. P012700050).

Acknowledgments: The authors would like to thank Jeong Min Bak, Korea for SEM technical advice.

Conflicts of Interest: The authors declare no conflict of interest.

References

1. Hansen, J.; Sato, M.; Ruedy, R.; Lacis, A.; Oinas, V. Global warming in the twenty first century: An alternative scenario. *Proc. Natl. Acad. Sci. U.S.A.* **2000**, *97*, 9875–9880. [CrossRef] [PubMed]
2. Xu, S.; He, H.; Luo, L. Status and prospects of municipal solid waste to energy technologies in China. In *Recycling of Solid Waste for Biofuels and Bio-chemicals*; Karthikeyan, O.P., Heimann, K., Muthu, S.S., Eds.; Springer: Berlin, Germany, 2016; pp. 31–54.
3. Ibrahim, H.A. Fouling in heat exchangers. In *MATLAB—A Fundamental Tool for Scientific Computing and Engineering Applications*; Katsikis, V., Ed.; InTechopen: London, UK, 2012; Volume 3, pp. 57–96.

4. Regueiro, A.; Patiño, D.; Granada, E.; Porteiro, J. Experimental study on the fouling behaviour of an underfeed fixed-bed biomass combustor. *Appl. Therm. Eng.* **2017**, *112*, 523–533. [CrossRef]

5. Energy Efficiency Benchmarking Study of Food Manufacturing Plants in Singapore. Available online: http://www.e2singapore.gov.sg/DATA/0/docs/Resources/Industry/FMBS%20Final%20Report%20v1.1.pdf (accessed on 18 February 2018).

6. Barma, M.C.; Saidur, R.; Raham, S.M.A.; Allouhi, A.; Akash, B.A.; Sadiq, M.S. A review on boilers energy use, energy savings, and emissions reductions. *Renew. Sustain. Energy Rev.* **2017**, *79*, 970–983. [CrossRef]

7. Gupta, R.D.; Ghai, S.; Jain, A. Energy efficiency improvement strategies for industrial boilers: A case study. *J. Eng. Technol.* **2011**, *1*, 52–56. [CrossRef]

8. Nussbaumer, T. Combustion and co-combustion of biomass: Fundamentals, technologies, and primary measures for emission reduction. *Energy Fuels* **2003**, *17*, 1510–1521. [CrossRef]

9. Gao, X.; Jiang, Z.; Gao, J.; Xu, D.; Wang, Y.; Pan, H. Boiler maintenance robot with multi-operational schema. In Proceedings of the 2008 IEEE International Conference on Mechatronics and Automation, Takamatsu, Japan, 5–8 August 2008.

10. Karell, M. Energy Saving Tips: Boiler Maintenance. Available online: http://www.energymanagertoday.com/energy-saving-tipsboiler-maintenance-091900 (accessed on 18 February 2018).

11. Nguyen, M.D.; Bang, J.W.; Kim, Y.H.; Bin, A.S.; Hwang, K.H.; Pham, V.H.; Kwon, W.T. Slurry spray coating of carbon steel for use in oxidizing and humid environments. *Ceram. Int.* **2018**, *44*, 8306–8313. [CrossRef]

12. Milad, F.; Amin, R.B.; Hossein, E.; Mahdi, S. Study on high-temperature oxidation behaviors of plasma-sprayed TiB$_2$-Co composite coatings. *J. Korean Ceram. Soc.* **2018**, *55*, 178–184.

13. Nguyen, M.D.; Bang, J.W.; Bin, A.S.; Kim, S.R.; Kim, Y.H.; Hwang, K.H.; Pham, V.H.; Kwon, W.T. Novel polymer-derived ceramic environmental barrier coating system for carbon steel in oxidizing environments. *J. Eur. Ceram. Soc.* **2017**, *37*, 2001–2010. [CrossRef]

14. Wang, K.; Unger, J.; Torrey, J.D.; Flinn, B.D.; Bordia, R.K. Corrosion resistant polymer derived ceramic composite environmental barrier coatings. *J. Eur. Ceram. Soc.* **2014**, *34*, 3597–3606. [CrossRef]

15. Günthner, M.; Kraus, T.; Dierdorf, A.; Decker, D.; Krenkel, W.; Motz, G. Advanced coatings on the basis of Si(C)N precursors for protection of steel against oxidation. *J. Eur. Ceram. Soc.* **2009**, *29*, 2061–2068. [CrossRef]

16. Chen, M.; Li, W.; Shen, M.; Zhu, S.; Wang, F. Glass-ceramic coatings on titanium alloys for high temperature oxidation protection: Oxidation kinetics and microstructure. *Corros. Sci.* **2013**, *74*, 178–186. [CrossRef]

17. Li, W.; Chen, M.; Wang, C.; Zhu, S.; Wang, F. Preparation and oxidation behavior of SiO$_2$-Al$_2$O$_3$-glass composite coating on Ti-47Al-2Cr-2Nb alloy. *Surf. Coat. Technol.* **2013**, *218*, 30–38. [CrossRef]

18. Günthner, M.; Schütz, A.; Glatzel, U.; Wang, K.; Bordia, R.K.; Greißl, O.; Krenke, W.; Motz, G. High performance environmental barrier coatings, Part I: Passive filler loaded SiCN system for steel. *J. Eur. Ceram. Soc.* **2011**, *31*, 3003–3010. [CrossRef]

19. Moreau, F.; Durán, A.; Munoz, F. Structure and properties of high Li$_2$O-containing aluminophosphate glasses. *J. Eur. Ceram. Soc.* **2009**, *29*, 1895–1902. [CrossRef]

20. Shan, X.; Wei, L.Q.; Liu, P.; Zhang, X.M.; Tang, W.X.; Qian, P.; He, Y.; Ye, S.F. Influence of CoO glass–ceramic coating on the anti-oxidation behavior and thermal shock resistance of 200 stainless steel at elevated temperature. *Ceram. Int.* **2014**, *40*, 12327–12335. [CrossRef]

21. *ASTM D3359 Standard Test Methods for Measuring Adhesion by Tape Test*; ASTM International: West Conshohocken, PA, USA, 2017.

22. *ASTM D4541 Standard Test Method for Pull-Off Strength of Coatings Using Portable Adhesion Testers*; ASTM International: West Conshohocken, PA, USA, 2017.

23. Wibberley, L.J.; Wall, T.F. Alkali-ash reactions and deposit formation in pulverized-coal-fired boilers: The thermodynamic aspects involving silica, sodium, sulphur and chlorine. *Fuel* **1982**, *61*, 87–92. [CrossRef]

24. Lee, S.H.; Themelis, N.J.; Castaldi, M.J. High-temperature corrosion in waste-to-energy boilers. *J. Therm. Spray Technol.* **2007**, *16*, 104–110. [CrossRef]

25. Melissari, B. Ash related problems with high alkali biomass and its mitigation experimental evaluation. *Memoria Investigaciones en Ingeniería* **2014**, *12*, 31–44.

26. Nielsen, H.P.; Frandsen, F.J.; Dam-Johansen, K.; Baxter, L.L. The implications of chlorine-associated corrosion on the operation of biomass-fired boilers. *Prog. Energy Combust. Sci.* **2000**, *26*, 283–298. [CrossRef]

27. Yan, Z.Q.; Xiong, X.; Xiao, P.; Chen, F.; Zhang, H.B.; Huang, B.Y. Si–Mo–SiO$_2$ oxidation protective coatings prepared by slurry painting for C/C–SiC composites. *Surf. Coat. Technol.* **2008**, *202*, 4734–4740. [CrossRef]

28. Barker, M.G.; Wood, D.J. The corrosion of chromium, iron, and stainless steel in liquid sodium. *J. Less Common Met.* **1974**, *35*, 315–323. [CrossRef]

29. Wall, T.F.; Bhattacharya, S.P.; Zhang, D.K.; Gupta, R.P.; He, X. The properties and thermal effects of ash deposits in coal-fired furnaces. *Prog. Energy Combust. Sci.* **1993**, *19*, 487–504. [CrossRef]

30. Baxter, L.L. Influence of ash deposit chemistry and structure on physical and transport properties. *Fuel Proc. Technol.* **1998**, *56*, 81–88. [CrossRef]

Article

Size Distribution of Contamination Particulate on Porcelain Insulators

Ming Zhang, Rumeng Wang, Lee Li and Yunpeng Jiang *

International Joint Research Laboratory of Magnetic Confinement Fusion and Plasma Physics, State Key Laboratory of Advanced Electromagnetic Engineering and Technology, School of Electrical & Electronic Engineering, Huazhong University of Science & Technology, Wuhan 430074, China; zhangming@hust.edu.cn (M.Z.); wangrumeng@hust.edu.cn (R.W.); leeli@hust.edu.cn (L.L.)
* Correspondence: jiangyunpeng@hust.edu.cn

Received: 19 July 2018; Accepted: 28 August 2018; Published: 25 September 2018

Abstract: The characteristics of contamination on the insulation medium surface play an important role in the surface flashover, especially size distribution of contaminated particles. After measuring the size of contaminated particles on the porcelain insulator surface, obvious size distribution characteristics of particles were found. To study the reason for these statistical characteristics, the movement of particles was analyzed in detail combining with fluid mechanics and collision dynamics. Furthermore, an adhesion model was established in this paper. In addition, the influences of different factors on the adhesion were studied. The results showed that the size of adhered particles on the porcelain insulator surface was easy to focus on a specific range, and the influences of relative humidity and wind speed were remarkable. However, the influences of electric field type, electric field strength, and aerodynamic shape were relatively weak. This research was significant and valuable to the study of artificial contamination simulation experiments, and the influence of particles size distribution on pollution flashover.

Keywords: particle size; distribution characteristics; contaminated particle; porcelain insulator; adhesion simulation model

---

1. Introduction

The research on insulator contamination characteristics is the fundamental research of external insulation in power systems, and it is of great significance to master the contamination characteristics of the insulator for the design, operation, and maintenance of external insulation [1–6]. Given the existence of the contamination, the insulation performance of insulators has changed greatly [7–9]. Contamination degree, leakage current, and pollution flashover voltage are important parameters for evaluating insulation performance of the insulator, and researchers have carried out a lot of research on them [10–14]. However, these researches were based on the surface that has been adhered to the contamination. The contribution on reduction of contamination accumulation and the physical properties of contamination is limited. Therefore, it is necessary to carry out research on related characteristics regarding accumulating contamination particles.

In recent years, researchers have carried out a lot of research on contamination accumulation characteristics through a variety of ways, including theoretical research, natural contamination accumulation experiments, artificial simulation experiments, and simulation analysis; especially in the research of the movement characteristics and deposition characteristics of contaminated particles from the microscopic point of view [15–27]. In theoretical research, Wang et al. [15] have analyzed the forces of particles moving around the insulator in detail, where they found that polarization force has little effect on the movement of particles, and fluid drag force plays an important role under high wind speed. Horenstein et al. [16] considered that the electric field force has a great influence on the movement of particles, especially for particles with a size less than 10 μm, after a wind tunnel simulation experiment and theoretical deduction. However, when the particle size is greater than 10 μm, the effect of fluid drag force on the particle trajectory will gradually increase. At low wind speed, Li et al. [17] thought that the energy loss was mainly caused by the inelastic collision of particles, and the friction loss caused by the relative motion of particles and surface during the collision. However, at high wind speed, the particles will experience elastic-plastic deformation, and thus generate energy loss [18].

Furthermore, researchers have also used fluid dynamics simulation software to study contamination accumulation characteristics. Jiang et al. [19] found that under the condition of horizontal wind, the collision coefficient of the contaminated particles on the upper surface of the insulator increases with the increase of particle size and wind speed, but the collision coefficient of the bottom surface is always at a low level. Nan et al. [20] also obtained the same conclusion through an artificial simulation experiment and simulation analysis. In addition, they found that the change of wind direction had obvious influence on the contamination distribution of double umbrella insulators. Wang et al. [21] suggested that the collision mass of particles increases rapidly with the increase of wind speed, and the shape of the umbrella skirt at the bottom surface has an important influence on the collision characteristics. Zhang et al. considered that the contamination degree of windward and leeward of the insulator increased with the increase of wind speed, whilst the contamination degree of crosswind side showed a decreasing trend. As a result, a great deal of contamination accumulates on the windward and leeward, but the crosswind surfaces are relatively clean [22]. Lv et al. [23–25] thought that under the same wind speed, the contamination degree in the DC (direct current) electrical field is much more than that in the AC (alternating current) and no electrical field. Furthermore, under the conditions of AC and no electrical field, the deposition amount of contamination at low wind speed was relatively low. Even under the condition of high wind speed, the increase amount of contamination was very limited.

In general, the current literature shows that researchers paid more attention to the amount of contamination accumulated from contaminated particles. However, researchers seldom studied and analyzed the size distribution characteristics of adhered particles. With the increase of the frequency of haze and other microclimate phenomena in recent years, the difference of adhered particle size distribution on the insulator surface will have some influence on the contamination characteristics, pollution flashover characteristics, etc. [26]. Therefore, it is necessary to study size distribution characteristics of contaminated particles on the insulator surface.

After measuring many contamination samples, it was found that the size distribution of contaminated particles had some significant statistical characteristics. To explore the reasons of these distribution characteristics, a physical model of collision and adhesion between particles and surface was established. In addition, the influences of different factors on the adhesion were studied respectively. The research of this paper can effectively explain that the size distribution characteristics of adhered particle are easy to focus on a specific range. Furthermore, the work of this paper can provide theoretical support for more accurate external insulation researches in the future, such as artificial contamination simulation experiments and the influence of particle size distribution on pollution flashover.

2. Size Distribution Characteristics of Adhered Particle

2.1. Measurement Method

To study microscopic topography and particle size distribution of the particles adhered to the insulator surface, scanning electron microscopy (SEM) and laser particle size analyzer were used to study contamination. To observe microscopic topography of the particles in the contamination sample, a small amount of the contamination sample was adhered by conductive tape from the insulator surface, and samples were placed in a scanning electron microscope (VEGA TS5136XM, TESCAN Corporation, Brno, Czech) for observation. In addition, when collecting the particles, the force should be uniform, and the contaminated sample should not be squeezed to avoid destroying the shape and size of the particles. To measure particle size distribution of contamination samples, the contamination on the insulator surface was collected into clean sealed bags with a clean brush, and the numbers were recorded. Mastersizer 2000 laser particle size analyzer (Malvern Panalytical Corporation, Malvern, UK) was used to analyze the particle size of contamination samples. The measurement range of the Mastersizer 2000 laser particle size analyzer is from 20 nm to 2000 μm, and the reappearance rate is better than 0.5%, and the accuracy is better than 1%, which effectively met the measurement requirements. In addition, some of contamination was easily soluble in water, easily causing measurement errors. Therefore, alcohol was used as substrate, and the method of wet dispersion measurement was carried out.

2.2. Typical Particle Size Distribution Characteristics

The Hami South-Zhengzhou (Ha-Zheng Line) ±800 kV ultra-high voltage (UHV)-DC transmission line and the Southeastern Shanxi-Nanyang-Jingmen (Chang-Nan Line) 1000 kV UHV-AC transmission line pass through Henan, China. Additionally, these two lines have a short distance to transmit power in parallel. Therefore, this was a good experimental environment to compare the difference of contamination accumulation characteristics of UHV-DC and UHV-AC transmission lines, under the same natural conditions. During the outage of these two UHV transmission lines in December 2014, 16 samples of natural contamination on the surface of the XP-160 insulator were collected from 4 towers of No. 249 and No. 251 (±800 kV, Ha-Zheng Line), and No. 3937 and No. 3938 (1000 kV Chang-Nan Line). In addition, the particle size was measured using a laser particle size analyzer and scanning electron microscope. The scanning electron microscope graph and particle size distribution graph of the contaminated samples, which were under the conditions of 1000 kV (AC), +800 kV and −800 kV (DC), are shown in Figures 1 and 2, respectively. The accumulation period of these 16 samples was from 21 March 2014 to 21 December 2014. Experiment towers were in the same area (distance less than 2 km), and the main pollution sources were farmland and chemical plants.

It can be seen from Figures 1 and 2, that the size of contaminated particles on the insulator surface is mainly concentrated in a specific range, which is 5–50 μm. Specifically, the average size was 19.76 μm, and the size of 90% of the particles was larger than 2.89 μm and less than 46.37 μm. Based on the above measurement results, this paper proposed a new conclusion, that there are obvious statistical characteristics. In other words, the size distribution of contaminated particles on the porcelain insulator surface was basically in logarithmic normal distribution, and the median size of adhered particles was about 20 μm, and the distribution of particles with sizes larger than 50 μm and less than 5 μm was rare.

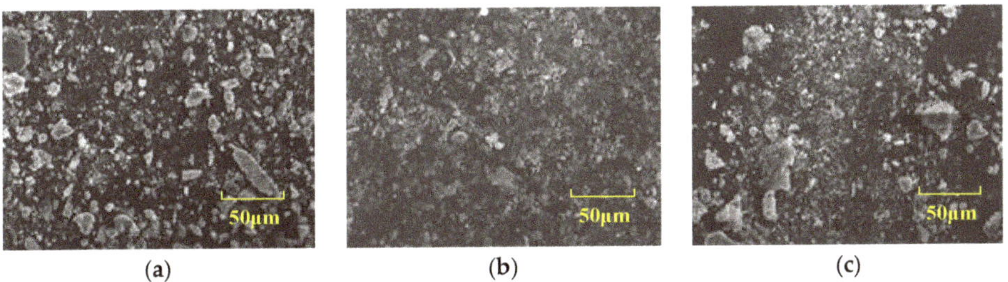

(a)   (b)   (c)

Figure 1. Scanning electron microscopy diagram of typical contaminated particle samples: (**a**) +800 kV (DC); (**b**) −800 kV (DC); and (**c**) 1000 kV (AC).

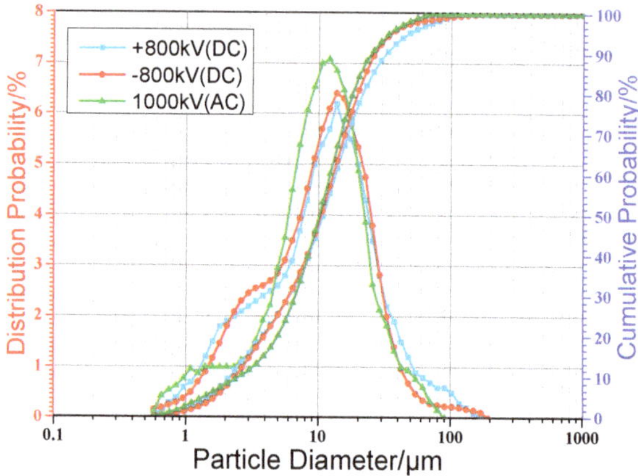

Figure 2. Size distribution diagram of typical contaminated particle samples.

2.3. Particle Size Distribution Characteristics

To verify whether the size distribution characteristics of contaminated particles, proposed in the previous section is generally applicable, 34 samples of contamination on the porcelain insulator surface were collected from 27 provinces and cities in China, and their particle size was measured using a laser particle size analyzer. The measurement results and essential information are shown in Table 1. In the samples, the voltage levels included AC (110, 220, 500, 750, and 1000 kV) and DC ($\pm$500, $\pm$660, and $\pm$800 kV). The results showed that the average particle size was mainly distributed in the range of 5–25 μm. It can be considered that these measurement results supported the particle size distribution characteristics proposed in this paper.

Table 1. Measurement results of particle size and essential information of contamination samples.

Province/Line	Voltage Level (kV)	Insulator Type	SPS Level *	Average Relative Humidity (%)	Average Wind Speed (m/s)	D_{10} (μm)	D_{50} (μm)	D_{90} (μm)	Contamination Accumulation Year
Heilongjiang/Xin-fu Line	110	XP-7	B	64	2.9	5.87	15.74	38.48	1
Hunan/Jin-ming Line	500	XWP-160	E	77	1.2	7.54	17.04	51.48	2
Shanghai/Si-du Line	500	XP-160	E	75	2.3	9.94	25.94	49.49	1
Fujian/Xing-zeng Line	220	XWP-70	D	72	2.4	4.15	16.57	35.71	4
Anhui/Hua-xia Line	110	XP-70	B	70	2.4	14.39	27.46	48.38	2
Ningxia/Wa-long Line	110	XP-70	D	55	2.2	15.67	36.09	59.45	1
Jilin/Yu-long Line	220	XP-70	E	62	3.5	3.87	11.11	34.87	1
Beijing/Li-da Line	110	XWP-7	C	54	2.3	3.62	10.04	37.14	1
Chongqin/Bai-tian Line	110	XP-7	C	80	1.4	2.27	9.88	21.61	2
Tianjin/Hou-hua Line	110	XWP-70	D	61	2.4	10.36	23.14	43.45	7
Jiangxi/Shang-ding Line	110	XP-7	B	75	2.3	8.46	20.35	40.19	10
Shanxi/Xi-shi Line	500	XP-160	E	61	3.1	13.21	33.37	58.74	3
Gansu/Zhang-tan Line	110	XP-70	C	48	2.1	4.48	13.72	27.39	8
Hubei/Guang-xian Line	500	XP-160	D	75	1.5	5.74	22.56	37.67	2
Zhejiang/Fang-tang Line	110	XP-70	D	73	1.8	2.81	13.83	29.74	11
Hebei/Shi-liu Line	220	XP-70	E	60	2.0	6.14	26.83	34.59	2
Henan/Jia-xiang Line	500	XP-160	C	62	2.0	4.92	16.83	37.71	4
Sichuan/Na-da Line	110	XWP-70	B	66	1.5	7.48	20.77	44.74	10
Liaoning/An-hong Line	220	XWP-100	C	65	2.6	5.93	17.18	38.38	2
Henan/Chang-nan Line	1000	XP-160	C	62	2.0	7.39	23.55	45.78	1
Henan/Chang-nan Line	1000	XP-160	C	62	2.0	6.87	17.18	41.90	1
Hunan/Jiang-cheng Line	$\pm$500	XP-160	E	78	1.9	9.73	23.14	52.47	2
Hunan/Jiang-cheng Line	$\pm$500	XP-160	E	78	1.9	11.35	20.64	41.90	2
Anhui/Yi-hua Line	$\pm$500	XP-70	D	76	2.8	6.10	16.63	38.76	1
Anhui/Long-zheng Line	$\pm$500	XP-70	C	75	1.9	6.34	18.80	38.86	2
Shanxi/Yin-dong Line	$\pm$660	XP-210	E	51	2.8	5.65	19.02	35.74	2
Shanxi/Yin-dong Line	$\pm$660	XP-210	D	51	2.8	9.59	22.84	48.86	2

Table 1. *Cont.*

Province/Line	Voltage Level (kV)	Insulator Type	SPS Level *	Average Relative Humidity (%)	Average Wind Speed (m/s)	D_{10} (µm)	D_{50} (µm)	D_{90} (µm)	Contamination Accumulation Year
Hebei/Yin-dong Line	±660	XP-210	C	60	2.0	8.71	16.12	47.29	1
Hebei/Yin-dong Line	±660	XP-210	D	60	2.0	6.46	23.57	52.43	1
Hubei/Jin-su Line	±800	XZP-210	D	77	1.4	5.74	17.08	36.72	1
Hubei/Fu-feng Line	±800	XZP-210	D	74	2.8	8.59	19.56	38.17	2
Henan/Tian-zhong Line	±800	XP-160	C	62	2.0	3.46	11.06	21.19	1
Henan/Tian-zhong Line	±800	XP-160	B	62	2.0	8.47	23.20	33.48	1
Henan/Tian-zhong Line	±800	XP-160	B	62	2.0	4.92	15.15	24.91	1

* SPS (site pollution severity) level is used to evaluate contamination accumulation degree in a region defined by IEC 60815 [27].

2.4. Particle Size Measurment Results of References

The measurement results of other researchers also support the conclusion that the size distribution of contaminated particles on the insulator surface is mainly concentrated in a specific range. Xu et al. [28] measured the particle size of contamination collected from a porcelain insulator surface, under charged and non-charged conditions. The measurement results showed that the average particle size was mainly concentrated in the range of 17.97–24.64 µm. At the same time, 90% of the particle size was greater than 6.55 µm, and 90% of the particle size was less than 41.9 µm. Su et al. [29] measured the contamination collected on the porcelain insulator, which worked in a natural contamination test station and converter station. The results showed that 50% of the particle size was less than 15 µm, and 90% of the particle size was less than 50 µm. Tu et al. [30] measured the particle size of the contamination on the porcelain insulator surface under the condition of haze and fog, and found that particle size was in logarithmic normal distribution, 90% of the particle size was less than 14.6 µm.

3. Motion Characteristics of Particle

There are two processes in which particles move from the air to the insulator surface. The first one is that the particles move toward the surface with the effect of fluid drag force, electric force, gravity, and other forces. The second one is that particles with a certain initial velocity, collide with the surface and adhere to the surface. To research the motion characteristics of particles more clearly, it is necessary to carry out detailed research on these two physical processes [31].

3.1. Aerodynamic Characteristics of Particle

In this paper, the Euler method was used to analyze the motion of airflow. However, for the particles, its concentration is relatively sparse, and its molecular weight is much greater than that of gas molecules. Thus, the Lagrange method was used to calculate the motion trajectory of the particles. In addition, the effect of the airflow phase on the particle phase was considered, but the effect of the particle phase on the airflow phase was neglected.

3.1.1. Mathematical Model of Airflow Phase

Given the complex structure of the suspension insulator, the airflow around the insulator will experience severe bending. If the standard k–ε model was used to calculate, it would produce some errors. However, the RNG (Renormalization-group) k–ε model has an advantage in dealing with airflow with low Reynolds number and serious streamline bending. Thus, the RNG k–ε model was used [19].

The N–S equation and the continuous equation are

$$\nabla U = 0 \tag{1}$$

$$\frac{\partial U}{\partial t} + \rho U \cdot \nabla U = -\nabla p + \mu \nabla^2 U \tag{2}$$

where, U is the average wind speed, m/s; ρ is the air density, kg/m^3; p is the average pressure, Pa; μ is dynamic viscosity coefficient of air.

The turbulent kinetic k equation and the dissipation rate ε equation of the RNG k–ε model are

$$\rho \frac{Dk}{Dt} = \frac{\partial}{\partial x_i}\left(\sigma_k \mu_{eff} \frac{\partial k}{\partial x_i}\right) + \mu_{eff} S^2 - \rho \varepsilon \tag{3}$$

$$\rho \frac{D\varepsilon}{Dt} = \frac{\partial}{\partial x_i}\left(\sigma_\varepsilon \mu_{eff} \frac{\partial \varepsilon}{\partial x_i}\right) + C_{1\varepsilon}\frac{\varepsilon}{k}\mu_t S^2 - C_{2\varepsilon}\rho\frac{\varepsilon^2}{k} - R_\varepsilon \tag{4}$$

$$R_\varepsilon = \frac{C_\mu \rho \varphi^3 (1 - \varphi/\varphi_0)\varepsilon}{k\left(1 + \beta\varphi^3\right)} \tag{5}$$

where, σ_k and σ_ε are Prandtl numbers corresponding to turbulent kinetic energy k and dissipation rate ε, respectively, $\sigma_k = \sigma_\varepsilon = 1.393$; μ_{eff} is effective dynamic viscosity coefficient of air, $\mu_{eff} = \mu + \mu_t$; μ_t is turbulent viscosity coefficient of air, $\mu_t = \rho C_\mu k^2/\varepsilon$; $C_\mu = 0.0845$; S is the modulus of the mean rate of strain tensor; $C_{1\varepsilon} = 1.42$, $C_{2\varepsilon} = 1.68$; $\varphi = Sk/\varepsilon$, $\varphi_0 = 4.38$, $\beta = 0.012$.

3.1.2. Mathematical Model of Particle Phase

The particles moving in the air are subjected to a variety of forces, including viscous resistance force, pressure gradient force, gravity, air buoyancy, virtual mass force, Brownian force, Basset force, Magnus lifting force, Saffman lifting force, thermophoresis forces, fluid drag force, electric field force, etc. [15]. Among them, the effects of gravity, fluid drag force, and electric field force on the movement of particles are significant. Therefore, this paper mainly considered these three forces. The motion equations of particle in the Lagrange coordinate system can be calculated after analyzing the forces of the particle.

$$m\frac{dv}{dt} = F_D + G + F_q \tag{6}$$

where, m is particle quality, kg; v is particle velocity, m/s; F_D is fluid drag force, N; G is gravity, N; F_q is electric field force, N.

Fluid drag force (F_D). In the mathematical model, the particles are assumed to be spherical, and their radius is R. The fluid drag force is calculated using the Stokes Equation [32].

$$F_D = \frac{18\mu}{\rho_p R^2} m(u - v) \tag{7}$$

where, u is the wind speed, m/s; ρ_p is the density of particle, kg/m^3; R is particle radius, m.

Electric field force (F_q). If the charge of particles is q, the electric field force is

$$F_q = qE \tag{8}$$

where, E is the electric field strength near the insulator, V/m; q is the particle charge, C.

3.2. Collision Process between Particles and Surface

The physical process of collision between particles and the insulator surface (hereinafter referred to as the surface) can be divided into three stages, namely, injection stage, collision deformation stage, and ejection stage, as respectively shown in Figure 3. In Figure 3, the injection stage is I→II→III, and the collision deformation stage is II→III→IV, and the ejection stage is III→IV→V. Then these three stages are analyzed in detail.

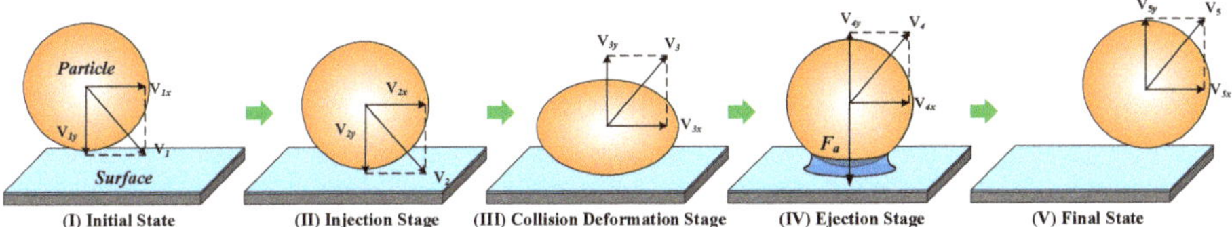

Figure 3. Sketch diagram of collision process between particle and surface.

3.2.1. Injection Stage

At this stage, the particles fly toward the surface with the initial velocity of V_1, in which V_{1x} is the tangential component of V_1, and V_{1y} is the normal component of V_1. When the particles move toward the surface, it will be affected by the water molecular layer attached to the surface [33], and then its velocity will change to V_2. However, the measurement results by Asay et al. [34] showed that the thickness of the water molecule layer varies only in the range of 0.5–2.5 nm, under different relative humidity. Compared with the particle size (1–100 μm), there is a great difference in magnitude. At the same time, the action distance of this process is too short, and the effect on the particles is so small that it can be neglected. Therefore, it can be considered that the particles hit the surface directly at the injection stage.

3.2.2. Collision Deformation Stage

The porcelain surface can be considered that it will not experience deformation during collision, due to its material properties. The particles will experience non-complete elastic deformation, and its velocity will change to V_3 after deformation recovery, and the direction of its velocity is outward along the surface normal. The theoretical model of Johnson collision recovery coefficient was used to analyze the velocity of particles in this paper, as outlined in [17]. The recovery coefficient e is:

$$e = \frac{V_3}{V_2} = 3.8 \left(\frac{\sigma_s}{E*} \right)^{1/2} \left(\frac{mV_2^2}{2\sigma_s R^3} \right)^{-1/8} \tag{9}$$

$$\frac{1}{E*} = \frac{1 - \lambda_1}{E_1} + \frac{1 - \lambda_2}{E_2} \tag{10}$$

where, σ_s is yield limit, σ_s = 200 N/mm^2; $E*$ is the effective elasticity modulus, GPa; E_1 is elastic modulus of particle, GPa; E_2 is the elastic modulus of surface, GPa; λ_1 and λ_2 are the Poisson's ratios of particles and surface, respectively.

3.2.3. Ejection Stage

Particles at this stage are mainly affected by the adhesion force F_{ad} produced by surface and liquid bridge, and the direction of adhesion force is downward along the surface normal. If the adhesion force is too weak, the particles cannot be adhered, and its velocity will change to V_5. If the adhesion force is strong, the particles will be adhered to the surface. After this stage, the collision process between particle and surface is concluded.

The adhesion force between particle and surface includes Van der Waals force, capillary force, electrostatic force, chemical bond force, and so on [35]. Among them, Van der Waals force (F_{vdw}) and capillary force (F_{cap}) play an important role in the adhesion force (F_{ad}). The contact model diagram between particle and surface is shown in Figure 4. The adhesion force can be expressed as the following series of equations, as described in [35,36].

$$F_{ad} = F_{vdw} + F_{cap} \tag{11}$$

$$F_{vdw} = \frac{H_1 R}{6D^2} + \frac{(H_2 - H_1)R}{6D^2}\left\{\frac{1}{[1+h/D]^2}\right\} \tag{12}$$

$$h = r_k(\cos\theta_1 + \cos(\beta + \theta_2)) \tag{13}$$

$$F_{cap} = \frac{2\pi R(2cr_k - D)\gamma_W}{r_k} \tag{14}$$

$$r_k = -\frac{\gamma_W V_0}{R_g T \ln(p/p_0)} = -\frac{0.53 \times 10^{-9}}{\ln(c_{RH})} \tag{15}$$

$$c = [\cos(\theta_1) + \cos(\theta_2)]/2 \tag{16}$$

where, H_1 and H_2 are the Hamaker constant, and the magnitudes of these values are related to the medium: in the air medium $H_1 = 10.38 \times 10^{-20}$ J, in the water medium $H_2 = 1.90 \times 10^{-20}$ J [35]. D is the distance between particle and surface, m; h is the height of the liquid bridge, m; r_k is the Kelvin radius, m; θ_1 and θ_2 are the contact angles of the bottom liquid bridge and upper liquid bridge, respectively; β is liquid bridge angle of the particle; c is contact angle coefficient; γ_w is the surface tension of water, $\gamma_w = 0.073$ N/m; V_0 is the molar volume of water, $V_0 = 18 \times 10^{-6}$ m^3/mol; R_g is the gas constant, $R_g = 8.31$ J/(mol K); T is the absolute temperature, $T = 290$ K; p is vapor pressure, Pa; p_0 is saturated vapor pressure, Pa; c_{RH} is relative humidity.

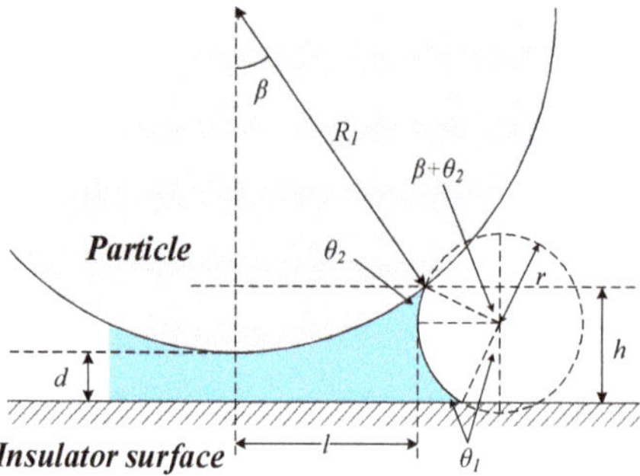

Figure 4. Contact model between particle and surface.

3.3. Adhesion Criterion of Particles

The energy loss of particles during collision is mainly composed of two parts: the collision energy loss caused by non-complete elastic deformation and the adhesion energy loss caused by adhesion force. The details are as follows.

When particles collide with the surface, the non-complete elastic deformation occurs, and the velocity of the particles will change to V_3.

$$V_3 = eV_2 = eV_1 \tag{17}$$

At the ejection stages, the work done (W_1) by adhesion force is

$$W_1 = \int_{a_{\min}}^{a_{\max}} F_{vdw}(D)\mathrm{d}D + \int_0^h F_{cap}(D)\mathrm{d}D \tag{18}$$

where, $a_{\max}$ is the maximum effect distance of Van der Waals force, $a_{\max} = 0.4$ nm; $a_{\min}$ is the minimum effect distance of Van der Waals force, $a_{\min} = 0.165$ nm [17].

At first, the particles fly toward the surface with the initial velocity V_1, and then through three stages of injection, collision deformation, and ejection, the final velocity V_5 becomes

$$V_5 = \sqrt{V_3^2 - \frac{2W_1}{m}} = \sqrt{(eV_1)^2 - \frac{2W_1}{m}} \tag{19}$$

In Equation (18), if $(eV_1)^2 - 2W_1/m > 0$, it can be considered that the particles cannot be adhered to the surface. However, if $(eV_1)^2 - 2W_1/m < 0$, it can be considered that the particles will be adhered to the surface.

4. Simulation Model

To analyze the reason why size distribution of contaminated particles on the porcelain insulator surface is concentrated in a specific range, a physical model of collision, rebound, and adhesion between particles and surface was built, and the adhesion of particles was simulated by COMSOL Multi-physics simulation software® (5.2a). In the simulation model, four types of insulators were considered, including bell type insulator XP-160, aerodynamic type insulator XMP-160, double umbrella type insulator XWP-160, and the three-umbrella type insulator XSP-160. The structure and parameter of these four kinds of insulators are shown in Table 2. In Table 2, H, D and L, respectively, represent height, umbrella skirt size and leakage distance.

In the simulation model, three pieces porcelain insulators were established to study the adhesion of contaminated particles. The top of the insulator string was set as the grounding terminal, and its potential was 0 kV; the bottom of the insulator string was set as the high voltage terminal, and its potential was 30 kV. The material of the umbrella skirt was set to porcelain and its relative dielectric constant was set to 6. The material of the fittings was set to steel and its relative dielectric constant was set to 10^{12}. In each simulation test, 9000 particles were released from the left side of the insulator. Among them, 3000 particles carried positive charges, its charge-mass ratio was 1.58×10^{-4} C/kg; 3000 particles carried negative charges, and its charge-mass ratio was -3.04×10^{-4} C/kg [37]; 3000 particles had no charge. Previous studies have shown that $CaSO_4$ is the major component of contamination [38], so the particle density was set to 2960 kg/m^3.

Table 2. Parameter and structure of insulator.

Type	Parameter			Shape
	H (mm)	**D (mm)**	**L (mm)**	
XP-160	170	280	405	
XMP-160	155	425	385	
XWP-160	170	340	525	
XSP-160	170	330	545	

5. Influence of Different Factors on Particle Adhesion

The adhesion process of particles is affected by a variety of complex factors, including relative humidity, wind speed, precipitation, particle properties, electric field type, electric field strength, aerodynamic shape, material, and so on. The existing literature shows that the influences of relative humidity, wind speed, electric field type, electric field strength, and aerodynamic shape on the adhesion are obvious [15,17,19,21]. Therefore, this paper carried out a series of studies on the influences of these five factors. To highlight the influences of relative humidity, wind speed, electric field type, and electric field strength, the paper took the XP-160 insulator as the research object. In addition, four kinds of insulators were used to study the influence of aerodynamic shape.

5.1. Influence of Relative Humidity

Historical meteorological data shows that annual average relative humidity of most cities in China is in the range of 50%–70%. Thus, the adhesion of particles was studied under relative humidity at 30%, 40%, 50%, 60%, 70%, and 80%, and the results are shown in Figure 5. The data points are connected by a B-Spline curve. In the simulation model, the conditions were set as a positive DC electric field, $v = 4$ m/s and $U = 30$ kV.

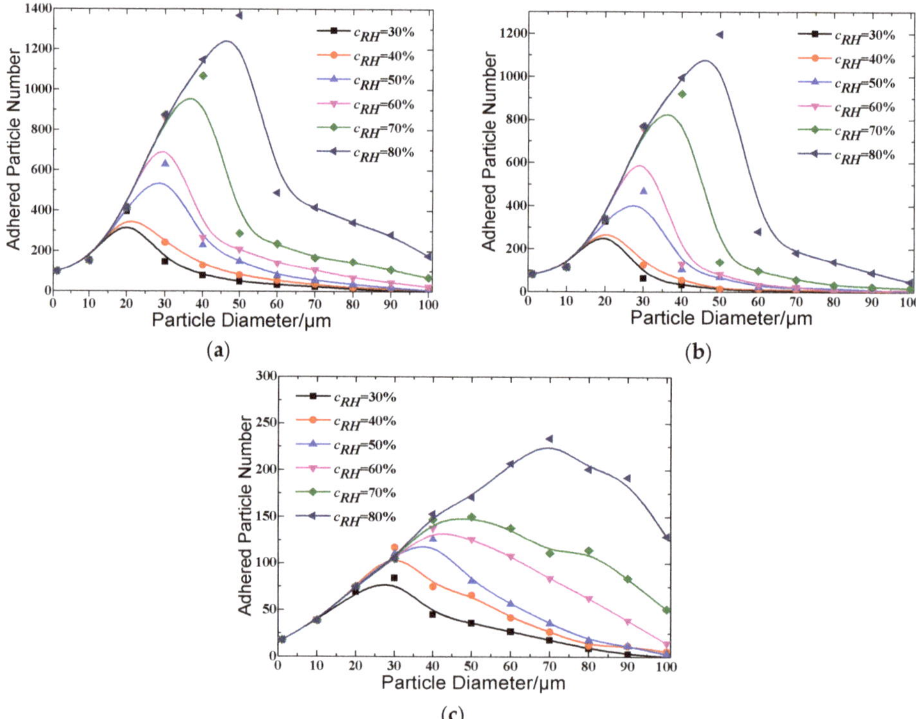

Figure 5. Adhesion number of particles with different size under different relative humidity: (**a**) all surface; (**b**) upper surface; and (**c**) bottom surface.

Figure 5 shows that the higher the relative humidity, the easier the large particles are adhered, and the more the number of adhered particles. Specifically, at low relative humidity (c_{RH} = 30% and 40%), particles with sizes in the range of 10–30 μm were easily adhered, and the D_{50} of adhered particles were 19.84 μm and 21.52 μm, respectively. With high relative humidity (c_{RH} = 70% and 80%), the particles with sizes in the range of 25–70 μm were easily adhered, and the D_{50} of the adhered particles were 48.76 μm and 37.42 μm, respectively. With normal relative humidity (c_{RH} = 50% and 60%), the particles with sizes in the range of 15–40 μm were easily adhered, and the D_{50} of the adhered particles were 29.47 μm and 30.14 μm, respectively. The measurement results were consistent with the statistical characteristics obtained above. In addition, it could also be found that the size distribution of adhered particles on the upper surface was similar to that of on all surface, and there was a small amount of adhered particles on the bottom surface. Moreover, the influence of relative humidity on the adhesion number of particles was relatively limited when the particle size was less than 15 μm and greater than 90 μm. However, when the particle size was in the range of 20–80 μm, the influence of relative humidity on the adhesion number of particles was quite significant.

In Section 3.2, Equations (11)–(16) show that when relative humidity increases, the capillary force F_{cap} will increase accordingly, and then the adhesion loss will also increase. Finally, the particles will be easier to adhere to the insulator surface with the same initial kinetic energy. For small particle (size ≤ 20 μm), the effect of fluid drag force is more obvious, and the trajectory of the particle is more likely to follow the change of wind direction. Therefore, it is easy to follow the movement of airflow, and bypass the insulator surface. So, collision and adhesion are difficult to happen. Although the small particles are easily adhered after collision, the number of adhered particles is rare due to the lower collision probability. For larger particles (size ≥ 80 μm), the effect of fluid drag force is remarkably weak, and the trajectory of particles cannot quickly follow the change of wind direction. Thus, the particles find it easy to pass through the boundary layer and achieve the collision. However, the energy loss during the collision process is so limited that the particles are not easily adhered, so there is a small number of adhered particles. However, for particles with sizes in the range of 20–80 μm, the order of magnitude of their initial kinetic energy and energy loss in collision are similar, so the adhesion is greatly affected by other external parameters. As relative humidity increases, the adhesion loss will increase correspondingly, which will cause the particles to be easily adhered to the surface. Therefore, the relative humidity has a significant influence on the adhesion number of particles, especially for particles with sizes in the range of 20–80 μm.

5.2. Influence of Wind Speed

In view of the fact that annual average wind speed in most cities of China is about 4m/s, the adhesions of the particles under wind speed of 2, 4, 6, 8 and 10 m/s were studied in this paper, respectively. The results are shown in Figure 6. B-Spline curve is used to connect data points, and the simulation conditions are set as positive DC electric field, c_{RH} = 60% and U = 30 kV.

As shown in Figure 6, the influence of wind speed on adhesion of particles is significant. At low wind speed (v = 4 m/s), the particles with greater size were easily adhered to the surface, and the size of adhered particles was mainly distributed in the range of 30–70 μm, and the D_{50} is 49.22 μm. At high wind speed (v = 10 m/s), the particles with smaller size were easily adhered, and the size of adhered particles was mainly distributed in the range of 10–30 μm, and the D_{50} is 20.14 μm. When wind speed was in the range of 2–6 m/s, there were obvious changes of the size distribution of adhered particles. However, when wind speed was in the range of 6–10 m/s, the size distribution of the adhered particles showed little change, and it showed saturation. Therefore, for the area in which annual average wind speed is about 4 m/s, the particles with sizes in the range of 20–40 μm are more likely to be adhered. These simulation results support the statistical characteristics of the particle size distribution obtained from the above measurement results. In addition, the size distribution of adhered particles on the upper surface was similar to that of on the all surface, and there were a small number of adhered particles on the bottom surface.

According to Figure 6, there is a certain concentration of the size distribution of adhered particles. A thin boundary layer will be formed near the insulator surface when airflow moves around the insulator [11]. In the boundary layer, there is a significant gradient change of force in the direction of the normal vertical surface. The order of magnitude of viscous force increases remarkably and reaches an order of magnitude which is similar to that of the inertial force [17]. Therefore, for the smaller particles, their trajectories tend to vary with the direction of the airflow due to the significant viscous force, so it is difficult to collide with the surface, and the number of adhered particles will be greatly reduced. However, for the larger particles, the inertia force is greater than the viscous force, and it plays a major role in the forces acting on the particles. Therefore, the change of airflow has little influence on its trajectory, which makes it easier to pass through the boundary layer and realize collision. Whereas, due to the larger initial kinetic energy and less energy loss during the collision, it is easier to experience rebound and fail to complete adhesion.

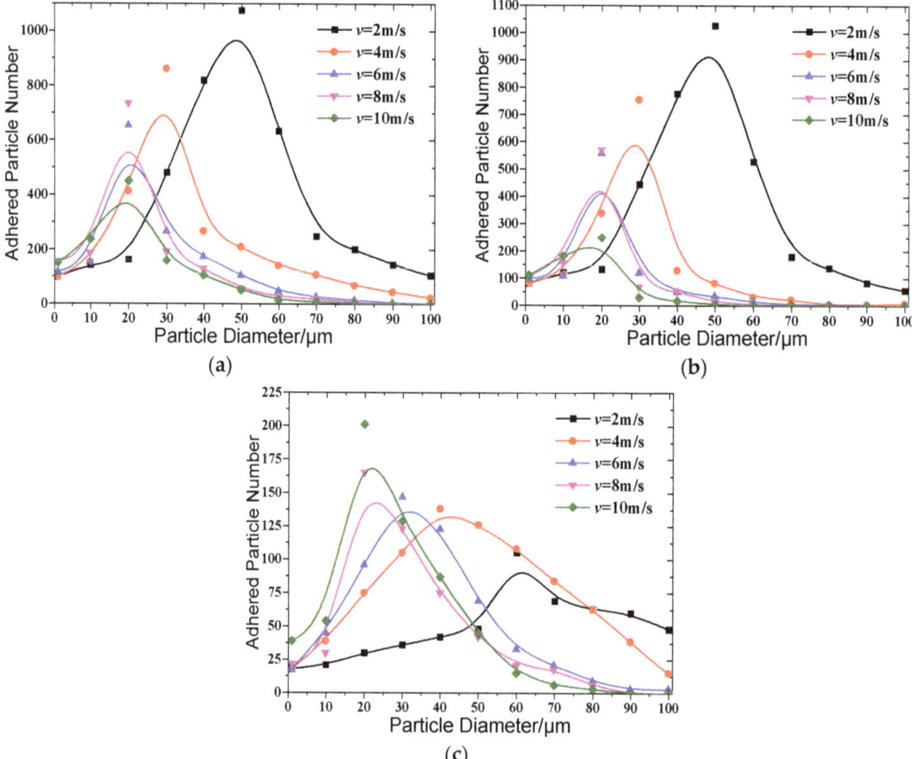

Figure 6. Adhesion number of particle with different size under different wind speed: (**a**) all surface; (**b**) upper surface; and (**c**) bottom surface.

Especially for the bottom surface, due to the existence of the umbrella skirt, the turbulent flow around the bottom surface is remarkable, and it will greatly reduce the speed of the airflow. At the same time, the velocity of particles will also reduce. Finally, it causes the large particles to be easily adhered. As shown in Figures 5c and 6c, there is a large amount of adhesion of larger particles on the bottom surface.

5.3. Influence of Electric Field Type

Adhesion of particles under four different electric field types were studied, including positive DC electric field, negative DC electric field, AC electric field, and no electric field. The voltages were set to +30 kV, −30 kV, 30sin (100πt) kV, and 0 kV, respectively, and the results are shown in Figure 7. In the simulation model, the simulation conditions were set to $c_{RH} = 60\%$, $v = 4$ m/s.

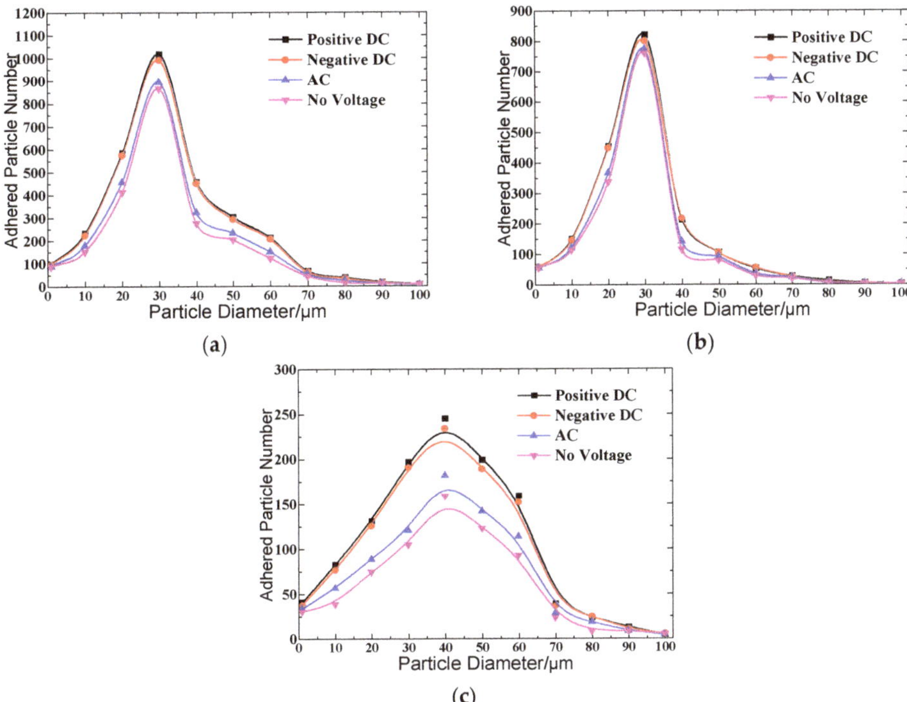

Figure 7. Adhesion number of particles with different size under different electric field types: (**a**) all surface; (**b**) upper surface; and (**c**) bottom surface.

In Figure 7, it shows that the influence of electric field type on adhesion is relatively weak. The difference of adhesion number curves under different electric field types is not obvious. The adhesion numbers of the particles with the same size from high to low, are positive DC electric field, negative DC electric field, AC electric field, and no electric field. The reason for these results is that the AC electric field changes periodically, which leads to the periodic change of the electric force acting on the particle, and it cannot achieve the continuous effect. Finally, the trajectory of particles is less affected. Under the condition of the DC electric field, the particles will move toward the surface with the effect of electric field force, because the electric field gradient near the insulator is perpendicular to the surface [15]. At the same time, due to the continued effect of electric field force, the collision number of particles will show an obvious rise. Therefore, it leads to a higher adhesion number of particles under the DC electric field than that under the AC electric field and no electric field.

5.4. Influence of Electric Field Strength

The adhesion of particles at different voltage levels were studied, including 10, 20, 30, 40, 50, and 60 kV, and the results are shown in Figure 8. In the simulation model, the conditions were positive electric field, $c_{RH} = 60\%$, $v = 4$ m/s.

Figure 8a shows that the greater the electric field strength, the more particles that are adhered to the insulator surface. The adhesion number of particles on the all surface reaches peak value when size is about 30 μm. In Figure 8b,c, it shows that the adhesion number of particles on the upper surface is greater than that on the bottom surface, and the size distribution of adhered particles on the upper and bottom surface is different. The adhesion number of particles reaches peak value when the size is about 40 μm on the bottom surface, but the adhesion number of particles reaches peak value when the size is about 30 μm on the upper surface.

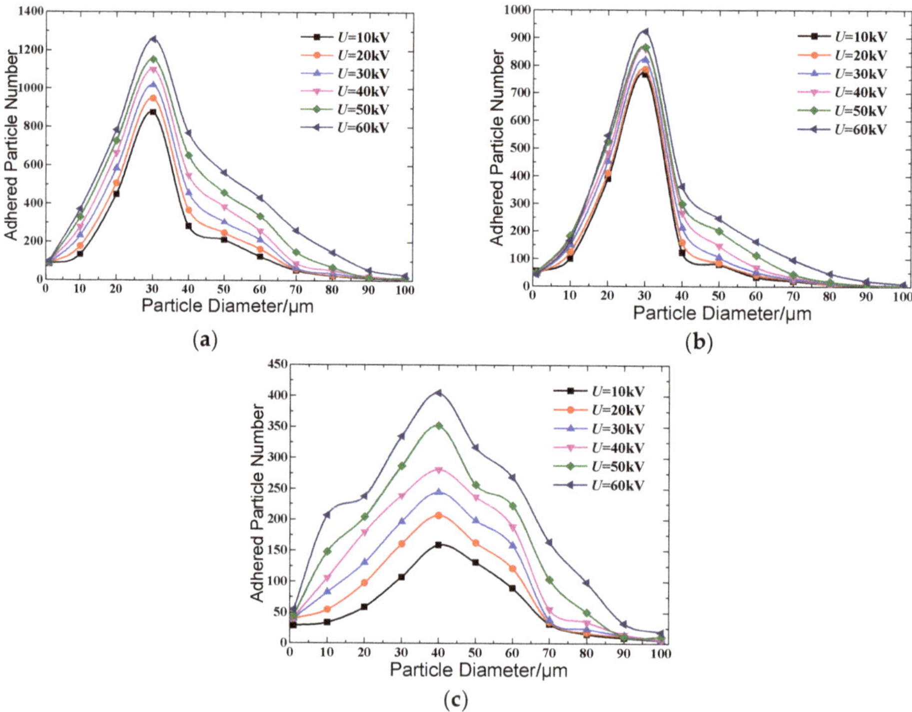

Figure 8. Adhesion number of particles with different size under different electric field strength: (**a**) all surface; (**b**) upper surface; and (**c**) bottom surface.

Furthermore, it was also found that when the particle size was in the range of 10–70 μm, the influence of electric field strength on the adhesion number was relatively obvious. However, when the particle size was less than 10 μm and greater than 70 μm, the influence of electric field strength on the adhesion number was very limited. This phenomenon can be explained by that the greater the electric field strength, the greater the electric field force. The electric field force causes more particles to move toward the insulator surface [15], thereby increasing the number of adhered particles. For small particles, the influence of electric field force is relatively weak due to less electric charge. At the same time, the influence of fluid drag force was stronger compared with electric field force, so the change of electric field strength showed little influence on the adhesion number. For large particles, its charge was greater. The increase of electric field strength will increase the colliding number of particles, but it will also increase the velocity of the particles when collision happens, resulting in a decrease in the number of adhered particles. Therefore, the influence of electric field strength is limited.

5.5. Influence of Aerodynamic Shape

In order to verify whether the above statistical characteristics are universally applicable, the adhesion of particles under the conditions of different aerodynamic shapes were studied, including the bell type insulator XP-160, aerodynamic type insulator XMP-160, double umbrella type insulator XWP-160 and the three-umbrella type insulator WSP-160. The results are shown in Figure 9, and the airflow field diagrams for these four kinds of aerodynamic models are shown in Figure 10. In addition, the same parameter conditions were set, including c_{RH} = 60%, v = 4 m/s, positive DC electric field, and U = 30 kV.

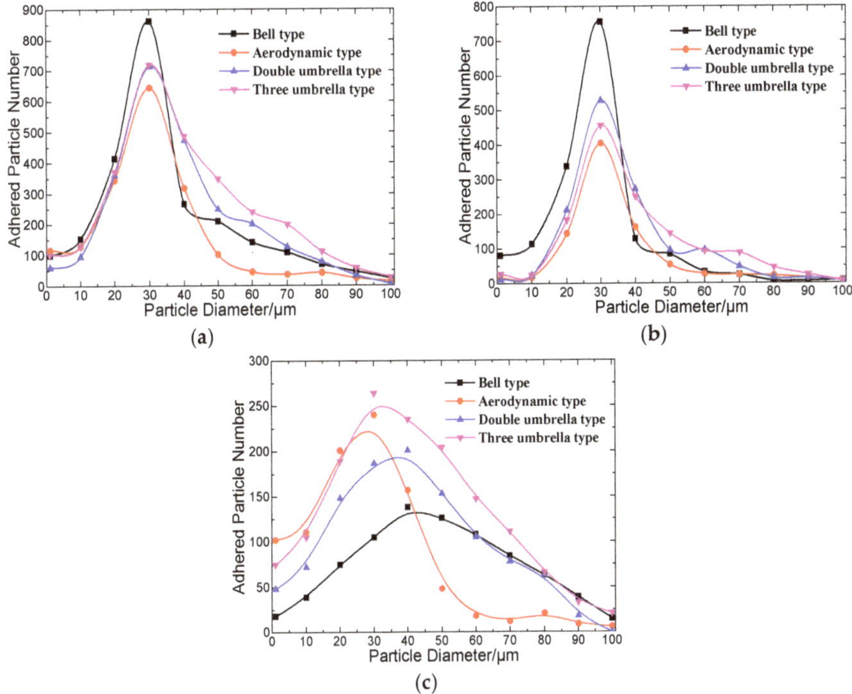

Figure 9. Adhesion number of particles with different size under different electric field strength: (**a**) all surface; (**b**) upper surface; and (**c**) bottom surface.

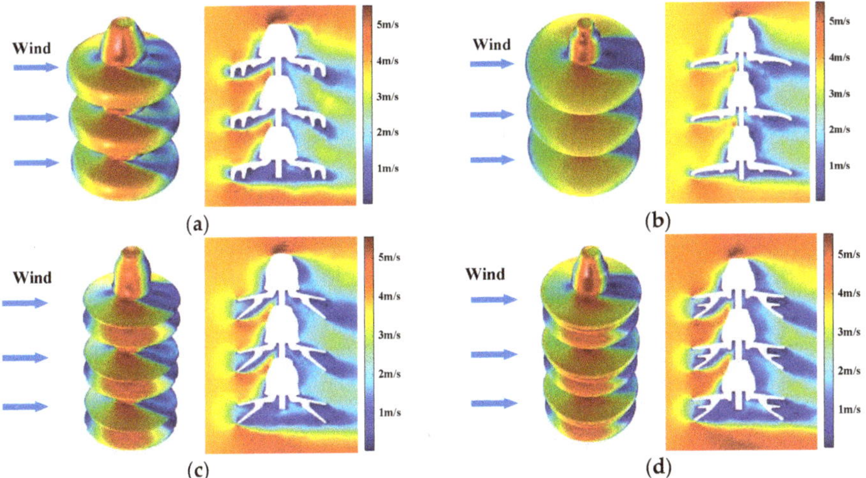

Figure 10. Airflow field diagram of different aerodynamic shapes under 4 m/s wind speed: (**a**) bell type insulator; (**b**) aerodynamic type insulator; (**c**) double umbrella type insulator; and (**d**) three umbrella type insulator.

According to the Figure 10, it can be found that there is no significantly low speed area around the bottom surface of the aerodynamic type insulator. The airflow is less disturbed because the structure of its umbrella skirts is relatively simple. However, for the bell type insulator, double umbrella type insulator, and the three-umbrella type insulator, there was obviously a low speed area around the bottom surface. The umbrella skirt structure of these three kinds of insulators is relatively complex, so the airflow is greatly disturbed (the blue part shown in the Figure 10).

As can be seen from Figure 9, the influence of aerodynamic shape on the adhesion is not significant. In general, the size distribution of adhered particles on the four kinds of insulators with different aerodynamic shape were similar, and the adhered particles were mainly concentrated in the range of 20–40 µm, and the adhesion number of particles reached peak value when size was about 30 µm. In particular, the adhesion number of particles on the bell type, double umbrella type, and three-umbrella type insulators was greater than that of the aerodynamic type insulator, especially for particles with sizes greater than 20 µm. This difference is attributed to the difference of the umbrella skirt structures of these four kinds of insulators. More particles can be adhered to the surface of the bell type, double umbrella type, and three-umbrella type insulators, due to the obvious low speed area around the insulators' surface. In addition, this difference is more remarkable, especially on the bottom surface. In the Figure 9c, it can be found that the adhesion number curve of greater particles ($R \geq 45$ µm) on the aerodynamic type insulator surface is the lowest.

6. Conclusions

Many contamination samples collected from porcelain insulator surfaces were measured, and obvious size distribution characteristics were found. Furthermore, an adhesion model was established to analyze the movement of particles, and the influences of different factors on adhesion were also studied. The following conclusions were obtained.

- The size distribution of adhered particles on the porcelain insulator surface is basically in logarithmic normal distribution, and the D_{50} is about 20 µm, and the distribution of particles ($R \leq 5$ µm and $R \geq 50$ µm) is rare.
- For small particles, their trajectory is easily affected by the fluid drag force, and it is difficult to experience collision and adhesion. For large particles, it is difficult to adhere to the surface due to great initial kinetic energy. Thus, there are significant size distribution characteristics of contaminated particles on the porcelain insulator surface.
- In the process of adhesion, the influences of relative humidity and wind speed on the adhesion were remarkable, whilst the influences of electric field type, electric field strength, and aerodynamic shape were relatively weak.

In addition, it can be considered that the size distribution of contaminated particles on the glass insulator surface is similar to that of the porcelain insulator, because glass and porcelain all belong to a rigid medium. However, the surface of a composite insulator will experience micro-elastic deformation during collision, resulting in a difference of the size distribution, so more detailed research about adhesion will continue to be carried out.

Author Contributions: Conceptualization, M.Z.; Methodology, R.W.; Software, Y.J.; Validation, M.Z., L.L. and Y.J.; Investigation, L.L.; Resources, L.L.; Data Curation, L.L.; Writing—Original Draft Preparation, R.W. and Y.J.; Writing—Review & Editing, M.Z. and L.L.; Visualization, Y.J.; Supervision, L.L.; Project Administration, M.Z.; Funding Acquisition, M.Z.

Funding: This research was funded by the New Theory, New Technology and Application Demonstration of Artificial Rain or Snow Catalyzed by Charged Particles (Project 2016YFC0401002), the Science Technology Program of State Grid Corporation of China (52170216000A), the Fundamental Research Funds for the Central Universities (2016YXZD069) and the National Natural Science Foundation of China (51777082).

Acknowledgments: We acknowledge the support given by Ming Lu and Xiaohui Yang, who work in State Grid Henan Electric Power Research Institute (Power Transmission Line Galloping Prevention and Control Technology Laboratory of State Grid, Zhengzhou 450052, China).

Conflicts of Interest: The authors declare no conflict of interest.

References

1. Ren, A.; Liu, H.; Wei, J.; Li, Q. Natural contamination and surface flashover on silicone rubber surface under haze-fog environment. *Energies* **2017**, *10*, 1580. [CrossRef]
2. Hu, J.; Sun, C.; Jiang, X.; Yang, Q.; Zhang, Z.; Shu, L. Model for predicting DC flashover voltage of pre-contaminated and ice-covered long insulator strings under low air pressure. *Energies* **2011**, *4*, 628–643. [CrossRef]
3. Jiang, Z.; Jiang, X.; Guo, Y.; Hu, Y.; Meng, Z. Pollution accumulation characteristics of insulators under natural rainfall. *IET Gener. Transm. Distrib.* **2017**, *11*, 1479–1485. [CrossRef]
4. Li, L.; Li, Y.; Lu, M.; Liu, Z.; Wang, C.; Lv, Z. Quantification and comparison of insulator pollution characteristics based on normality of relative contamination values. *IEEE Trans. Dielectr. Electr. Insul.* **2016**, *23*, 965–973. [CrossRef]
5. Banik, A.; Dalai, S.; Chatterjee, B. Autocorrelation aided rough set based contamination level prediction of high voltage insulator at different environmental condition. *IEEE Trans. Dielectr. Electr. Insul.* **2016**, *23*, 2883–2891. [CrossRef]
6. Jiang, Y.; Mcmeekin, S.G.; Reid, A.J.; Nekahi, A.; Judd, M.D.; Wilson, A. Monitoring contamination level on insulator materials under dry condition with a microwave reflectometer. *IEEE Trans. Dielectr. Electr. Insul.* **2016**, *23*, 1427–1434. [CrossRef]
7. Cherney, E.A.; Hackam, R.; Kim, S.H. Porcelain insulator maintenance with RTV silicone rubber coatings. *IEEE Trans. Power Deliv.* **1991**, *6*, 1177–1181. [CrossRef]
8. Sundararajan, R.; Gorur, R.S. Role of non-soluble contaminants on the flashover voltage of porcelain insulators. *IEEE Trans. Dielectr. Electr. Insul.* **2015**, *3*, 113–118. [CrossRef]
9. Venkataraman, S.; Gorur, R.S. Prediction of flashover voltage of non-ceramic insulators under contaminated conditions. *IEEE Trans. Dielectr. Electr. Insul.* **2015**, *13*, 862–869. [CrossRef]
10. Richards, C.N.; Renowden, J.D. Development of a remote insulator contamination monitoring system. *IEEE Trans. Power Deliv.* **1997**, *12*, 389–397. [CrossRef]
11. Chandrasekar, S.; Kalaivanan, C.; Cavallini, A.; Montanari, G.C. Investigations on leakage current and phase angle characteristics of porcelain and polymeric insulator under contaminated conditions. *IEEE Trans. Dielectr. Electr. Insul.* **2009**, *16*, 574–583. [CrossRef]
12. Karady, G.G.; Shah, M.; Brown, R.L. Flashover mechanism of silicone rubber insulators used for outdoor insulation-I. *IEEE Trans. Power Deliv.* **1995**, *10*, 1965–1971. [CrossRef]
13. Matsuoka, R.; Shinokubo, H.; Kondo, K.; Mizuno, Y.; Naito, K.; Fujimura, T.; Terada, T. Assessment of basic contamination withstand voltage characteristics of polymer insulators. *IEEE Trans. Power Deliv.* **1996**, *11*, 1895–1900. [CrossRef]
14. Dong, B.; Jiang, X.; Hu, J.; Shu, L.; Sun, C. Effects of artificial polluting methods on ac flashover voltage of composite insulators. *IEEE Trans. Dielectr. Electr. Insul.* **2012**, *19*, 714–722. [CrossRef]
15. Wang, J.; Liang, X.; Chen, L.; Liu, Y. Combined effect of different fields on the motion characteristics of dust particles around the insulators. In Proceedings of the 2011 Annual Report Conference on Electrical Insulation and Dielectric Phenomena, Cancun, Mexico, 16–19 October 2011; pp. 373–376. [CrossRef]
16. Horenstein, M.N.; Melcher, J.R. Particle contamination of high voltage DC insulators below corona threshold. *IEEE Trans. Dielectr. Electr. Insul.* **1978**, *EI-14*, 297–305. [CrossRef]
17. Li, H.; Lai, J.; Lei, Q.; Deng, S.; Liu, G.; Li, L. Collision and adsorption of pollution particles on the surface of electrical insulator. *High Volt. Eng.* **2012**, *38*, 2596–2603. [CrossRef]
18. Lv, F.; Liu, H.; Wang, F.; Yang, S.; Ma, J. Deposit criterion of pollution particles on composite insulators surface under high speed aerosol. *Trans. Chin. Electrotech. Soc.* **2017**, *32*, 206–213. [CrossRef]
19. Jiang, X.; Li, H. Application of computational fluid dynamics to analysis of contamination depositing characteristics of insulators. *High Volt. Eng.* **2010**, *36*, 329–334. [CrossRef]
20. Nan, J.; Xu, T.; Wan, X.; Liang, X.; Wang, J.; Li, H.; Hu, L.; Huo, F. Fast contamination depositing characteristics of insulators under artificial horizontal wind field conditions. *Proc. CSEE* **2017**, *37*, 3323–3330. (In Chinese) [CrossRef]
21. Wang, L.; Liu, T.; Mei, H.; Xiang, Y. Research on contamination deposition characteristics of post insulator based on computational fluid dynamics. *High Volt. Eng.* **2015**, *41*, 2741–2749. [CrossRef]

22. Zhang, Z.; Zhang, W.; You, J.; Jiang, X.; Zhang, D.; Bi, M.; Wu, B.; Wu, J. Influence factors in contamination process of XP-160 insulators based on computational fluid mechanics. *IET Gener. Transm. Distrib.* **2016**, *10*, 4140–4148. [CrossRef]

23. Lv, Y.; Li, J.; Zhang, X.; Pang, G.; Liu, Q. Simulation study on pollution accumulation characteristics of XP13-160 porcelain suspension disc insulators. *IEEE Trans. Dielectr. Electr. Insul.* **2016**, *23*, 2196–2206. [CrossRef]

24. Lv, Y.; Li, J.; Zhao, W.; Zhang, X.; Liu, Q. Simulation study on dynamic contamination characteristics of porcelain suspension disc insulator. In Proceedings of the 2016 IEEE International Conference on High Voltage Engineering and Application (ICHVE), Chengdu, China, 19–22 September 2016. [CrossRef]

25. Lv, Y.; Zhao, W.; Li, J.; Zhang, Y. Simulation of contamination deposition on typical shed porcelain insulators. *Energies* **2017**, *10*, 1045. [CrossRef]

26. Guo, Y.; Jiang, X.; Liu, Y.; Meng, Z.; Li, Z. AC flashover characteristics of insulators under haze–fog environment. *IET Gener. Transm. Distrib.* **2016**, *10*, 3563–3569. [CrossRef]

27. IEC. *IEC TS 60815-1:2008 Selection and Dimensioning of High-Voltage Insulators for Polluted Conditions—Part 1: Definitions, Information and General Principles*; IEC: Geneva, Switzerland, 2008.

28. Xu, S.; Wu, C.; Li, S.; Bao, W.; Liu, Y.; Liang, X. Research on pollution accumulation characteristics of insulators during fog-haze days. *Proc. CSEE* **2017**, *37*, 2142–2150. (In Chinese) [CrossRef]

29. Su, Z.; Liu, Y. Comparison of natural contaminants accumulated on surfaces of suspension and post insulators with DC and AC stress in Northern China's inland areas. *Power Syst. Tech.* **2004**, *28*, 13–17. [CrossRef]

30. Tu, Y.; Sun, Y.; Peng, Q.; Wang, C.; Gong, B.; Chen, X. Particle size distribution characteristics of naturally polluted insulators under the fog-haze environment. *High Volt. Eng.* **2014**, *40*, 3318–3326. [CrossRef]

31. Lan, L.; Zhang, G.; Wang, Y.; Wen, X. Force and motion characteristics of contamination particles near the high voltage end of UHVDC insulator. *Energies* **2017**, *10*, 969. [CrossRef]

32. O'Neill, M.E. A sphere in contact with a plane wall in a slow linear shear flow. *Chem. Eng. Sci.* **1968**, *23*, 1293–1298. [CrossRef]

33. Asay, D.B.; Kim, S.H. Evolution of the adsorbed water layer structure on silicon oxide at room temperature. *J. Phys. Chem. B* **2005**, *109*, 16760–16763. [CrossRef] [PubMed]

34. Asay, D.B.; Kim, S.H. Effects of adsorbed water layer structure on adhesion force of silicon oxide nanoasperity contact in humid ambient. *J. Phys. Chem.* **2006**, *124*, 174712. [CrossRef] [PubMed]

35. Xiao, X.; Qian, L. Investigation of humidity-dependent capillary force. *Langmuir* **2000**, *16*, 8153–8158. [CrossRef]

36. You, S.; Man, P. Mathematical models for the van der Waals force and capillary force between a rough particle and surface. *Langmuir* **2013**, *29*, 9104–9117. [CrossRef] [PubMed]

37. Zhang, Y.; Zhao, S.; Chen, Z.; Yu, Z.; Dong, H.; Wang, S. Influence of suspended sand particles on potential and electric field distribution along long rod insulator. *High Volt. Eng.* **2014**, *40*, 2706–2713. [CrossRef]

38. Li, H.; Liu, G.; Li, L. Study on the natural contamination chemical composition of line insulator in Guangzhou area. In Proceedings of the 2013 Annual Report Conference on Electrical Insulation and Dielectric Phenomena, Shenzhen, China, 20–23 October 2013; pp. 323–326. [CrossRef]

Article

Fouling Release Coatings Based on Polydimethylsiloxane with the Incorporation of Phenylmethylsilicone Oil

Miao Ba [1,2], Zhanping Zhang [1,2,*] and Yuhong Qi [1,2]

[1]   Key Lab of Ship-Machinery Maintenance & Manufacture, Dalian Maritime University, Dalian 116026, China; bamiao90@126.com (M.B.); yuhong_qi@dlmu.edu.cn (Y.Q.)

[2]   Department of Materials Science and Engineering, Dalian Maritime University, Dalian 116026, China

*   Correspondence: zzp@dlmu.edu.cn; Tel.: +86-0411-8472-3556

Received: 2 February 2018; Accepted: 10 April 2018; Published: 24 April 2018

Abstract: In this study, phenylmethylsilicone oil (PSO) with different viscosity was used for research in fouling release coatings based on polydimethylsiloxane (PDMS). The surface properties and mechanical properties of the coatings were investigated, while the leaching behavior of PSO from the coatings was studied. Subsequently, the antifouling performance of the coatings was investigated by the benthic diatom adhesion test. The results showed that the coatings with high-viscosity PSO exhibited high levels of hydrophobicity and PSO leaching, while the high PSO content significantly decreased the elastic modulus of the coatings and prolonged the release time of PSO. The antifouling results indicated that the incorporation of PSO into coatings enhanced the antifouling performance of the coating by improving the coating hydrophobicity and decreasing the coating elastic modulus, while the leaching of PSO from the coatings improved the fouling removal rate of the coating. This suggests a double enhancement effect on the antifouling performance of fouling release coatings based on PDMS with PSO incorporated.

Keywords: polydimethylsiloxane; phenylmethylsilicone oil; viscosity; additive amount; leach; antifouling

---

1. Introduction

Marine biofouling refers to the colonization of submerged surfaces by marine micro- and macro-organisms, and is a worldwide problem affecting maritime and aquatic industries [1–5]. For a long time, painting antifouling coatings on seawater-impregnated substrates has been an effective way to inhibit the attachment of marine organisms. It has been widely applied, such as in shipping vessels, heat exchangers, offshore rigs, jetties, aquaculture cages, and other submerged structures in the marine environment [6,7]. Meanwhile, one statistical analysis indicated that the practice saves the global shipping industry an estimated 150 billion US dollars per year [8]. Among them, self-polishing antifouling coatings incorporating tributyltin-based compounds (TBT-based coatings) are the most efficient. Unfortunately, with the development of relevant research, these coatings have been revealed as toxicants toward non-targeted species [2]. Furthermore, TBT compounds have also been reported to accumulate in mammals and debilitate the immune defenses of fish [2,3]. Therefore, in 2001 the International Marine Organization (IMO) forbade the use of TBT-based coatings. Moreover, the biocide used in the marine environment is now under strict control in many countries, and it has become a driving force for the development of environmentally-friendly alternative coatings, which mainly include fouling-resistant coatings, fouling release coatings, and fouling degrading coatings [8–12].

It is known that marine biofouling involves a wide variety of organisms (more than 4000 species have been identified) [12]. It has the following characteristics: the fouling organisms change very quickly, attaching to the growth is very easy, there is no selective attachment to most of the matrix materials, and the environmental adaptability of organisms is very strong. Therefore, it is difficult to completely reject the adhesion of the fouling organisms on the substrate. For fouling release coatings, by minimizing the adhesion strength between fouling organisms and the material surface, fouling may be readily removed by simple mechanical cleaning or hydrodynamic stress during navigation [13]. These coatings use physical means to protect the substrate immersed in seawater, and are not causes of environmental pollution. They are now widely used as environmentally-friendly alternatives [8,13].

Polydimethylsiloxane (PDMS) has a linear structure and is the most common hydrophobic material. The molecular structure of PDMS is arranged in helixes, whereas the outwards-directed methyl groups provide hydrophobicity [14,15]. PDMS with low surface energy and low elastic modulus reduces the adhesion of marine organisms [16,17]. It has been prepared for fouling release coatings, and the first organic PDMS-based polymer antifouling coating came out in 1972 [18].

The incorporation of non-reactive silicone oil in fouling release coatings based on PDMS has been seen since 1977 [19]. Related studies have suggested that silicone oil additives in fouling release PDMS coatings could reduce the coefficient of friction and favor an easier release of fouling organisms [20,21]. Further studies showed that the increase of interfacial slippage with silicone oil leached from the coated surfaces improved the fouling release performance [22]. Although the potential of silicone oil to affect the ability to overcome cellular barriers has been pointed out [17], there is no statistical evidence to support these concerns [13,20]. The incorporation of silicone oil into fouling release coatings based on PDMS showed that the amount of silicone oil was negligible [22,23], and the low toxicity of silicone oil caused little harm in marine life [24–26].

The research on the incorporation of silicone oil into coatings is mainly focused on the evaluation of antifouling behavior on certain specific fouling organisms by the leaching of silicone oil [20–23]. There is no related research on the choice of silicone oil, including viscosity and added amount. A particularly significant research gap exists regarding the effect of these factors on the leaching speed and the leaching percent of silicone oil. In this experiment, the phenylmethylsilicone oil (PSO) with different viscosity and content was incorporated into fouling release coatings based on PDMS. The effect of the leaching amount of PSO on the antifouling performance of the coating is studied herein, and the leaching cycle of PSO is predicted by a reasonable measurement. The analysis of related properties can reveal the reasons for the improvement of the antifouling performance of the coating.

2. Materials and Methods

2.1. Materials

Hydroxyl-terminated polydimethylsiloxane (PDMS) was obtained from Dayi Chemical Industry Co., Ltd. (Yantai, China). The kinematic viscosity of PDMS was 10,000 mm^2/s and the relative molecular weight was about 60,000. Phenylmethylsilicone oil (PSO) was purchased from Shanghai Hualing Resin Co., Ltd. (Shanghai, China). The kinematic viscosity of PSO was 30 mm^2/s, 75 mm^2/s, and 100 mm^2/s for products known as Si-30, Si-75, and Si-100, respectively. Tetraethylorthosilicate (TEOS) was obtained from Tianjin Kemiou Chemical Reagent Co., Ltd. (Tianjin, China). Bismuth neodecanoate (BiND) was obtained from Deyin Chemical Co., Ltd. (Shanghai, China). Xylene and ethyl acetate were also analytical grade and supplied by Yongda Chemical Reagent Co., Ltd. (Tianjin, China).

2.2. Preparation of Coating Samples

The coating was composed of three parts. Pre-dispersed slurry included PDMS and PSO. TEOS and xylene were mixed to make the curing agent, and the mixture of BiND and ethyl acetate was prepared into the catalytic agent. The two-stage process of the preparation method was as follows: PDMS (100 g) and PSO were added into a 500 mL stirring tank at 2000 rpm for 30 min. Afterward, coatings were prepared in a 20:4:1 weight ratio of PDMS (from the pre-dispersed slurry):curing agent:catalytic agent. The coating was brushed on glass slides with dimensions of 75 mm × 25 mm × 1 mm, and also poured into a Teflon mold with dimensions of 150 mm × 150 mm × 2 mm for at least 8 h to form a cross-linked elastomer. The blank control sample without PSO was set as A, and other experimental samples were set as Ax-y, where x represents the viscosity of PSO and y represents the mass of PSO.

2.3. Experimental Procedure

The antifouling performance of the fouling release coating based on PDMS depends mainly on the surface and mechanical properties. Therefore, we analyzed the effect of the PSO on the above properties. In this study, the change in surface properties was due to the change of the chemical composition caused by the incorporation of PSO, which could be analyzed by Fourier transform infrared (FTIR) spectroscopy. The influence of the PSO on the crosslink density of the coating was beneficial to the analysis of the coating's mechanical properties. The focus of this study was on the observation of leaching PSO, including leaching amount and leaching percent. The leaching of the PSO with different viscosity and content needed to be characterized by reasonable parameters. Finally, the antifouling performance of the coating was analyzed. Error values are quoted as standard error of the mean (SEM), based on the number of samples analyzed (n).

2.4. Characterization

2.4.1. Surface Properties of the Coatings

Contact angle measurements were conducted using the sessile drop method on a JC2000C contact angle measurement system (Shanghai Zhongchen Co., Ltd., Shanghai, China). Three-microliter droplets of distilled H_2O and CH_2I_2 were placed on the sample surface using a syringe. Digital images of the droplet silhouette were captured with a charge-coupled device camera and the contact angles were evaluated using the measuring angle method. Six points for each sample were tested for the contact angle measurement. The surface free energies were calculated from the measured water contact angles and diiodomethane contact angles using Owens two-liquid method [27].

In order to analyze the effect of the chemical composition on the surface properties of the coating, FTIR spectra were recorded on a Frontier PerkinElmer infrared spectrometer (Thermo Fisher Scientific Inc., Waltham, MA, USA) within a scan range of 4000–650 cm^{-1} and a resolution of 2 cm^{-1}. For each sample, 32 scans were recorded and a high-performance diamond single-bounce Attenuated Total Reflection accessory was used. The spectral evaluation was performed using the Spectrum 10.3.6 software package from PerkinElmer.

2.4.2. Mechanical Properties of the Coatings

Tensile samples were prepared into strips with 150 mm × 20 mm × 2 mm and then stretched on a Labthink XLM auto tensile tester (Labthink Co., Ltd., Jinan, China) with the 1.15 software package. The tensile speed was at 50 mm/min. Elastic modulus was fitted with the data, which showed that the strain was less than 0.02 mm/mm. Three samples of each coating were prepared for the experiment. The stress–strain curves of the coatings were presented by selecting the data for which the elastic modulus value was close to the average value.

A HT220 shore hardness tester was applied to test the hardness of the coating. The thickness of the test sample must exceed 5 mm, and thus the prepared casting sample was folded repeatedly to meet the experimental requirement [28]. Similarly, three samples for each coating were used in the experiment.

In order to determine the origin of the mechanical properties of the coating, the crosslink density of the coating was measured by the equilibrium swelling method [29]. Generally, it can be expressed by the molecular weight between crosslink points (M_C). Toluene was used as the solvent which could dissolve the PSO in the coating, and then the swelling of the gel could be determined. The value of M_C was calculated by the Flory–Rhener relation,

$$M_C = \frac{-\rho V \left(v^{\frac{1}{3}} - v/2 \right)}{\ln(1 - v) + v + \chi_1 \times v^2} \tag{1}$$

where v represents the volume fraction of the polymer in the swollen specimen [29]; ρ represents the density of samples before swelling; V represents the molar volume of the solvent, and it is 106.125 cm^3/mol for toluene; χ_1 refers to the Flory–Huggins interaction parameter between the sample and toluene, and it is 0.45 in this experiment.

In order to ensure the accuracy of the experimental data, the weight of the sample was measured every 3 h by using a precision balance (Mettler Toledo Co., Ltd., Zurich, Switzerland) during the swelling process. When the difference of the adjacent measurement data did not exceed 0.1 mg, the sample was up to the equilibrium swelling state. The whole experiment was performed at 25 °C, while three samples for each coating were used in the experiment.

2.4.3. The Leaching Observation of the PSO

The morphology of the leached PSO on the sample surface was analyzed by an Olympus OLS4000 CLSM (OLYMPUS (China) Co., Ltd., Beijing, China) with a field of view 2575 µm × 2581 µm. The software version was 2.2.4.

The leaching percent of the PSO with exposure time was analyzed. Before brushing slide samples, the mass of the slide (m_0) was measured with a precision balance, and the mass of the slide sample (m_1) was also measured after 2 days of curing. Then, the mass of the PSO in the coatings could be calculated. Every t days, the surface of the coating was wiped with alcohol swabs, and the mass of the cleaned slide sample (m_t) was determined. The leaching percent of PSO (W_{PSO}) was calculated by Equation (2), where w denotes the mass percent of PSO in the coating.

$$W_{PSO} = \frac{(m_1 - m_t)}{w \times (m_1 - m_0)} \times 100\% \tag{2}$$

The M_C value of coating A could be calculated by the equilibrium swelling method, and then the PSO was used as the solvent to swell Coating A at 25 °C. The experimental method was as reported before. By Equation (1), the Flory–Huggins interaction parameter (χ) between Coating A and the PSO was also calculated [29], and could be used as a parameter to evaluate the leaching of the PSO with different viscosity. Densities of the samples and the PSO were measured by a MH-300A electron density meter (Kunshan Creator Testing Instrument Co., Ltd., Kunshan, China) and are shown in Table 1. In this experiment, the molar volume of the PSO was calculated by the relative molecular mass and density. The relative molecular mass of the PSO was researched via PL-GPC-50 normal temperature gel permeation chromatograph (POLYTECH Co., Ltd., Beijing, China) and are also listed in Table 1.

Table 1. The density and the relative molecular mass of experimental materials. PDMS: polydimethylsiloxane.

Materials	Coating A	Si-30	Si-75	Si-100	PDMS Resin
Density (g/cm^3)	0.975	0.996	1.001	1.007	-
Relative molecular mass	-	450	610	700	60,000

2.4.4. Benthic Diatom Adhesion Tests

Navicula sp. was used to measure the antifouling performance of the coatings, and was obtained from the Institute of Oceanology, Chinese Academy of Sciences (Qingdao, China). The influence of the PSO on benthic diatom adhesion tests of the coatings included two stages: the stage without the leaching of PSO and the stage with the leaching of PSO. Therefore, it was necessary to conduct benthic diatom adhesion tests on the coatings before and after the PSO leached. In this experiment, samples cured for 1 day with no leached PSO observed and samples cured for 30 days with leached PSO observed were selected. Before the benthic diatom adhesion tests, all samples were cleaned with alcohol to ensure that there was no dust or leached PSO.

Six coated slides of each coating were immersed in 10^5–10^6 cell/mL fresh benthic diatom suspension for 24 h, then three slides were rinsed lightly with sterile deionized water to remove unsettled diatoms, whereas the other three slides were washed with a CB-8L-C high-pressure water gun at 0.1 MPa for 20 s (Haishu Chebo Industry & Trade Co., Ltd., Ningbo, China). All slides were then ground into test tubes filled with 45 mL acetone solution (90%, v/v) mixed with 1 mL magnesium carbonate solution (1 wt %). The tubes were stored in a cold (8 °C), dark area. After 24 h, the supernatant was poured into a stoppered test tube and centrifuged (4000 rpm, 15 min). The resulting solution was poured into a quartz cuvette with a path length of 1 cm, and the absorbance of the solution was determined at fixed wavelengths (750 nm, 663 nm, 645 nm, and 630 nm) to calculate the concentration of chlorophyll-a using an ultraviolet–visible spectrophotometer (Labtech UV-2000, Labtech Co., Ltd., Beijing, China). The fouling removal rate based on benthic diatom adhesion to the coating samples was calculated by the formula in Equation (3), where W_r, W_w, and R represent the concentration of chlorophyll-a for rinsed samples, the concentration of chlorophyll-a for washed samples, and the removal rate of the coating, respectively:

$$R = \frac{W_r - W_w}{W_r} \times 100\% \tag{3}$$

3. Results and Discussion

3.1. Surface Properties of the Coatings

Determining the water contact angles and surface free energies can provide essential information on the surface properties of the coatings. Table 2 showed the contact angle results and calculated surface free energies of the coatings. As observed from Table 2, the water contact angles increased with the increase of the PSO content and viscosity. Meanwhile, PSO incorporated into the coatings also decreased the surface free energies of the coatings, which were calculated by the Owens two-liquid method. However, unlike the change in water contact angle, the effect of low viscosity PSO on the reduction of the surface free energies was more obvious. In terms of the functional group, the hydrophobicity of the phenyl group is better than that of the methyl group. Therefore, the addition of PSO enhanced the hydrophobicity of the coatings. Results also revealed the change of the chemical composition of the coatings with the incorporation of PSO. The more phenyl groups in the coating, the stronger the hydrophobicity of the coating. The FTIR results (Figure 1) also revealed that the strength of the featured absorption peaks of the phenyl group in coatings increased by increasing the viscosity of the PSO, indicating that the increase in the water contact angle (the featured absorption peak at 725 cm^{-1} was interfered by the featured absorption peak of Si–CH$_3$ at 790 cm^{-1}). Related studies also showed that the surface free energies of the PSO (7–100 mm^2/s) were about 22.2 to 24.1 mJ/m^2 [30]. Thus, low-viscosity PSO could significantly reduce the surface free energies of the coatings.

Table 2. The contact angle and surface free energies of coating samples (±SEM (standard error of the mean), $n = 6$).

Sample	Contact Angle (°)		Surface Free Energy (mJ/m^2)
	Water	Diiodomethane	
A	107.1 ± 0.31	65.4 ± 0.11	25.8 ± 0.41
A30-5	113.9 ± 0.23	73.4 ± 0.27	21.6 ± 0.22
A30-10	114.7 ± 0.17	74.3 ± 0.31	21.1 ± 0.18
A30-15	115.3 ± 0.08	75.3 ± 0.28	20.6 ± 0.33
A30-20	116.0 ± 0.22	76.5 ± 0.21	19.9 ± 0.28
A75-20	116.9 ± 0.15	72.1 ± 0.21	22.9 ± 0.19
A100-20	118.5 ± 0.44	70.7 ± 0.33	24.2 ± 0.38

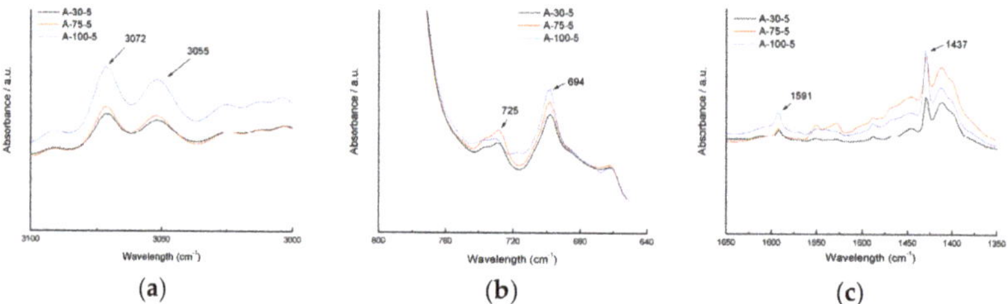

Figure 1. Fourier transform infrared (FTIR) spectra of the phenyl group in coating samples: (**a**) the stretching vibration peaks of =C–H in the phenyl group at 3072 and 3055 cm^{-1}; (**b**) the out-of-plane bending vibration peaks of =C–H in the phenyl group at 725 and 694 cm^{-1}; (**c**) the stretching vibration peaks of C=C in the phenyl group at 1591 and 1488 cm^{-1}.

3.2. Mechanical Properties of the Coatings

The mechanical properties of the coatings were evaluated via tensile test. The fitting elastic modulus and the measured shore hardness are presented in Figure 2. Experimental determination showed that the elastic modulus and the shore hardness decreased with the increase of the PSO content. For the coatings with different viscosity of the PSO, these properties were basically the same. The tensile curve of the sample is also shown in Figure 3. The tensile curves of the coatings with different viscosities of PSO were basically coincident under the same additive amount. By increasing the content of the PSO, the stress of the coating decreased at the same strain. This result was in agreement with the results of Truby [25]. In order to analyze the reason for the decrease of the mechanical properties of the coating, the crosslink density of the coating was measured. Crosslink density is defined as the moles of effective network chain per cubic centimeter [29]. The crosslink density of a polymer is an important parameter affecting its mechanical properties. Polymer with high crosslink density has better mechanical properties. Generally, it can be expressed by the number average molecular weight between two adjacent crosslinks (M_C), and a larger value of M_C indicates a lower polymer crosslink density [31,32]. For this experiment, the M_C value of the coating is shown in Table 3. It is indicated that the value of M_C increased with the increase of the PSO content, revealing the decrease of the crosslink density. The results also showed that the changing of the PSO viscosity did not change the crosslink density of the coatings. Non-reactive PSO was stored in the gap between the molecular chains of the coating, reducing the density of the polymer network chain. Therefore, the crosslink density of the coating was decreased. In general, the PSO with low molecular weight and high additive amount could have a more pronounced effect. The relative molecular mass of the PSO is shown in Table 1. Although the three kinds of PSO had different relative molecular weights, the difference of the relative molecular weight between the PSOs could be ignored compared with the relative molecular weight of the PDMS resin. Therefore, the viscosity of the PSO had little effect on the mechanical properties of the coatings, while the effect of the PSO content on the mechanical properties was more obvious. With the increase of the PSO content, the crosslink density of the coatings decreased, causing the decrease of the mechanical properties.

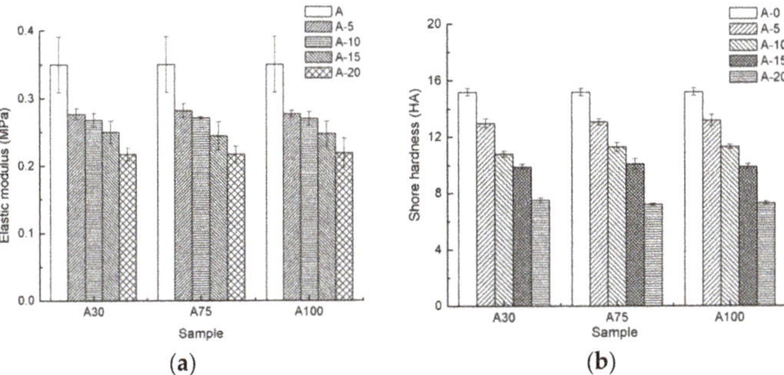

Figure 2. The mechanical properties of coating samples: (**a**) elastic modulus; (**b**) shore hardness.

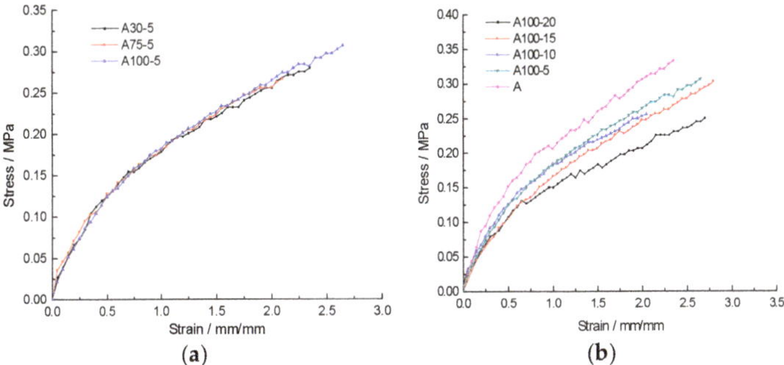

Figure 3. The tensile stress of samples (**a**) with different phenylmethylsilicone oil (PSO) viscosities and (**b**) with different additive amounts of PSO (Si-100).

Table 3. The M_C value of samples. ($\pm$SEM, $n = 6$).

Sample	A	A30-5	A30-10	A30-15	A30-20	A75-20	A100-20
M_C	11,728 $\pm$ 246	19,840 $\pm$ 374	22,495 $\pm$ 277	24,336 $\pm$ 406	32,799 $\pm$ 178	32,710 $\pm$ 323	32,706 $\pm$ 301

3.3. Leaching Observation of PSO

Samples with dimensions of 150 mm × 150 mm × 2 mm were used to observe the leaching of the PSO. The morphology of the sample is shown in Figures 4 and 5 with 1 day and 30 days of exposure. After 1 day of exposure, no PSO was leached on the coating surface. With the extension of exposure time, PSO was gradually leached. It was also shown that the coating incorporating high viscosity and high additive amount of the PSO leached more PSO in the same amount of time. The leaching percent–time curve is shown in Figure 6. It basically follows the Higuchi diffusion equation (Figure 6c). By fitting the curve, the time needed for the complete leaching of the PSO could be simulated (Table 4). Although the increase of the PSO content resulted in the increase in the leaching amount of the PSO from the coatings at the same elapsed time, it also showed that the leaching percent of the PSO decreased with increasing PSO content. Coatings with low-viscosity PSO had a longer release time. For the coating A30-20, the time required for the complete leaching of the PSO exceeded 5890 days. In contrast, the coating of A100-5 released all the PSO in only 350 days.

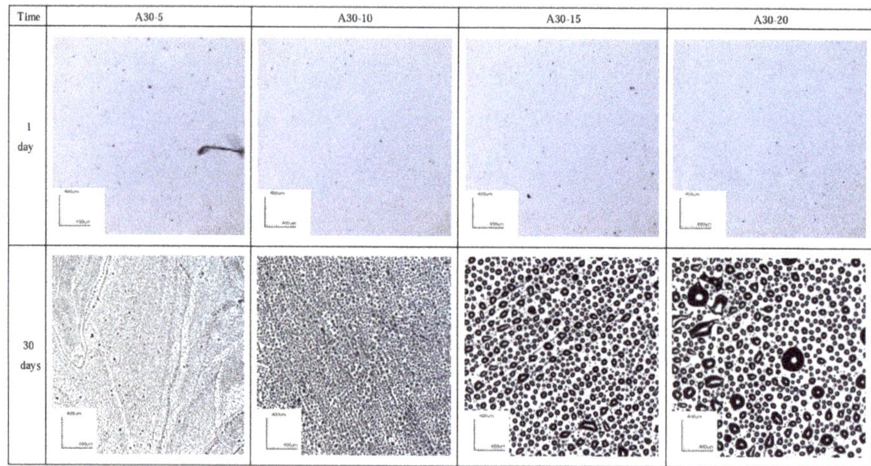

Figure 4. Morphologies of different amounts of PSO leached on samples.

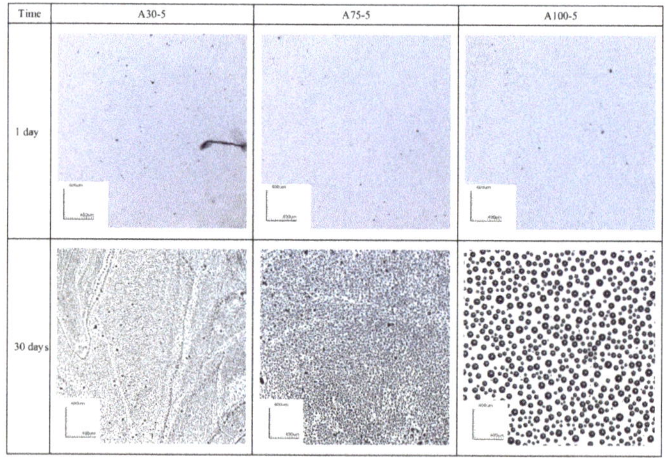

Figure 5. Morphologies of different viscosities of PSO leached on samples.

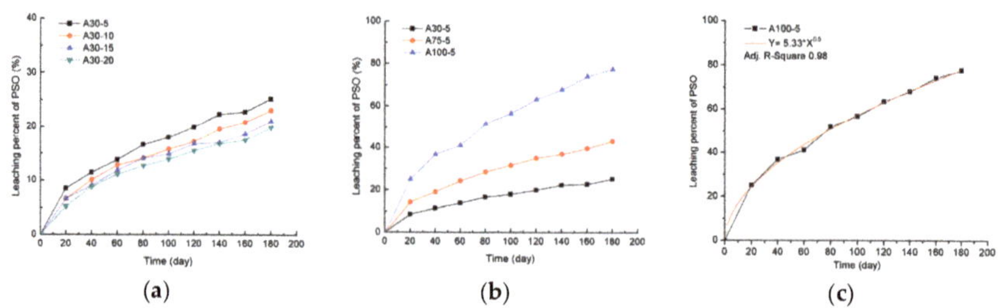

Figure 6. The leaching percent–time curve: (**a**) with different viscosities of PSO; (**b**) with different PSO contents; (**c**) Higuchi fitting curve of A100-5.

Table 4. The estimated time for leaching PSO completely.

Sample	A30-5	A30-10	A30-15	A30-20	A75-5	A100-5
Time (day)	2250	3810	5000	5890	760	350

Studies [5,20–22] on the incorporation of silicone oil into the fouling release coatings show that the reason for the leaching behavior of silicone oil from the coatings was the compatibility difference between silicone oil and the coatings, and note that compatibility should be characterized by a reasonable parameter. The leaching phenomenon can be considered as the phase separation of silicone oil and the coatings. Correspondingly, the Flory–Huggins interaction parameter (χ) between the coating A and the PSO is shown in Table 5. With the increase of the PSO viscosity, the value of χ increased. The high viscosity of the PSO contained more phenyl groups (Figure 1). This would lead to the increase in the molecular structure difference between the coating and the PSO. Therefore, with the increase of the PSO viscosity, the compatibility between the PSO and the coating decreased, causing more obvious leaching behavior of the PSO from the coatings. In addition, with the increase of the PSO content, the crosslink density of the coating decreased. The decrease of the crosslink density caused the decrease of the resistance to PSO leaching. Therefore, the PSO from the coatings with low crosslink density could be leached easily. This revealed that the leaching amount of the PSO was affected by the Flory–Huggins interaction parameter (χ) between the coating and the PSO, as well as the crosslink density of the coating. Although the leaching amount of the PSO at the same time increased, the increase of the PSO content eventually resulted in the decrease of the PSO leaching percent at the same time.

Table 5. Flory–Huggins interaction parameter between samples and PSO. ($\pm$SEM, $n = 6$).

Materials	Si-30	Si-75	Si-100
χ	1.037 $\pm$ 0.0297	1.434 $\pm$ 0.0266	1.779 $\pm$ 0.0177

3.4. Benthic Diatom Adhesion Tests

The antifouling performance of the coatings was examined by conducting benthic diatom adhesion tests, and the results on the coating samples in the stage without leached PSO are shown in Table 6. The coatings incorporated with high-viscosity PSO had lower chlorophyll-a concentration, suggesting low levels of benthic diatom adhesion and high levels of the antifouling property. With the increase of the PSO content, the fouling removal rate of the coating increased. For the fouling release coating based on PDMS, the coating with high hydrophobicity showed a good ability to inhibit the adhesion of fouling organisms [1,3]. In addition, the low elastic modulus could effectively reduce the energy consumption required for the removal of fouling organisms, increasing the fouling removal rate of the coating [8]. The effect of the incorporation of PSO on the hydrophobicity and the elastic modulus of the coating was analyzed as above. Accordingly, the improvement of the antifouling performance of the coating could also be demonstrated by benthic diatom adhesion tests. The coating with high PSO content exhibited better hydrophobicity and lower elastic modulus, reducing the adhesion amount of benthic diatoms and increasing the fouling removal rate. Therefore, the corresponding antifouling performance was enhanced.

Subsequently, the antifouling performance of the coatings in the stage with leached PSO was also examined. The pretreatment ensured that there was no leached PSO on the surface of experimental samples. However, the leaching of the PSO was a continuous process, and PSO could still be released in fresh benthic diatom suspension. The experimental results are shown in Table 7. Comparison of Tables 6 and 7 indicated that the concentration of chlorophyll-a for rinsed samples in the different stages were basically the same; however, the concentration of chlorophyll-a for washed samples in the stage with the leached PSO decreased obviously, meaning better antifouling performance. Further, the removal rate of samples in the stage with the leached PSO generally increased, and the values of some samples exceeded 90%. With the leaching of the PSO, the removal of fouling organisms was not only dependent on the elastic modulus of the coatings. The leaching PSO formed continuous oil films to block the adhesion of the benthic diatom to samples, or reduced the adhesion strength between the benthic diatom and the sample surface [22,23]. After washing, most of the benthic diatoms were removed. Section 3.3 showed that the coating with high PSO viscosity and high PSO content could leach more PSO in the same time, while it also showed that the coatings with high leached PSO could exhibit high levels of the removal rate. Therefore, benthic diatom adhesion tests of the coating indicated that the enhancement of the coating hydrophobicity effectively inhibited the adhesion of fouling organisms. In addition, the decrease of the elastic modulus of the coating as well as the PSO leached onto the coating surface increased the removal rate of the coating. The cause of this phenomenon was due to the incorporation of PSO.

Table 6. Experimental results of the benthic diatom adhesion test on the stage of no leaching of PSO. ($\pm$SEM, $n = 3$).

Sample	A	A30-5	A75-5	A100-5	A100-10	A100-15	A100-20
Concentration of chlorophyll-a for rinsed samples (mg/m^2)	1735 ± 56	1567 ± 27	1445 ± 19	1375 ± 28	830 ± 23	719 ± 31	460 ± 12
Concentration of chlorophyll-a for washed samples (mg/m^2)	747 ± 27	622 ± 42	565 ± 32	534 ± 28	273 ± 33	197 ± 40	106 ± 20
Fouling removal rate (%)	56.5 ± 4.71	60.3 ± 3.89	61.7 ± 2.77	60.9 ± 4.14	67.1 ± 1.80	72.6 ± 2.97	76.7 ± 4.36

Table 7. Experimental results of the benthic diatom adhesion test on the stage of leaching of PSO. ($\pm$SEM, $n = 3$).

Sample	A	A30-5	A75-5	A100-5	A100-10	A100-15	A100-20
Concentration of chlorophyll-a for rinsed samples (mg/m^2)	1774 ± 48	1615 ± 37	1485 ± 22	1365 ± 42	847 ± 33	707 ± 12	471 ± 22
Concentration of chlorophyll-a for washed samples (mg/m^2)	786 ± 41	326 ± 19	293 ± 43	205 ± 27	115 ± 19	66 ± 25	37 ± 15
Fouling removal rate (%)	55.7 ± 3.21	79.8 ± 3.11	80.3 ± 2.16	85.0 ± 5.10	86.4 ± 3.67	90.7 ± 2.77	92.1 ± 3.21

4. Conclusions

This research investigated the influence of PSO with different viscosity and content on the antifouling performance of fouling release coatings based on PDMS. With the incorporation of PSO, the hydrophobicity of the coating was enhanced, and the mechanical properties of the coating, especially the elastic modulus, decreased. The leaching behavior of the PSO also indicated that the PSO with high viscosity and large content was more easily leached. By fitting the leaching percent–time curve of PSO, the leaching cycle of PSO could be correctly predicted. Due to the incorporation of PSO, the change of the above properties improved the antifouling performance of the coating. Therefore, the coating showed a low adhesion of the benthic diatom. The higher viscosity and the greater the content of PSO incorporated in the coating, the better the antifouling performance of the coating.

Author Contributions: Miao Ba, Zhanping Zhang, and Yuhong Qi designed and performed the experiments; Miao Ba and Zhanping Zhang analyzed the data; Zhanping Zhang and Yuhong Qi contributed reagents/materials/analysis tools; Miao Ba and Zhanping Zhang wrote the paper.

Acknowledgments: This work was supported by the National Natural Science Foundation of China (51079011, 51179018), the application basic research fund of the Ministry of Transport (2013329225330). The authors gratefully acknowledge for financial support.

Conflicts of Interest: The authors declare no conflict of interest.

References

1. Callow, J.A.; Callwo, M.E. Trends in the development of environmentally friendly fouling-resistant marine coatings. *Nat. Commun.* **2011**, *2*, 244. [CrossRef] [PubMed]
2. Chen, R.R.; Li, Y.K.; Tang, L.; Yang, H.C.; Lu, Z.T.; Wang, J.; Liu, L.H.; Takahashi, K. Synthesis of zinc-based acrylate copolymers and their marine antifouling application. *RSC Adv.* **2017**, *7*, 40020–40027. [CrossRef]
3. Yang, W.J.; Neoh, L.G.; Kang, E.T.; Teo, L.M.; Rittschof, D. Polymer brush coatings for combating marine biofouling. *Prog. Polym. Sci.* **2014**, *39*, 1017–1042. [CrossRef]
4. Nurioglu, A.G.; Esteves, A.C.C.; de With, G. Non-toxic, nonbiocide-release antifouling coatings based on molecular structure design for marine applications. *J. Mater. Chem. B* **2015**, *3*, 6547–6570. [CrossRef]
5. Krishnan, S.; Weinman, C.J.; Ober, C.K. Advances in polymers for anti-biofouling surfaces. *J. Mater. Chem.* **2008**, *18*, 3405–3413. [CrossRef]
6. Stafslien, S.J.; Christianson, D.; Daniels, J.; VanderWal, L.; Chernykh, A.; Chisholm, B.J. Combinatorial materials research applied to the development of new surface coatings XVI: Fouling-release properties of amphiphilic polysiloxane coatings. *Biofouling* **2015**, *31*, 135–149. [CrossRef] [PubMed]
7. Loriot, M.; Linossier, I.; Vallee-Rehel, K.; Fay, F. Influence of biodegradable polymer properties on antifouling paints activity. *Polymers* **2017**, *9*, 36. [CrossRef]
8. Hellio, C.; Yebra, D.M. *Advances in Marine Antifouling Coatings and Technologies*; Woodhead Publishing Ltd.: Cambridge, UK, 2009; pp. 1–15.
9. Voulvoulis, N.; Scrimshaw, M.D.; Lester, J.N. Alternative antifouling biocides. *Appl. Organomet. Chem.* **1999**, *13*, 135–143. [CrossRef]
10. Omae, I. General aspects of tin-free antifouling paints. *Chem. Rev.* **2003**, *103*, 3431–3448. [CrossRef] [PubMed]
11. Matyjaszewski, K.; Tsarevsky, N.V. Nanostructured functional materials prepared by atom transfer radical polymerization. *Nat. Chem.* **2009**, *1*, 276–288. [CrossRef] [PubMed]
12. Chambers, L.D.; Stokes, K.R.; Walsh, F.C.; Wood, R.J.K. Modern approaches to marine antifouling coatings. *Surf. Coat. Technol.* **2006**, *201*, 3642–3652. [CrossRef]
13. Lejars, M.; Margillan, A.; Bressy, C. Fouling release coatings: A nontoxic alternative to biocidal antifouling coatings. *Chem. Rev.* **2012**, *112*, 4347–4390. [CrossRef] [PubMed]
14. Heilen, W. *Silicone Resins and Their Combinations*; Vincentz Network GmbH & Co.: Hannover, Germany, 2005; p. 28.
15. Buyl, F.D. Silicone sealants and structural adhesives. *Int. J. Adhes. Adhes.* **2001**, *21*, 411–422. [CrossRef]
16. Patwardhan, S.V.; Taori, B.P.; Hassan, M.; Agashe, N.R.; Franklin, J.E.; Beaucage, G.; Mark, J.E.; Clarson, S.J. An investigation of the properties of poly(dimethylsiloxane)-bioinspired silica hybrids. *Eur. Polym. J.* **2006**, *42*, 167–178. [CrossRef]
17. Eduok, U.; Faye, O.; Szpunar, J. Recent developments and applications of protective silicone coatings: A review of PDMS functional materials. *Prog. Prg. Coat.* **2017**, *111*, 124–163. [CrossRef]
18. Mueller, W.J.; Nowacki, L.J. Ship's Hull Coated with Antifouling Silicone Rubber. U.S. Patent 3,702,778, 14 November 1972.
19. Selvig, T.A.; Leavitt, R.I.; Powers, W.P. Anti-Fouling Marine Compositions. U.S. Patent 4,025,693, 24 May 1977.
20. Shivapooja, P.; Cao, C.Y.; Orihuela, B.; Levering, V.; Zhao, X.H.; Rittschof, D.; Lopez, G.P. Incorporation of silicone oil into elastomers enhances barnacle detachment by active surface strain. *Biofouling* **2016**, *32*, 1017–1028. [CrossRef] [PubMed]
21. Howell, C.; Vu, T.L.; Lin, J.J.; Kolle, S.; Juthani, N.; Watson, E.; Weaver, J.C.; Alvarenga, J.; Aizenberg, J. Self-replenishing vascularized fouling-release surfaces. *ACS Appl. Mater. Interfaces* **2014**, *6*, 13299–13307. [CrossRef] [PubMed]
22. Galhenage, T.P.; Hoffman, D.; Silbert, S.D.; Stafslien, S.J.; Daniels, J.; Milikovic, T.; Finlay, J.A.; Franco, S.C.; Clare, A.S.; Nedved, B.T.; et al. Fouling-release performance of silicone oil-modified siloxane-polyurethane coatings. *ACS Appl. Mater. Interfaces* **2016**, *8*, 29025–29036. [CrossRef] [PubMed]

23. Stein, J.; Truby, K.; Wood, C.D.; Stein, J.; Gardner, M.; Swain, G.; Kavanagh, C.; Kovach, B.; Schultz, M.; Wiebe, D.; et al. Silicone foul release coatings: Effect of the interaction of oil and coating functionalities on the magnitude of macrofouling attachment strengths. *Biofouling* **2003**, *19*, 71–82. [CrossRef] [PubMed]

24. Ware, C.S.; Smith-Palmer, T.; Peppou-Chapman, S.; Scarratt, L.R.; Humphries, E.M.; Balzer, D.; Neto, C. Marine antifouling behavior of lubricant-infused nanowrinkled polymeric surfaces. *ACS Appl. Mater. Interfaces* **2018**, *10*, 4173–7182. [CrossRef] [PubMed]

25. Truby, K.; Wood, C.D.; Stein, J.; Cella, J.; Carpenter, J.; Kavanagh, C.; Swain, G.; Wiebe, D.; Lapota, D.; Meyer, A.; et al. Evaluation of the performance enhancement of the silicone biofouling-release coatings by oil incorporation. *Biofouling* **2000**, *15*, 141–150. [CrossRef] [PubMed]

26. Afsar, A.; De Nys, R.; Steinberg, P. The effect of foul-release coatings on the settlement and behavior of cyprid larvae of the barnacle balanus amphitrite darwin. *Biofouling* **2003**, *19*, 105–110. [CrossRef] [PubMed]

27. Owens, D.K.; Wendt, R.C. Estimation of the surface free energy of polymers. *J. Appl. Polym. Sci.* **1969**, *13*, 1741–1747. [CrossRef]

28. Ba, M.; Zhang, Z.P.; Qi, Y.H. The dispersion tolerance of micro/nano particle in polydimethylsiloxane and its influence on the properties of fouling release coatings based on polydimethylsiloxane. *Coatings* **2017**, *7*, 107. [CrossRef]

29. Jain, S.R.; Sekkar, V.; Krishnamurthy, V.N. Mechanical and swelling properties of HTPB-based copolyurethan networks. *J. Appl. Polym. Sci.* **1993**, *48*, 1515–1523. [CrossRef]

30. Lai, G.Q.; Xin, S.M. *Synthetic Process and Application of Organosilicon*; Chemical Industry Press: Beijing, China, 2010; p. 420. (In Chinese)

31. Flory, P.J.; Rehner, J.J. Statistical mechanics of cross-linked polymer networks II. swelling. *J. Chem. Phys.* **1943**, *11*, 521–526. [CrossRef]

32. Oikawa, H.; Murakami, K. Some comments on the swelling mechanism of rubber vulcanizates. *Rubber Chem. Technol.* **1987**, *60*, 579–590. [CrossRef]

Article

SiO$_2$@TiO$_2$ Coating: Synthesis, Physical Characterization and Photocatalytic Evaluation

A. Rosales [1], A. Maury-Ramírez [2], R. Mejía-De Gutiérrez [3], C. Guzmán [1] and K. Esquivel [1,*]

[1] Graduate and Research Division, Engineering Faculty, Universidad Autónoma de Querétaro, Cerro de las Campanas, Santiago de Querétaro 76010, Mexico; alicia.ros.perez@hotmail.com (A.R.); cgm1909@hotmail.com (C.G.)

[2] Civil and Industrial Engineering Department, Engineering Faculty, Pontificia Universidad Javeriana Cali, calle 18 #118-250, Cali 760031, Colombia; anibal.maury@javerianacali.edu.co

[3] Group of Composite Materials, School of Materials, Faculty of Engineering, Universidad del Valle, Calle 13 #100-00, Cali 76001, Colombia; ruby.mejia@correounivalle.edu.co

* Correspondence: karen.esquivel@uaq.mx; Tel.: +52-442-192-12-00 (ext. 65401)

Received: 2 February 2018; Accepted: 19 March 2018; Published: 24 March 2018

Abstract: Use of silicon dioxide (SiO$_2$) and titanium dioxide (TiO$_2$) have been widely investigated individually in coatings technology, but their combined properties promote compatibility for different innovative applications. For example, the photocatalytic properties of TiO$_2$ coatings, when exposed to UV light, have interesting environmental applications, such as air purification, self-cleaning and antibacterial properties. However, as reported in different pilot projects, serious durability problems, associated with the adhesion between the substrate and TiO$_2$, have been evidenced. Thus, the aim of this work is to synthesize SiO$_2$ together with TiO$_2$ to increase the durability of the photocatalytic coating without affecting its photocatalytic potential. Therefore, synthesis using sonochemistry, synthesis without sonochemistry, physical characterization, photocatalytic evaluation, and durability of the SiO$_2$, SiO$_2$@TiO$_2$ and TiO$_2$ coatings are presented. Results indicate that using SiO$_2$ improved the durability of the TiO$_2$ coating without affecting its photocatalytic properties. Thus, this novel SiO$_2$@TiO$_2$ coating shows potential for developing long-lasting, self-cleaning and air-purifying construction materials.

Keywords: hydrophobic; photocatalytic; sonochemistry; coating; mortar

---

1. Introduction

Current environmental problems observed in big cities, such as air pollution and associated infrastructure deterioration, encourage research for the development of new technologies and products that mitigate these modern, urban threats. Among the different environmentally-friendly technologies, heterogeneous photocatalytic oxidation using TiO$_2$ has become an interesting technology due to its durability and high photocatalytic activity [1]. Recently, the incorporation of TiO$_2$ (e.g., coatings or additives) into construction materials used in urban infrastructure, such as concrete and mortars, has been an interesting approach to reduce NO$_x$ and VOCs (volatile organic compounds) at outdoor concentrations using sunlight as the only energy source; these are the so-called air purifying properties. TiO$_2$ under UV-A light irradiation can generate oxidative (·OH) and reductive (·O$_2$) species, which are able to degrade different organic and inorganic compounds [2–4]. Furthermore, exposure to UV-A light enhances the superhydrophilic effect on the TiO$_2$ surface, which makes it easier to remove the fouling substances on TiO$_2$ loaded surfaces; this is the so-called self-cleaning ability [5–7]. However, recent applications of photocatalytic building materials in urban pilot projects have demonstrated that maintaining the durability of the air-purifying and self-cleaning properties remains challenging, especially for the application of photocatalytic building materials under outdoor conditions [6]. Among other environmental factors, dust and oil accumulation have been reported as major factors affecting the properties of photocatalytic construction materials at an urban scale [7].

On the other hand, hydrophobic surfaces have also received attention for their self-cleaning, anti-flogging, anti-adherent and anti-polluting properties. The natural model for the design of superhydrophobic synthetic films is the lotus plant, which is known for its self-cleaning properties that allow the capture of air under water droplets that contribute to the rolling water droplet, a characteristic of well-designed superhydrophobic surfaces [8]. Due to the nano-manufacturing technologies that have been established for silicon substrates, silicon has been widely used for producing superhydrophobic surfaces; moreover, this kind surfaces, for instance, promotes durability in structures by avoiding the incrustation of corrosive salts (Cl^- and SO_4^-) that promote cracking or surface erosion [9]. To make superhydrophobic surfaces of intrinsically hydrophilic materials, a two-step process is usually required, i.e., first, make a rough surface and second, modify it with a coating of chemicals, such as organosilane, which may offer low surface energy after binding to the rough surface [9,10]. This is the case for polydimethylsiloxane (PDMS), which can be easily processed to make a hydrophobic surface with a rough texture and reduced free surface energy [11,12]. The methods to create hydrophobic surfaces have very long reaction times and strict chemical conditions. A method that uses sonochemistry has smaller reaction times, is more likely to undergo a complete chemical reaction and more ordered crystallization. Sonochemistry is a process of cavitation that refers to the rapid growth and collapse of implosion bubbles in a liquid in an unusual reaction environment [13,14]. Therefore, this article reports the development of a $SiO_2@TiO_2$ coating applicable to cement based materials, such as mortars and glass. The SiO_2 matrix, based on PDMS (polydimethylsiloxane), has the potential to increase the adherence of TiO_2 particles and to improve their photocatalytic efficiency [15].

2. Materials and Methods

As a strategy to develop an efficient $SiO_2@TiO_2$ coating, pure TiO_2 and pure SiO_2 coatings that used the same precursors, proportions, and two different synthesis methods were evaluated.

2.1. Synthesis of $SiO_2@TiO_2$ Coating Coupled with Sonochemistry

For TiO_2 sol production, titanium Iso-propoxide (97%, Sigma Aldrich, St. Louis, MO, USA) was added dropwise into an organic solvent (isopropyl alcohol, 99%, Sigma Aldrich), previously stirred under an inert nitrogen atmosphere for 5 min.

For SiO_2 synthesis, sonotrode equipment (Hielscher Ultrasound Technology UP200Ht, Teltow, Germany) was used working at 100% cavitation and 20% amplitude. A solution of distillated water, absolute ethyl alcohol and oxalic acid, in a 5/5/0.1 molar relation was prepared and stirred sonochemically for 15 min. Afterwards, tetraethyl orthosilicate was added dropwise and the mixture was stirred sonochemically for 3 min. Next, polydimethylsiloxane was added dropwise and continuously stirred for 3 min.

Finally, the titanium dioxide sol and the silicon dioxide sol were mixed. Beforehand, sonotrode working conditions were modified from the initial conditions to 100% cavitation and 60% amplitude. Immediately after mixing, 10 mL of distilled water was added and mixed continuously using sonotrode conditions for 20 min. The resultant mixture was applied on glass and mortar surfaces and left to dry at room temperature.

Synthesis of $SiO_2@TiO_2$ Coating without Sonochemistry

For TiO_2 sol preparation, after sol formation, a hydrolysis process was carried out with the addition of distilled water, added dropwise. The resulting solution was filtered, washed with distilled water and dried at room temperature for 18 h. Finally, a calcination process was carried out at 450 °C for 3 h.

For SiO_2, a solution of distillated water, absolute ethyl alcohol and oxalic acid, in a 5/5/0.1 molar relation was prepared and stirred for 15 min. Afterwards, tetraethyl orthosilicate was added dropwise and stirred for 3 min. Then, polydimethylsiloxane was added dropwise and stirred continuously for 3 min.

Finally, the titanium dioxide particles and the silicon dioxide sol were mixed for 20 min. The resultant mixture was applied on mortar surfaces and dried at room temperature for at least 1 h.

2.2. Preparation of Mortar Samples

The mortar samples were manufactured using a previous mix design (Table 1), the same local materials, and the ASTM method C192/C192M [16–18].

Table 1. Mortar components and proportions for 1 m^3.

Material	Proportions Related to Cement Content	Mass (kg)	Absolute Volume (dm^3)
Water	0.59	324.00	324.00
Cement	1.00	549.00	186.70
Aggregate	2.66	1421.40	489.30

2.3. Physical Characterization of SiO_2@TiO_2 Coating

The microstructures of the materials were examined by transmission electron microscopy (TEM) using a JEOL JEM-1010 (Tokyo, Japan), operating at a voltage of 200 kV. The crystallinity of the SiO_2@TiO_2 coating was determined by X-ray diffraction (XRD) using Bruker D8 equipment (Billerica, MA, USA) with a sealed copper tube to generate Cu–Kα radiation (λ = 1.15406 Å) with angles of $10 < 2\theta < 80°$ in a pitch of $0.01°$. To verify the crystallinity, the structures of the obtained samples were characterized using Raman spectroscopy with the LabRAM HR equipment (Horiba Scientific, Kyoto, Japan), which used an NdYGA laser (λ = 532 nm). The samples were analyzed using a microscope with an objective of $10\times$ at a power of 6 mW over a circle 1.5 μm in diameter. The optical transmittance of the glass substrates coated with SiO_2@TiO_2 was measured with a Cary 5000 UltraViolet-Visibe-Near-Infra-Red spectrophotometer (Agilent, Santa Clara, CA, USA) at wavelengths ranging from 350 to 800 nm. Water contact angle was measured using an optical tensiometer (Analyzer-DSA100W Krüss, Hamburg, Germany), which produces water droplets with a volume adjusted to 10 μL using a needle (stainless steel, model NE60).

2.4. Photocatalytic Evaluation

2.4.1. Evaluation of Photocatalytic Properties

The photocatalytic performance of the coating was evaluated using the method (UNI 11259-2016) based on Rhodamine B (RhB) degradation on the sample exposed to UV-A irradiation [19,20]. To monitor dye removal, the mortar samples were divided into three parts, in which TiO_2, SiO_2 and SiO_2@TiO_2 coatings were applied, as shown in Figure 1. Using a pipette, an RhB solution with a concentration of 50 ppm was evenly applied to 3 standardized positions on the samples, and they were left to dry overnight.

Then, the dye-contaminated samples were exposed to UV-A irradiation for 26 h using the UV reactor shown in Figure 2. In this reactor, UV-A irradiation was provided by an Electrolux T8 20 W/BLB. This lamp type emits light with a peak wavelength of 360 nm and an intensity of 10.3 $W \cdot m^{-2}$ at a distance of 5 mm. Finally, changes in color at 0, 4, and 26 h were measured using a portable X-rite Ci60 spectrophotometer (Photometric Solutions International, Victoria, Australia). Measurements were reported in the L^*, a^*, b^* colorimetric coordinates of the CIE LAB system (32/64 bit software), which corresponds to the white and black color range, red and green color range and yellow and blue color range, respectively, where the a^* coordinate is the comparison parameter. Based on these measurements, the following parameters were calculated, as shown in Equations (1) and (2). Where R stands for color removal at time 0, 4, and 26 h of UV light exposition.

$$R_4 = \frac{(R_4 - R_0)}{R_0} \times 100 \tag{1}$$

$$R_{26} = \frac{(R_{26} - R_0)}{R_0} \times 100 \tag{2}$$

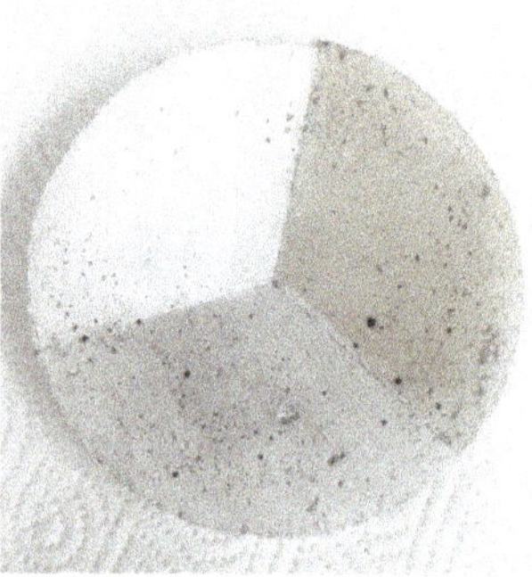

Figure 1. Mortar surface coated with TiO_2, SiO_2 and $SiO_2@TiO_2$ coatings.

Figure 2. UV-A reactor using Electrolux T8 20W/BLB, λ = 360 nm, intensity = 10.3 W·m^{-2}.

2.4.2. Water Contact Angle Measurements

To evaluate the water behavior on the TiO_2, SiO_2 and $SiO_2@TiO_2$ coated samples, a preliminary test using the "rising drop" method was employed, using a camera and digital measurements, before and after the UV A light exposure (0, 4, and 26 h).

2.5. Durability

To assess the durability of the SiO_2, TiO_2 and $SiO_2@TiO_2$ coated samples, an adherence test was performed following the ASTM 3359 [21] and the appropriate references [22–27]. In this case, the corresponding test for thin films with thicknesses less than or equal to 2 mm was selected. To perform this test, a grid of 1 mm × 1 mm with eleven cuts of $\frac{3}{4}$ (20 mm) in length was drawn on top of the coated mortar sample. Subsequently, a piece of scotch tape, three inches long, was placed in the center of the grid and soft pressed with an eraser. A change in color of the tape indicated complete contact. Then, the scotch tape was removed from the opposite end of the application, forming a 180° angle. Next, the coated area was compared with the patterns presented in Table 2 [22–25]. In addition, the adherence test was carried out before and after UV light exposure (0, 4 and 26 h), to determine the photocatalytic activity of the $SiO_2@TiO_2$ coating.

Table 2. Detachment patterns and classification of different coated surfaces after the adherence test (Modified from ASTM-3359-02 classification chart).

Classification	Area Removed (%)	Cross-Cut Surface Area with Adhesion Range by Percent
5B	None 0%	
4B	<5%	
3B	5–15%	
2B	15–35%	
1B	35–65%	
0B	>65%	

3. Results and Discussion

3.1. Physical Characteristics

TEM images (10×) of a power sample from the $SiO_2@TiO_2$ coating are shown in Figure 3. From these, it can be seen that layered agglomerates are formed by amorphous silicon dioxide while titanium dioxide is not visible. In general, the reported agglomerates vary in shape and size, ranging from 20–600 nm. In addition, it was observed that the morphology of the $SiO_2@TiO_2$ coating was not affected by the use of sonochemistry.

Figure 4 shows the SEM micrographs of the $SiO_2@TiO_2$ coating. It can be observed that the surface was rugged and the morphology was not uniform due to the formation of denser particles and their agglomeration. Additional cross-selection elemental mappings, combined with EDS analysis for SiO$_2$@TiO$_2$, showed the presence of Si, Ti and O as elements (Figure 5) as expected.

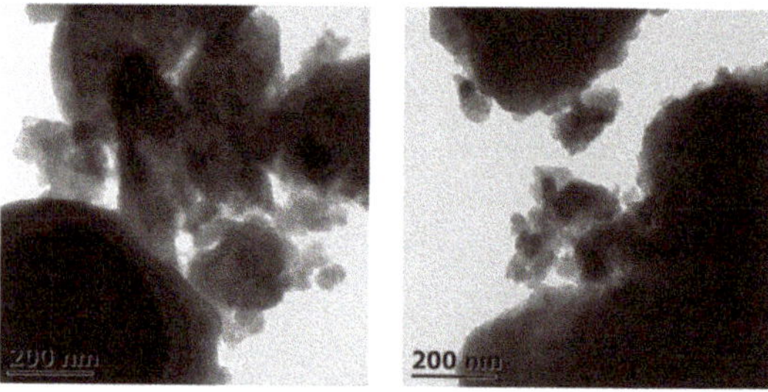

Figure 3. TEM images of the SiO$_2$@TiO$_2$ powder sample (10×).

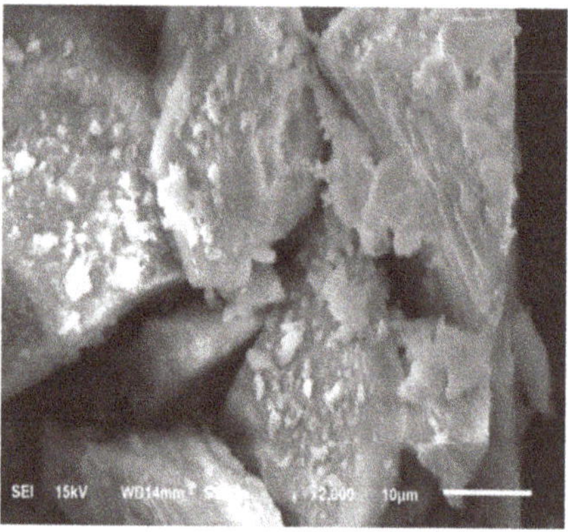

Figure 4. SEM micrographs of the SiO$_2$@TiO$_2$ coating in a power sample.

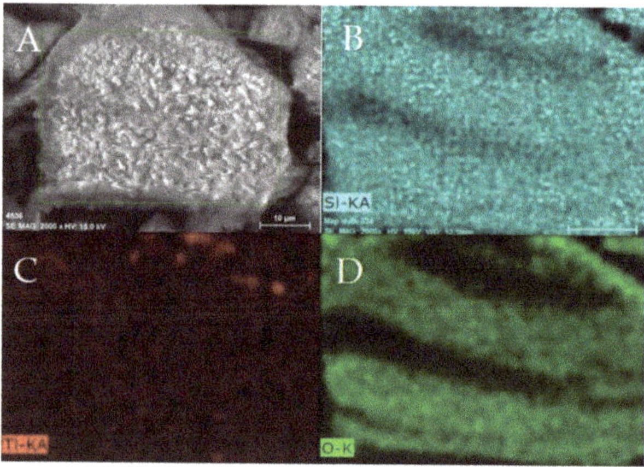

Figure 5. EDS analysis, elemental mapping images of the SiO$_2$@TiO$_2$ power sample: (**A**) EDS area; (**B**) Silicon; (**C**) Titanium and (**D**) Oxygen.

Figure 6 shows an X-ray diffractogram of the $SiO_2@TiO_2$ coating. It shows the high intensity peak of silicon dioxide at 24°, characteristic of the amorphous SiO_2 phase. In addition, signals observed at 12.8° and 22.6° are characteristic of the PDMS compound. On the other hand, signals of titanium dioxide that presented at peaks of 27.3° (110) and 55.5° (220) corresponded to the rutile crystalline phase [24] and the peaks of 25.3° (101), 38.6° (004) and 48.08° (200) corresponded to the anatase crystalline phase [25]. Additionally, it was observed that the diffractogram of the $SiO_2@TiO_2$ coating was not affected by the use of sonochemistry on a macro scale. Previous information was obtained using the standard XRD pattern (JCPDS FILES No. 21-1272). Moreover, the reflections corresponding to the silicon covered up the other signals of the TiO_2 phases and PDMS compound. The crystallite size was obtained by Scherrer's equation [26], which obtained a crystal size of 13 nm, being an amorphous compound due to its matrix of silicon dioxide.

The Raman spectra of TiO_2, SiO_2, SiO_2-PDMS and $SiO_2@TiO_2$ are presented in Figure 7. The Raman spectrum of TiO_2 contained a strong peak at 143 cm^{-1} and weak peaks at 395 cm^{-1}, 515 cm^{-1} and 638 cm^{-1}. The Raman spectrum of SiO_2 contained a strong peak at 450 cm^{-1} and weak peaks at 80 cm^{-1}, 90 cm^{-1} and 980 cm^{-1}. These peaks can be attributed to the bending of O–Si–O and Si–O–Si symmetric bond stretching. The Raman spectrum of SiO_2-PDMS exhibited peaks at 680 cm^{-1}, 816.1 cm^{-1}, 830.1 cm^{-1} and 882.4 cm^{-1}, these peaks are characteristic of the PDMS compound [27].

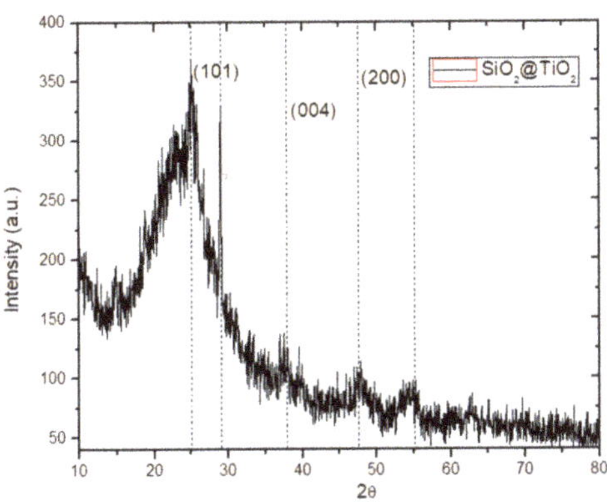

Figure 6. XRD pattern of the $SiO_2@TiO_2$ coating.

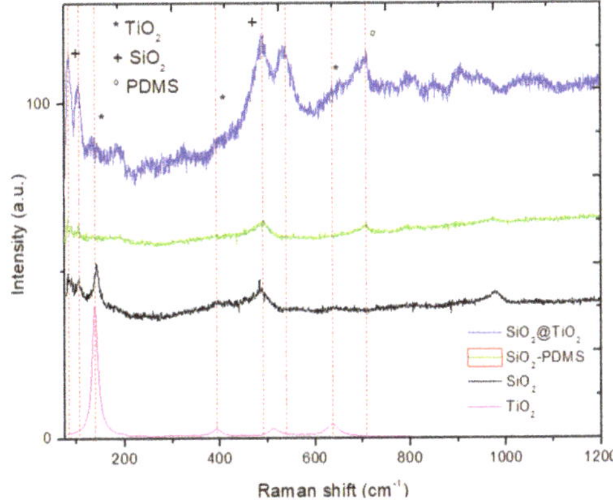

Figure 7. Raman spectra in the range of 80–1200 cm^{-1} from TiO_2, SiO_2 and $SiO_2@TiO_2$.

The Raman spectrum of the $SiO_2@TiO_2$ nanocomposite exhibited a decrease in the highest intensity peak of titanium dioxide while the other peaks were inhibited. This can be attributed to the highly dispersed titanium dioxide. Furthermore, the signals of silicon dioxide decreased due to the presence of PDMS that modifies the crystallinity and makes noise (fluorescence) on the $SiO_2@TiO_2$ coating. The Raman spectrum of the $SiO_2@TiO_2$ coating, without the application of sonochemistry, showed a lower crystallinity for the composite. Furthermore, the TiO_2 signals were decreased and not even located by the Raman Spectroscopy.

Figure 8 shows the UV-visible spectrum of a glass sample coated with $SiO_2@TiO_2$ with reference to a blank (uncoated glass). The glass substrate had a transmittance of 92–93% (black line). After placing the coating on the glass, the transmittance of sample (blue line) was 85%. This result shows that the coating of $SiO_2@TiO_2$ has high transparency over a wide wavelength range.

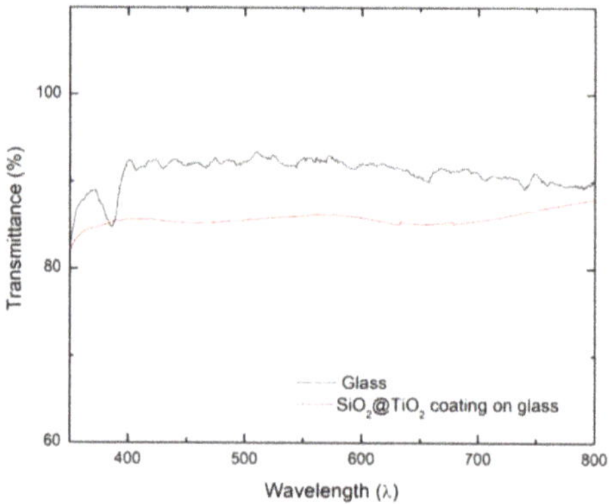

Figure 8. UV-Vis transmittance spectra of $SiO_2@TiO_2$ coated on glass (red) and glass (black).

3.2. Photocatalytic Evaluation

By measuring the RhB degradation before (0 h) and after UV-A irradiation (4 and 26 h) as shown in Figure 9, the TiO_2, SiO_2 and $SiO_2@TiO_2$ coated mortar samples were evaluated (Figure 10). With RhB removal of $R_4 = 25\%$ and $R_{26} = 55\%$, the developed $SiO_2@TiO_2$ coating satisfies the boundaries as to what can be considered photocatalytic material ($R_4 > 20\%$ and $R_{26} > 50\%$) [19]. However, the use of sonochemistry showed an improvement in the efficiencies of degradations of $R_4 = 30.4\%$ and $R_{26} = 70.5\%$. Similar values have been also reported by a photocatalytic coating applied on mortar samples [28].

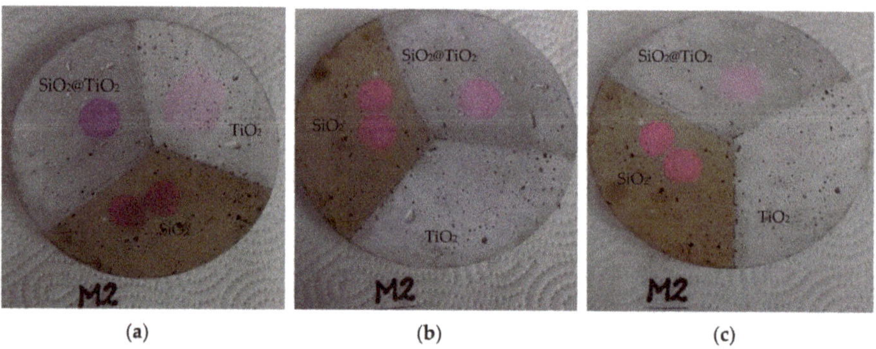

Figure 9. Comparison of the degradation of RhB, before and after UV-A irradiation. (**a**) 0 h; (**b**) 4 h; (**c**) 26 h.

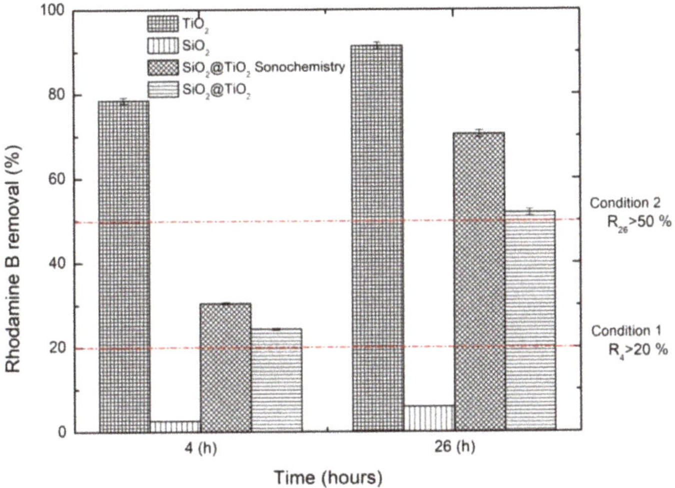

Figure 10. Rhodamine B removal efficiencies of the TiO_2, SiO_2 and $SiO_2@TiO_2$ coated mortar samples under UV-A irradiation (4 and 26 h).

On the other hand, as expected, TiO_2 coated samples displayed the best activity with $R_4 = 79\%$ and $R_{26} = 92\%$ [29]. In contrast, the SiO_2 coated samples exhibited a significantly lower degradation efficiency ($R_4 = 0.5\%$, $R_{26} = 8\%$). As there was no photocatalytic material present, the RhB removal was associated with dye photolysis, as previously reported [30]. According to the physico-chemical characterization of the $SiO_2@TiO_2$ composite previously described, the synthesis coupled with sonochemistry showed a non-significant difference in performance. Nevertheless, in the photocatalytic activity, the use of the sonochemical assisted synthesis helped to improve the Rhodamine B removal. This could be attributed to a better TiO_2 dispersion over the SiO_2-PDMS matrix and a higher anatase phase appearance without any thermal treatment as is used with the conventional sol-gel $SiO_2@TiO_2$ composite synthesis. However, this effect must be examined in further experiments by an extensive XPS analysis and by modifying the sonochemical synthesis parameters to achieve a macroscopic change in the physico-chemical characterization.

The water contact angle measurements of TiO_2, SiO_2 and $SiO_2@TiO_2$ coated mortar samples before, during and after UV irradiation (0, 4 and 26 h) are shown in Figure 11. As expected, TiO_2 exhibited hydrophilic behavior with values around 10°. On the contrary, the coated sample with $SiO_2@TiO_2$ presented water contact angles varying between 100° and 105° after UV-A irradiation. Previous research reports similar contact angles of around 114°–111° for a coating of TiO_2-SiO_2-PDMS [31]. Meanwhile the SiO_2 coated samples remained constant (around 98°) because silicon dioxide has hydrophobic properties.

Table 3 shows the adherence test results of the coated mortar samples using TiO_2, SiO_2 and $SiO_2@TiO_2$. In the case of the $SiO_2@TiO_2$, a 10% detachment was quantified using the grid which classifies as 3B according to the ASTM D3359-02 [22,32]. On the other hand, the TiO_2 presented with 40% detachment, which classifies as 1B. Finally, SiO_2 had the highest adherence of the tested coatings and presented with 5% detachment and classifies as 4B.

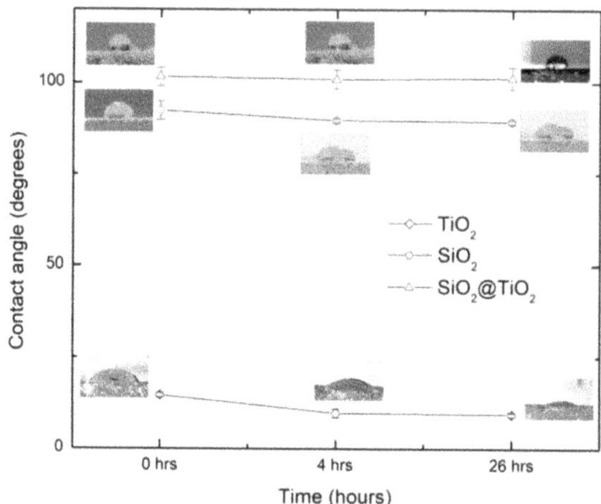

Figure 11. Water contact angles of TiO$_2$, SiO$_2$ and SiO$_2$@TiO$_2$ coated mortar samples before (0 h) and under UV-A irradiation (4 and 26 h).

Table 3. Test of adherence results of the coated mortar samples using TiO$_2$, SiO$_2$ and SiO$_2$@TiO$_2$.

Coated Sample	% Detachment	Image	Classified under ASTM D3359-02
TiO$_2$	40%		1B
SiO$_2$	5%		4B
SiO$_2$@TiO$_2$	10%		3B

After the evaluation of durability using the adherence test, Rhodamine B removal and water contact angle measurements were evaluated again on the coated samples. Results indicated that TiO$_2$ decreased its photocatalytic activity to $R_4 = 44\%$ and $R_{26} = 70\%$. For the SiO$_2$@TiO$_2$ coating no difference was noticed, while SiO$_2$ did not show a change in its photocatalytic activity. The contact angle was also maintained for all the tested materials. The results were 5° for the TiO$_2$ 90° SiO$_2$ and 100° SiO$_2$@TiO$_2$. Further experiments will be needed to find out the effects of extreme weather conditions on the durability of the coat.

4. Conclusions

In the present study, a hydrophobic and photocatalytic $SiO_2@TiO_2$ coating for mortar and glass protection was synthesized through a sol-gel with and without sonochemistry assistance. The completed analysis of Scanning Microscopy (SEM), Elemental Analysis (EDS), Transmission Electron Microscopy (TEM), X-ray diffraction (XRD) and Raman Spectroscopy of the $SiO_2@TiO_2$ coating revealed their composition and microstructure. The TEM images made it possible to observe agglomerates of the composite without a regular shape. But, by mapping the EDS analysis the main elements were found over the entire surface in a homogeneous way. The use of XRD enabled the visualization of the TiO_2 phases formed using sonochemistry. These phases are the rutile and anatase phase. Additionally the SiO_2 remained amorphous. Further, the Raman spectroscopy signals can be attributed to the bending and stretching of the O–Si–O and Si–O–Si symmetric bonds and without the application of sonochemistry a lower crystallinity of the composite and the TiO_2 signals was observed. Finally, according to the physico-chemical characterization of $SiO_2@TiO_2$, the coating displayed a high transparency over a wide wavelength range.

In addition, the application of sonochemistry in the sol-gel synthesis promoted the photocatalytic phase of the titanium dioxide and improved the removal of the Rhodamine B dye. The transparency of the titanium dioxide coating was around 85% of that compared to glass without a cover.

The photocatalytic activity of the $SiO_2@TiO_2$ coating showed an RhB removal of $R_4 = 25\%$ and $R_{26} = 55\%$ establishing itself as a photocatalytic material, while the $SiO_2@TiO_2$ coating coupled with sonochemistry showed $R_4 = 30.4\%$ and $R_{26} = 70.5\%$ indicating a major photocatalytic activity. The adherence test was used to study the durability, indicating a 3B type adhesion of the $SiO_2@TiO_2$, in accordance with the ASTM D3359-02 scale. Additionally, the $SiO_2@TiO_2$ composite after the durability tests showed no photocatalytic activity loss in contrast with the pure TiO_2 coating. These results show the potential of the developed $SiO_2@TiO_2$ coating for self-cleaning and air-purifying applications.

Acknowledgments: A. Rosales thanks CONACyT for the scholarship granted and also thanks to UniValle and PUJ Cali for the facilities in the international mobility exchange. C. Guzmán and K. Esquivel thanks to Luis A. Ortiz-Frade from CIDETEQ for the SEM and EDS analysis, to QFB. Lourdes Palma from UNAM for the TEM images and to Luis Escobar-Alarcón from ININ for the Raman analysis.

Author Contributions: K. Esquivel and A. Maury-Ramírez conceived and designed the experiments; A. Rosales performed the experiments; all the authors analyzed the data; R. Mejía-De Gutiérrez and C. Guzmán contributed reagents/materials/analysis tools; A. Rosales, A. Maury-Ramírez and K. Esquivel wrote the paper.

Conflicts of Interest: The authors declare no conflict of interest.

References

1. Creelman, L.W.; Pešinová, V.; Cohen, Y.; Winer, A.M. Intermedia transfer factors for modeling the enviromental impact of air pollution from industrial sources. *J. Hazard. Mater.* **1994**, *37*, 15–25. [CrossRef]

2. Hoffmann, M.R.; Martin, S.T.; Choi, W.; Bahnemann, D.W. Environmental applications of semiconductor photocatalysis. *Chem. Rev.* **1995**, *95*, 69–96. [CrossRef]

3. Ghosh, P.K.; Jasra, R.V.; Shukla, D.B.; Bhatt, A.K.; Tayade, R.J. Photocatalytic Auto-Cleaning Process of Stains. U.S. Patent No. 8,343,282, 1 January 2013.

4. Miyauchi, M.; Kieda, N.; Hishita, S.; Mitsuhashi, T.; Nakajima, A.; Watanabe, T.; Hashimoto, K. Reversible wettability control of TiO_2 surface by light irradiation. *Surf. Sci.* **2002**, *511*, 401–407. [CrossRef]

5. Natarajan, K.; Natarajan, T.S.; Bajaj, H.C.; Tayade, R.J. Photocatalytic reactor based on UV-LED/TiO_2 coated quartz tube for degradation of dyes. *Chem. Eng. J.* **2011**, *178*, 40–49. [CrossRef]

6. Boonen, E.; Beeldens, A.; Dirkx, I.; Bams, V. Durability of cementitious photocatalytic building materials. *Catal. Today* **2017**, *287*, 196–202. [CrossRef]

7. Etxeberria, M.; Guo, M.-Z.; Maury-Ramirez, A.; Poon, C.S. Influence of dust and oil accumulation on effectiveness of photocatalytic concrete surfaces. *J. Environ. Eng.* **2017**, *143*, 04017040. [CrossRef]

8. Cheng, Y.-T.; Rodak, D.E. Is the lotus leaf superhydrophobic? *Appl. Phys. Lett.* **2005**, *86*, 144101. [CrossRef]

9. Wang, D.; Hou, P.; Zhang, L.; Yang, P.; Cheng, X. Photocatalytic and hydrophobic activity of cement-based materials from benzyl-terminated-TiO_2 spheres with core-shell structures. *Constr. Build. Mater.* **2017**, *148*, 176–183. [CrossRef]

10. Stanton, M.M.; Ducker, R.E.; MacDonald, J.C.; Lambert, C.R.; Grant McGimpsey, W. Super-hydrophobic, highly adhesive, polydimethylsiloxane (PDMS) surfaces. *J. Colloid Interface Sci.* **2012**, *367*, 502–508. [CrossRef] [PubMed]

11. Ma, M.; Hill, R.M. Superhydrophobic surfaces. *Curr. Opin. Colloid Interface Sci.* **2006**, *11*, 193–202. [CrossRef]

12. Swart, M.; Mallon, P.E. Hydrophobicity recovery of corona-modified superhydrophobic surfaces produced by the electrospinning of poly(methyl methacrylate)-graft-poly(dimethylsiloxane)hybrid copolymers. *Pure Appl. Chem.* **2009**, *81*, 495–511. [CrossRef]

13. Mettin, R.; Cairós, C.; Troia, A. Sonochemistry and bubble dynamics. *Ultrason. Sonochem.* **2015**, *25*, 24–30. [CrossRef] [PubMed]

14. Rae, J.; Ashokkumar, M.; Eulaerts, O.; von Sonntag, C.; Reisse, J.; Grieser, F. Estimation of ultrasound induced cavitation bubble temperatures in aqueous solutions. *Ultrason. Sonochem.* **2005**, *12*, 325–329. [CrossRef] [PubMed]

15. Sangchay, W. The self-cleaning and photocatalytic properties of TiO_2 doped with SnO_2 thin films preparation by sol-gel method. *Energy Procedia* **2016**, *89*, 170–176. [CrossRef]

16. Sánchez de Guzmán, D. *Concrete and Mortar Technology (Tecnología del Concreto y del Mortero)*; Pontificia Universidad Javeriana: Bogotá, Colombia, 2001.

17. Medina Medina, A.F.; Torres Rojas, D.F.; Meza Girón, G.; Villota Grisales, R.A. System Design to Generate air Purification and Self-Cleaning Surfaces of the Tunnel of Colombia Avenue (Cali). (Diseño de un Sistema Para Generar Purificación del aire y Auto-Limpieza en las Superficies del Tunel de la Avenida Colombia (Cali)). Ph.D. Thesis, Pontificia Universidad Javeriana, Cali, Colombia, 2016.

18. ASTM International. *ASTM C192/C192M-16a Standard Practice for Making and Curing Concrete Test Specimens in the Laboratory*; ASTM International: West Conshohocken, PA, USA, 2016.

19. SAI Global. *UNI 11259: 2016 Determination of the Photocatalytic Activity of Hidraulic Binders-Rodammina Test Method*; SAI Global: Rome, Italy, 2016.

20. Maury-Ramirez, A.; Demeestere, K.; De Belie, N. Photocatalytic activity of titanium dioxide nanoparticle coatings applied on autoclaved aerated concrete: Effect of weathering on coating physical characteristics and gaseous toluene removal. *J. Hazard. Mater.* **2012**, *211–212*, 218–225. [CrossRef] [PubMed]

21. ASTM International. *ASTM D3359-02 Standard Test Methods for Measuring Adhesion by Tape Test*; ASTM International: West Conshohocken, PA, USA, 2002.

22. Fufa, S.M.; Labonnote, N.; Frank, S.; Rüther, P.; Jelle, B.P. Durability evaluation of adhesive tapes for building applications. *Constr. Build. Mater.* **2018**, *161*, 528–538. [CrossRef]

23. Bishop, C.A. *Vacuum Deposition onto Webs, Films and Foils*, 3rd ed.; William Andrew Publishing: Boston, MA, USA, 2015; pp. 197–208.

24. Cheng, P.; Zheng, M.; Jin, Y.; Huang, Q.; Gu, M. Preparation and characterization of silica-doped titania photocatalyst through sol–gel method. *Mater. Lett.* **2003**, *57*, 2989–2994. [CrossRef]

25. Haider, A.J.; AL–Anbari, R.H.; Kadhim, G.R.; Salame, C.T. Exploring potential Environmental applications of TiO_2 Nanoparticles. *Energy Procedia* **2017**, *119*, 332–345. [CrossRef]

26. Maury-Ramirez, A.; Nikkanen, J.-P.; Honkanen, M.; Demeestere, K.; Levänen, E.; De Belie, N. TiO_2 coatings synthesized by liquid flame spray and low temperature sol–gel technologies on autoclaved aerated concrete for air-purifying purposes. *Mater. Charact.* **2014**, *87*, 74–85. [CrossRef]

27. Al-Asbahi, B.A. Influence of anatase titania nanoparticles content on optical and structural properties of amorphous silica. *Mater. Res. Bull.* **2017**, *89*, 286–291. [CrossRef]

28. Guo, M.-Z.; Maury-Ramirez, A.; Poon, C.S. Photocatalytic activities of titanium dioxide incorporated architectural mortars: Effects of weathering and activation light. *Build. Environ.* **2015**, *94*, 395–402. [CrossRef]

29. Hashimoto, K.; Irie, H.; Fujishima, A. TiO_2 Photocatalysis: A Historical Overview and Future Prospects. *Jpn. J. Appl. Phys.* **2005**, *44*, 8269. [CrossRef]

30. Allen, N.S.; Binkley, J.P.; Parsons, B.J.; Phillips, G.O.; Tennent, N.H. Light stability and flash photolysis of azo dyes in epoxy resins. *Dyes Pigments* **1984**, *5*, 209–223. [CrossRef]

31. Kapridaki, C.; Maravelaki-Kalaitzaki, P. TiO$_2$–SiO$_2$–PDMS nano-composite hydrophobic coating with self-cleaning properties for marble protection. *Prog. Org. Coat.* **2013**, *76*, 400–410. [CrossRef]
32. Castañeda, A.; Rivero, C.; Corvo, F. Evaluación de sistemas de protección contra la corrosión en la rehabilitación de estructuras construidas en sitios de elevada agresividad corrosiva en Cuba. *Rev. Constr.* **2012**, *11*, 49–61. [CrossRef]

Article

Bacterial Biofilm Characterization and Microscopic Evaluation of the Antibacterial Properties of a Photocatalytic Coating Protecting Building Material

Thomas Verdier [1,*], Alexandra Bertron [1], Benjamin Erable [2] and Christine Roques [2]

[1] LMDC, Université de Toulouse, INSA/UPS Génie Civil, 135 Avenue de Rangueil, 31077 Toulouse Cédex 04, France; bertron@insa-toulouse.fr
[2] LGC, Université de Toulouse, UOS-CNRS Dép. BioSyM, 4 allée Emile Monso, 31029 Toulouse, France; benjamin.erable@ensiacet.fr (B.E.); roques730@aol.com (C.R.)
* Correspondence: tverdier@insa-toulouse.fr

Received: 2 February 2018; Accepted: 4 March 2018; Published: 5 March 2018

Abstract: Use of photocatalytic paint-like coatings may be a way to protect building materials from microbial colonization. Numerous studies have shown the antimicrobial efficiency of TiO_2 photocatalysis on various microorganisms. However, few have focused on easy-to-apply solutions and on photocatalysis under low irradiance. This paper focuses on (a) the antibacterial properties of a semi-transparent coating formulated using TiO_2 particles and (b) the microscopic investigations of bacterial biofilm development on TiO_2-coated building materials under accelerated growth conditions. Results showed significant antibacterial activity after few hours of testing. The efficiency seemed limited by the confinement of the TiO_2 particles inside the coating binder. However, a pre-irradiation with UV light can improve efficiency. In addition, a significant effect against the formation of a bacterial biofilm was also observed. The epifluorescence approach, in which fluorescence is produced by reflect rather than transmitted light, could be applied in further studies of microbial growth on coatings and building materials.

Keywords: coating; building materials; bacterial growth; proliferation; biofilm; antibacterial

---

1. Introduction

Since microorganisms are ubiquitous and dispersible from soil and water as well as from air, building materials are permanently exposed to them and may easily become targets for contamination and growth. The requirements' microorganism development, i.e., an energy source, carbon and water, can be satisfied by building materials (organic and/or inorganic) in a large variety of contexts [1–3]. The growth of microorganisms can have various detrimental consequences, including the biodeterioration of materials, which may concern structural and aesthetic properties, and/or the release of aerial contaminants that can be deleterious for human health. Their proliferation on building materials is of growing concern to the scientific community [3–6].

A substantial number of man-made constructions have been reported to be widely contaminated by microorganisms (mainly fungi, bacteria and algae). They include pipes carrying aggressive aqueous media such as sewers [3], moisture-damaged buildings [1,7,8], historical monuments [9], etc. Contaminated building materials are usually visually deteriorated by the activity of fungi and bacteria, which are known to be the most harmful organisms inducing physical and chemical changes in materials.

Microbial contamination of building materials in indoor environments has also been much studied. The degradation of indoor air quality is a growing public health concern worldwide and microorganisms or their components are major polluting agents [2]. Their contribution to the Sick

Building Syndrome (SBS) and Building Related Illness (BRI) is widely emphasized in the literature. Microorganisms may release aerial particles such as spores, allergens, toxins and other metabolites that can be harmful to human health [10–14]. Serious health troubles such as irritations and toxic effects, infections, allergies, etc. have been experienced by building occupants frequently exposed to microbial contaminants [15,16]. Several studies have already reported that indoor building materials can become major sites of microbial growth [13,15,17]. Actually, growth-promoting conditions, i.e., high humidity and presence of nutrients, are easily fulfilled in damp environments such as water-damaged, damp, and/or badly-insulated buildings. Microorganisms grow on interfacial zones (air–substrates, water–air, etc.). Under specific environmental conditions, they develop on surfaces in complex three-dimensional structures, in which adherent cells are embedded within a self-produced matrix of extracellular polymeric substance (EPS). The overall organization, commonly called a microbial biofilm, provides the microorganisms with particular resistance towards physical and chemical aggressions, be they environmental or from disinfection treatments [18–20].

Different solutions have been studied to prevent, stop or at least reduce microbial biofim development on building materials, including core treatments and coatings applied to surfaces. Antimicrobial products can be formulated using biocides [21], metal oxide nanoparticles [22–24] or bio-based products [25]. A substantial amount of literature has been published on the effect of photocatalytic TiO_2 nanoparticles on microorganisms [26–29]. Studies have shown high efficiency of photocatalysis on the viability of a wide variety of microorganisms and on microbial contaminants, including bacteria, endospores, fungi, algae, protozoa, viruses and prions (shown to be the organisms most resistant to disinfectants [30]) as well as on microbial toxins. Photocatalysis needs to be activated by a light irradiation of sufficient energy to create electron–hole pairs. The electron–hole pairs may react with electron donors or acceptors previously adsorbed on the photocatalyst, e.g., O_2, H_2O. The resulting redox reactions lead to the formation of oxide radicals that are highly reactive and degrade organic matter by non-selective action. Additionally, irradiated TiO_2 presents photo-induced superhydrophilicity [31] and may be incorporated in materials to provide self-cleaning properties.

The use of titanium dioxide as a self-cleaning agent in building materials is common for some years now. Photocatalytic concretes have been developed to prevent fouling and biofouling [32,33] or to purify ambient air (VOC, NO_x ...) [34–38]. Regarding bacterial and fungal disinfection, research has mainly focused on the efficiency of photocatalytic TiO_2 particles used alone, powdered or immobilized in the form of thin film coatings by complex processes (usually involving calcination). It seems that only few studies have investigated easy-to-apply devices such as paint-like coatings. In addition, one of the main advantages of such coatings are their passive aspect: once applied to a surface, the photocatalysis is activated by natural light. Furthermore, efficiencies of TiO_2 photocatalysis found in the literature are usually related to very high irradiances (>10 W/m^2) [28,29,39] that are far from the real-world conditions. Only recent studies have begun to investigate on the photocatalytic activity under low irradiances [40].

The main objectives of the present paper were (a) to focus on the antibacterial efficiency under low irradiance of a photocatalytic semi-transparent coating that is easy to apply on existing materials and (b) to carry out microscopic observations (epifluorescence and SEM) of bacterial biofilm development on a coated building materials under accelerated growth conditions.

2. Materials and Methods

2.1. Bacterial Cultivation

The microorganisms the most frequently detected on indoor building materials are (i) fungi genera *Cladosporium*, *Penicillium*, *Aspergillus* and *Stachybotrys*, and (ii) Gram negative bacteria and mycobacteria [2].

In this study, the model bacterium *Escherichia coli* was chosen for methodological reasons, i.e., its relative ease of culture and its high speed of growth. *E. coli* CIP 53126 was obtained from the Collection of Institute Pasteur (CIP), Paris, France. The strain was preserved at $-80\ ^\circ$C in Eugon medium (Biomérieux, Craponne, France) supplemented with 10% glycerol. The *E. coli* strain identification was checked by Gram staining and biochemical characterization (oxydase reaction and Analitical Profile Index 20E) (Biomérieux, Craponne, France). Before each experiment, bacterial cells were pre-cultured on a nutrient agar Petri dish. Colonies were then transferred to trypcase soy agar (TSA) (Biomérieux) and incubated at a temperature of 36 $^\circ$C for 16 to 24 h. A second culture on TSA was incubated at a temperature of 36 $^\circ$C for 16 to 20 h prior to the test. For testing, one plastic loop of bacteria was dispersed evenly in a small amount of sterile distilled water and the suspension for inoculation was adjusted to about 10^8 cells/mL using a spectrophotometer (640 nm). A 10-fold dilution of the cell suspension was then prepared to obtain the final concentration of the test suspensions, depending on the type of evaluation (antibacterial activity or resistance to biofilm formation).

2.2. Coatings and Preparation of Supports

2.2.1. Coatings

Two semi-transparent coatings were formulated using water and acrylic resin. The formulation contained 5 wt % TiO_2 dispersion (Kronos type 7454, trial product, KRONOS/Société Industrielle du Titane, Paris, France), 2 wt % acrylic resin and 93 wt % water. A control coating was made of 2 wt % acrylic resin and 98 wt % water. The TiO_2 dispersion contained TiO_2 particles (Kronos VLP7050), some physical characteristics of which are presented in Table 1.

Table 1. Physical characteristics of the VLP5070 (Kronos) TiO_2 particles

Description	VLP5070
TiO_2-Content	>85 %
Crystal modification	anatase
Crystallite size	approx. 15 nm
Specific surface area	>225 m^2/g

2.2.2. Nature and Preparation of Supports

Three types of support were used in this study: glass slides, polycarbonate membranes, and cementitious matrices.

Glass Slides

Preliminary measurements of the antibacterial activity were carried out on glass microscope slides ($26 \times 76\ mm^2$) covered by pipetting the TiO_2 coating or the control coating (without TiO_2). Regarding the coating of glass slides, while drying, the water left an even film of acrylic binder scattered with TiO_2 particles. Although the distribution of TiO_2 was not optimal on these supports, it was rather homogeneous over the entire surfaces of samples.

Polycarbonate Membranes

Isopore hydrophilic polycarbonate membranes (Millipore, 47 mm diameter, 0.4 μm filter pore size, 5%–20% porosity) were also used as supports for the coating. Membranes were placed in sterile Petri dishes, covered by pipetting the coatings and placed under a sterile flow hood for air drying. Then, samples were pre-irradiated with UV light (5 W/m^2) for different durations, from 0 h to 109 h. The role of the pre-irradiation was to increase the antibacterial activity of the coating. It is discussed later in the paper.

Cementitious Supports

Cement paste was chosen as reference building material because of its availability, its ease of casting, and because it is morphologically representative of many building materials (porosity and roughness). The cement paste samples were made of ordinary Portland cement CEM I 52.5R were cast with a 0.55 Water/Cement ratio in cylindrical molds (d = 2.8 cm, h = 1 cm). Samples were demolded after 24 h and stored for 27 days in a moisture chamber (100 %RH, room temperature). It is widely known that the pH of materials is a determining factor in microbial contamination. The pH of unaltered cementitious materials is around 13. In their work on an accelerated laboratory test to study fungal biodeterioration, Wiktor et al. [41,42] showed that, for cementitious materials, a low pH of the surface was essential to increase the bioreceptivity of samples. Recently-cast cementitious materials are thus not so subject to microbial colonization, probably because of their relatively high alkalinity incompatible with microbial growth. Usually, the bioreceptivity of these materials increases over time as carbonation tends to decrease the surface pH, but this phenomenon takes place very slowly (typically, a few weeks in natural conditions). Consequently, accelerated aging of the cementitious mortars was performed. After 28 days, samples were placed in a 2-liter-borosilicate reactor for a leaching operation, as follows: the reactor was continuously supplied with a leaching solution by a peristaltic pump (Masterflex L/S, Cole-Palmer, Vernon Hills, IL, USA). The flow rate was set to 1.4 mL/m in order to renew the reactor content in 24 h. The leaching solution was composed of water and acetic acid (CH_3COOH) 0.12 mol/L and the pH was adjusted to 4.7 with sodium hydroxide (NaOH) at 0.076 mol/L. The operation was carried out for 15 days. Then, samples were removed from the reactor and a pH-paper was applied on their wet surfaces. The pH measured was around 7 for all samples. After this leaching operation, samples were washed with distilled water and stored at 100% RH until tested.

Prior to the experiments, samples were covered with the coatings (TiO_2 coating or control coating) by pipetting 500 µL over the entire surface (6.16 cm^2). They were placed under a sterile flow hood for over 12 h for air drying. Then, samples coated with TiO_2 and control samples coated without TiO_2, were pre-aged by UV irradiation (5 W/m^2) for different durations, from 0 h to 109 h.

2.3. Photocatalysis Activation

The TiO_2 photocatalyst (anatase) is naturally activated by light irradiation in the UV-range. The light intensity, or irradiance, greatly influences photocatalysis efficiency. In the literature, intensity values are usually chosen between 10 and 500 W/m^2 [43,44]. These values are quite different from natural light intensity values, in indoor or outdoor conditions. For example, studies have reported outdoor intensity around 30 W/m^2 on sunny days [36,45,46] and between 5 and 10 W/m^2 on cloudy days [47,48]. The few studies that observed inactivation of microorganisms by photocatalysis with light intensities around 10^{-3} W/m^2 showed much lower inactivation rates [49], confirming the major impact of light intensity on the process. Regarding the resistance to microbial biofilm, a recent study also showed significant efficiency of TiO_2 coatings against the development of microbial biofilms under relatively low irradiance (13 W/m^2) closer to real-world light intensity [40].

UV light in the 325–400 nm range can strongly damage microorganisms and lead to their death. Standards usually recommend the use of low intensities for photocatalytic activity evaluation [50]. The position and the power of the lamp must be chosen so as activate the photocatalyst while making sure to avoid any lethal action of the UV light on bacterial cells. In our experiment, 8 W black light bulbs were chosen (peak wave 365 nm). The distance between lamps and samples was adjusted to obtain 2–4 W/m^2 at the bottom of the Petri dish (irradiance measured with a UV-A radiometer Gigahertz-Optik, GmbH, Türkenfeld, Germany, in the 310–400 nm range).

In order to assess the effect of UV irradiation on the survival of *Escherichia coli*, a drop of a bacterial suspension in phosphate buffer (about 1 mL of a 10^6 CFU/mL suspension) was deposited in a borosilicate Petri dish and irradiated for several hours. Different configurations were tested: irradiation with a borosilicate lid on the Petri dish, irradiation without the lid, and kept in the dark (control).

After the test, bacteria were collected with a recovery solution (9 mL of Soybean Casein Lecithin Polysorbate 80 medium, SCDLP broth, as described in [51]) and numerated (CFU) in TSA after a 24 h incubation at 36 °C . Results are presented in Figure 1.

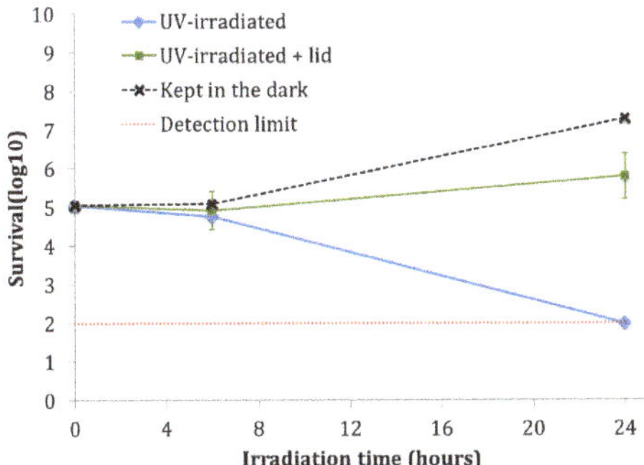

Figure 1. Direct effect of the time of UV irradiation on the survival of *Escherichia coli* CIP 53126 expressed as the log residual viable bacteria, light intensity $\approx$2.5 W/m^2.

For the three configurations, no significant evolution in the number of CFU was observed in the first 6 h of experiment. Control samples kept in the dark and irradiated in the presence of the borosilicate lid showed an increase of 2 log and 0.75 log, respectively, in 24 h, while no CFUs were detected on samples irradiated without the lid (limit of detection = 2 log). The borosilicate lid allowed UV light to penetrate while keeping the hygrometry constant in the Petri dish and preventing the drop from drying.

2.4. Evaluation of Antibacterial Activity

The antibacterial activity characterizes the ability of a substrate to inactivate bacteria cells, i.e., to kill them or to inhibit their growth. It is a quantitative test that provides a bacterial reduction in terms of Colony Forming Units (CFUs).

The experiment was based on the contact between an antibacterial surface and a bacterial suspension for a given period of time, according to standards JIS Z 2801:2010 [52] and ISO 27447:2009 [50]. The test was carried out on polycarbonate membranes covered with TiO$_2$ coating, or with control coating, and placed in Petri dishes under a sterile flow hood. Each membrane was inoculated with 0.4 ml of bacterial suspension (cell concentration adjusted between 8×10^4 and 2×10^5 CFU/mL). A transparent sterile film (9–10 cm^2) was carefully placed on the suspension in order to increase the contact area between bacterial cells and coatings [51]. Then, some membranes were UV-irradiated as described earlier (Section 2.3) and others were kept in the dark. The irradiation times were 2 h, 4 h, and 6 h.

After a certain time of contact (*t*), bacteria were recovered using the SCDLP broth. Quantitative evaluation was performed by CFU counting as described earlier (Section 2.4). The antibacterial activity was then calculated as the difference between the average logarithm of the number of viable bacteria on the control without TiO$_2$ and the average logarithm of the number of viable bacteria on samples coated with TiO$_2$:

$$A = \log \left(N_{TiO_2} \right) - \log \left(N_{control} \right) = \log \left(\frac{N_{TiO_2}}{N_{control}} \right),$$

(1)

with A: antibacterial activity; N_{TiO_2}: average number of CFU on TiO$_2$-coated samples; $N_{control}$: average number of CFU on control-coated samples.

2.5. Development of a Bacterial Biofilm under Accelerated Conditions

This experiment was designed to observe the effect of the photocatalytic coating on bacterial growth and spread once coated on a building material. Conditions favorable to microbial growth were chosen to obtain a biofilm very rapidly (24 h).

The test was carried out in aqueous media and in static conditions as previously described [53–55]. The media was chosen so that bacterial development was favored in the form of biofilm without planktonic growth. Direct observations of samples were carried out by epifluorescence microscopy and by scanning electron microscopy in order to highlight the distribution of bacteria.

2.5.1. Biofilm Nutrient Broth

The biofilm nutrient broth used to promote bacterial adhesion and biofilm formation was prepared as described in [53–55]. The composition of the broth is given in Table 2.

Table 2. Composition of the biofilm nutrient broth.

Components	Concentration (g/L)
$FeSO_4, 7H_2O$	0.0005
Na_2HPO_4	1.25
KH_2PO_4	0.5
$(NH_4)_2SO_4$	0.1
Glucose	0.05
$MgSO_4, 7H_2O$	0.02

2.5.2. Procedure

The test was carried out with the cementitious supports covered with TiO_2 coating or with control coating. Cement supports were placed in 6-well culture plates filled with approximately 6 mL per well of biofilm nutrient broth at room temperature. Then, each sample was inoculated with 300 μL of the bacterial test suspension adjusted to about 10^3 CFU/mL . The plates were covered with UV-transparent pyrex lids and incubated at 36 °C . One plate was irradiated with UV light and the other was kept in the dark.

Light intensity on the samples was settled around 3 W/m^2 . In order to avoid planktonic cells (non adherent or released from the biofilm), the wells were washed with sterile distilled water and the biofilm broth was renewed after 4 h, 6 h and 24 h of incubation. Rapid assessments with pH paper was carried out before and after the experiment. The broths of different samples were between 7 and 9. No jump of pH was detected, indicating that the pre-treatment (leaching) of the cementitious supports was efficient. After 24 h, the samples were washed and removed with sterile tweezers and placed in new culture plates. Then, 6 mL of phosphate buffer was added to the wells and the top surface of each sample was gently scraped with a steel spatula. After homogenization of the solution, 1 mL was collected and diluted in sterile distilled water.

2.5.3. Scanning Electron Microscopy

Microstrucutural observations and chemical analyses were performed using a scanning electron microscope (JSM-6380LV, Jeol, Tokyo, Japan) fitted with an Energy Dispersive Spectroscopy detector (XFlash® 3001, Röntec, Berlin, Germany).

2.5.4. Epifluorescence Microscopy

Epifluorescence microscopy was used for rapid assessment of the bacterial colonization of sample surfaces. Following the addition of phosphate buffer at the end of the experiment, 0.75 μL of Syto9 was used as a fluorochrome to stain the bacterial biofilm (Syto9 excitation range: 470–520 nm ; emission ange: 510–540 nm). After 10 min, samples were imaged with a Carl Zeiss Axio Imager-M2 microscope (Carl Zeiss, Oberkochen, Germany) equipped for epifluorescence with an HXP 200 C light source and the Zeiss 09 filter (excitor HP450r HP450200 C light source). Images were acquired with a digital camera (Zeiss AxioCam MRm) every 0.5 mm along the Z-axis and the set of images was processed with the Zen (Carl Zeiss) ® software. It should be emphasized that the Syto9 fluorochrome penetrates through damaged membranes as well as through whole membranes. Consequently, the method used here does not differentiate viable and non-viable cells but provides a visualization of global microbial colonization.

3. Results and Discussions

3.1. Antibacterial Activity

The results obtained from the analysis of TiO_2 antibacterial activity on coated glass are summarized in Figure 2.

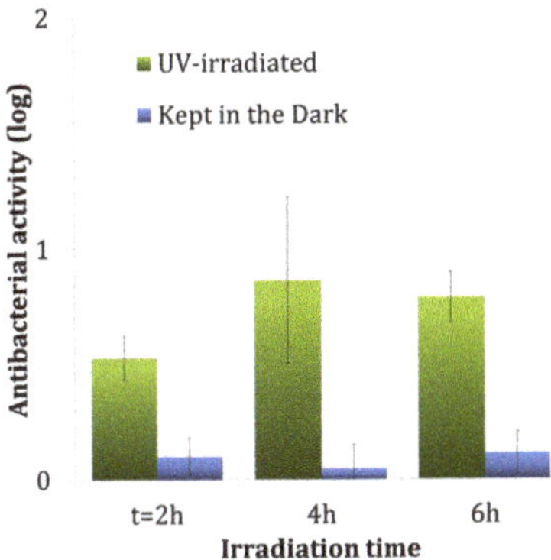

Figure 2. Antibacterial activity on *E. coli* of TiO_2-coating (glass slides), after 2 h, 4 h, and 6 h of contact. Mean $\pm$ s.e., $n = 3$.

The TiO_2 coating showed no antibacterial activity in the dark. Irradiation led to lower CFU values and activities were defined at 0.53 ± 0.1 log ($p = 0.241$), 0.96 ± 0.43 log ($p = 0.018$) and 0.79 ± 0.11 log ($p = 0.03$) after 2 h, 4 h, and 6 h of UV irradiation, respectively.

These inactivation values are relatively low when compared to the inactivation values found in the literature on photocatalytic TiO_2 coatings. Many factors strongly impact the microbial inactivation of TiO_2: the inoculum concentration, the nature of aqueous medium of the inoculum, the support material of the coating, the contact between TiO_2 particles and microbial cells, etc. [1–3]. However, two factors are quite often not highlighted: Irradiance and TiO_2 implementation. Regarding irradiance, in the frame of building engineering, it is desired to set up 'passive' antibacterial solutions, meaning that can be activated by low irradiances (daylight or conventional indoor lightning). This is why we chose to work under such values (<5 W/m^2). Regarding the photocatalyst implementation, its optimization often leads to expensive processes or processes that are impossible to implement on building (calcination process generally involved). The most interesting solutions are therefore inclusion in the mass (photocatalytic concrete) or paint-like coatings. One purpose of this study was to emphasize that such photocatalytic paint-like coating presents significant inactivation rates. Even if these values are far from those found in the literature, they are encouraging especially because they are observed in extreme contamination conditions (still far from the contamination that can occur in real conditions).

Nevertheless, these activities are quite a lot lower than those obtained from the commercial TiO_2 powder alone (VLP5070) alone in a previous study [51]: 2.62 ± 0.20 log ($p < 0.001$) after 4 h and 3.73 ± 0.24 log ($p = 0.001$) after 6 h of experiments. This can be explained by the presence of organic binder, which can interfere in the photocatalytic disinfection by directly reacting with TiO_2 particles. Various studies have shown that the presence of organic compounds could reduce photocatalytic efficiency [51,56–60]. Moreover, such results could be explained by the encapsulation of TiO_2 particles in the organic binder of the coating substantially reducing the antibacterial efficiency. The encapsulation of particles inside the acrylic binder might hamper the UV incidence on TiO_2 particles, which would reduce photocatalytic reactions.

Figure 3 presents the antibacterial activity of the photocatalytic coating (with TiO_2) on membrane after 4 h of experiment under UV irradiation, for several times of pre-irradiation of the coating (0 h, 48 h, 90 h, and 109 h prior to the test). The antibacterial activity increased with the pre-irradiation time. Samples that had not been pre-irradiated (0 h) showed an antibacterial activity of 0.20 ± 0.06 log ($p = 0.016$) after 4 h of experiment. The activities increased for samples that had been pre-irradiated for 48 h (0.68 ± 0.06 log) or for 90 h (0.75 ± 0.10 log) with the higher detected activity of 1.22 ± 0.13 log ($p < 0.001$) for a 109 h pre-irradiation.

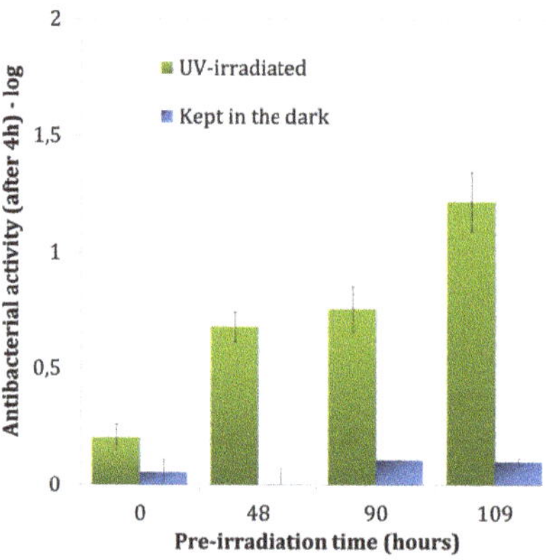

Figure 3. Antibacterial activity of TiO_2-coating on *E. coli* after 4 h of test, depending on the time of pre-irradiation of the coating (3 W/m^2). Mean $\pm$ s.e., $n = 3$.

After a few hours of UV-irradiation, the TiO_2 coating showed significant antibacterial activities that could reflect bactericidal action and/or growth inhibition (bacteriostatic action). Moreover, the antibacterial activity increased significantly after 48 h of pre-irradiation.

A possible explanation for these results may be the degradation of a first layer of the coating that covers most of TiO_2 particles. When samples were irradiated before the test (pre-irradiation), some of the UV beams may have reached encapsulated particles near the surface and activated the photocatalytic process. The organic coating would thus be partially degraded, generating additional porosity in the coating surface. The apparition of this new porosity would favor:

1. The access of UV beams for photocatalytic activation.
2. The access of the pollutants (here bacteria cells) to the particles.
3. Diffusion of the reactive radicals produced by photocatalysis toward pollutants, as shown in Figure 4.

Regarding this phenomenon of self-degradation of the coating, it is important to specify that this study follows previous work on photocatalytic coatings for air purifying [36] and for bacteria inactivation [51]. Coatings' formulation was initially engineered in the thesis of Martinez [48] on air purifying. In his work, TiO_2 coatings included a part of silicates as inorganic binder, promoting adhesion on building materials and sustainability of coatings. In the frame of bacterial inactivation, the presence of alkaline silicates (pH = 11–12) has a major impact on bacteria survival. In order to prevent overlapping between pH and photocatalysis effects, we decided to use a 'partial' coating, silicate free. Regarding the silicate-based coatings, no degradation phenomenon of the coating was observed over time (same samples were tested after several weeks). It seems possible that only the surface layer of the coating is degraded and that TiO_2 particles will subsequently react with external pollutants or other molecules from ambient air. Further work on pre-irradiation times greater than 109 h would be useful to evaluate if the activity increases, is maintained or decreases.

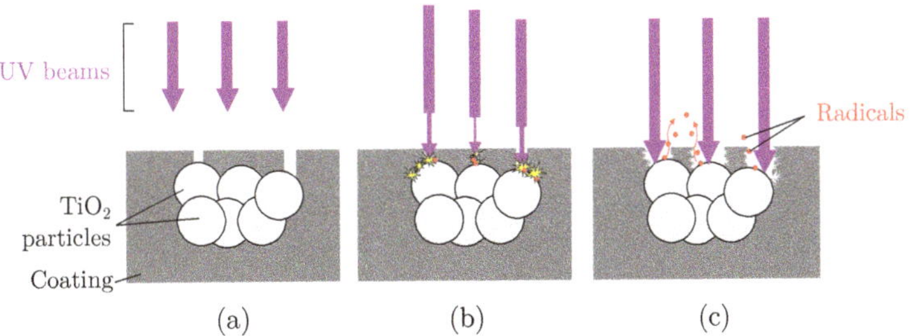

Figure 4. Schematic illustration of the disencapsulation mechanism of TiO_2 particles from acrylic coating. (**a**) starting UV pre-irradiation; (**b**) part of the UV beams reach particles near the surface, activating the photocatalytic reaction: the coating starts to degrade around particles; (**c**) apparition of additional porosity, favoring access of UV beams, diffusion of radicals outside the coating and diffusion of pollutants inside the coating.

The antibacterial activity experiment, carried out under low UV irradiation (2.5–3 W/m^2) confirmed that the efficiency on the bacterial inactivation of the TiO_2 coating increased after 48 h of pre-irradiation. The combined effect of photoinduced hydrophilicity and antibacterial activity of TiO_2 coating suggests an antibacterial action that is effective immediately on contact, providing resistance to bacterial adhesion.

3.2. Developement of a Bacterial Biofilm under Accelerated Conditions

Epifluorescence observations were carried out on the surface of samples after 24 h of experiment. Epifluorescence observation gave the best images of the spread of microbial colonization on samples. Figure 5 shows epifluorescence images of the surface of samples after 24 h of experiment. A light blue color was chosen to represent the coating and red to highlight the bacteria. Images taken on control coating either stored in the dark or exposed to UV irradiation showed bacterial colonization spread over the entire surface of samples (Figure 5a,b).

As can be seen from Figure 5c-1, TiO_2-coated samples that had been irradiated for 24 h showed areas of bacterial colonization spread over the entire surface but with very low density. Only a few red spots were visible in the near-center area of samples, meaning that bacterial cells were somewhat isolated from each other. Figure 5c-2 was taken on the edge area of the sample surface. The figure shows high concentrations of bacterial cells (red spots) surrounding TiO_2 coating (blue areas). It can be clearly seen that bacteria have developed mainly around the coating (directly on cementitious support) and very little on it. This 'repulsive' effect seems to confirm the results from antibacterial activity of the coating. Even under very severe conditions of microbial contamination, the coating was still active and prevented the biofilm development on its surface. These results are also in accordance with a recent study that showed a significant antibacterial effect of TiO_2 coatings against the formation of microbial biofilms under low irradiance [40].

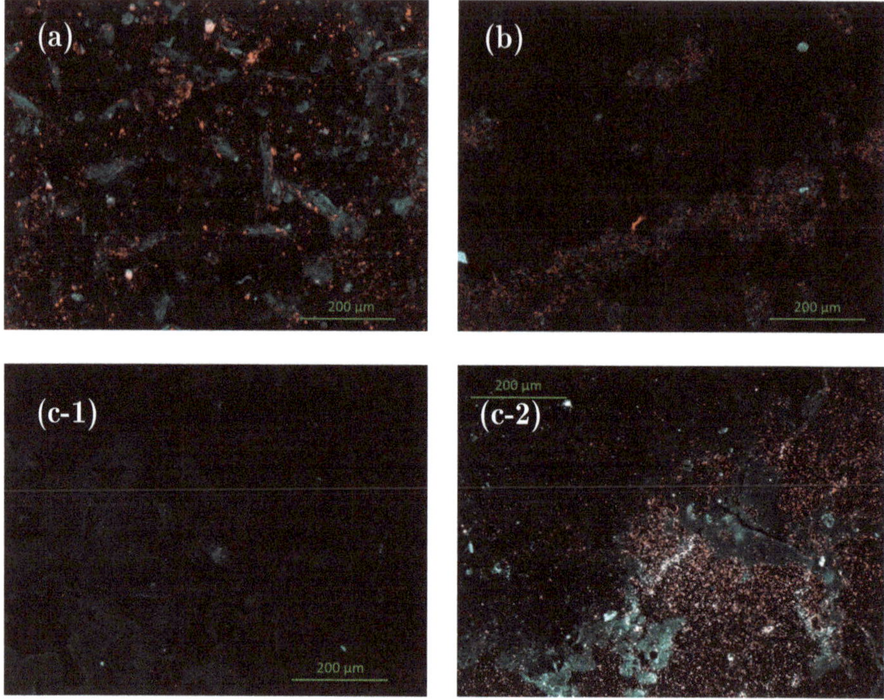

Figure 5. Epifluorescence pictures of surface samples after 24 h of experiment. (**a**) control-coated sample kept in the dark; (**b**) control-coated sample under UV irradiation; (**c**) TiO$_2$-coated samples under UV irradiation. For TiO$_2$-coated samples, one image was taken at the center of the sample (**c-1**) and another at the edge (**c-2**).

More tests would be required to validate the methodology. However, the presented images of early trials encourages the use of such device to study microbial growth on coated building materials. Moreover, further experimental investigations could be carried out with bacteria genetically modified to emit their own fluorescence, which would allow the evolution of the microbial colonization of material to be observed in real time, using a time lapse method [61,62]. In addition, the use of a fluorochrome such as propidium iodide, which is able to penetrate through damaged membrane only, would enable live cells, damaged membrane and dead cells to be quantified as previously described by Gregori et al. [63].

Figure 6 shows SEM observations of control-coated samples irradiated with UV (Figure 6a,b). Other samples that have been kept in the dark, i.e., control-coated samples and TiO$_2$-coated samples, presented the same colonization pattern (as did replicate samples, not shown). On Figure 6a,b, rounded protuberances can be observed, which gather to form a cohesive substance. This substance was not detected on control samples that had not been contaminated (Figure 6c,d). It was likely to be organic-based and produced by bacteria. A similar matrix was observed in the form of a dense network of curli fibers organized as a 'basket'-like structure around the cells [64]. On a smaller scale (Figure 6b), many bacterial cells were visible evolving near the surface or encased by the dense curli network.

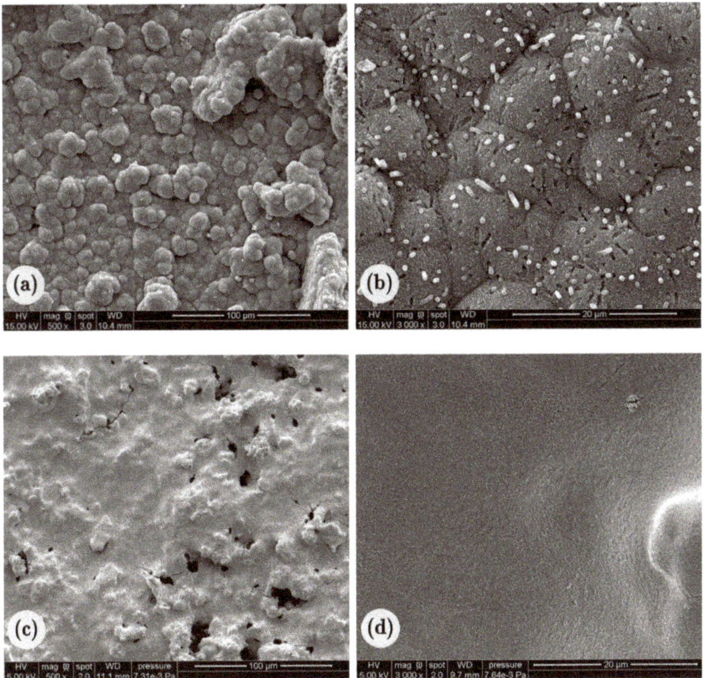

Figure 6. SEM images (secondary electron mode) of samples after 24 h of experiment. (**a**,**b**) SEM images of control samples without TiO$_2$ ($\times$500 and $\times$3000) showing an organic matrix and ovoid bacteria encased by or near surface of this dense network of curli fibers. (**c**,**d**) SEM images of TiO$_2$-coated samples that have not been inoculated ($\times$500 and $\times$3000) showing the surface of coating bacteria-free.

Observations of TiO$_2$-coated samples irradiated with UV light for 24 h of experiment showed colonization patterns. As can be seen from Figure 7, bacteria also colonized the coating in some areas. Figure 7c,d also show that parts of the coating were almost completely covered by the oragnic matrix network.

Damaged bacteria were also detected on the photocatalytic coating that had been UV-irradiated for 24 h. Figure 8a,b show intact bacterial cells from control samples kept in the dark. The natural shape of the bacillus, the classic physical appearance of *E. coli*, is quite visible. Figure 8c,d,e show damaged bacteria that were detected on UV-irradiated TiO$_2$-coated samples. Several similar clusters of damaged bacteria were found on the coating. The cells were always gathered in clusters. In addition, it can be seen from Figure 8d,e that small aggregates were attached on the cell walls of some bacteria. These particles were obviously submicrometer sized. They were detected only on visibly damaged bacteria. Moreover, such clusters of damaged bacteria were not detected on TiO$_2$-coated samples kept in the dark nor on control-coated samples. These aggregates thus might be TiO$_2$ particles that may have participated in the degradation of the cells.

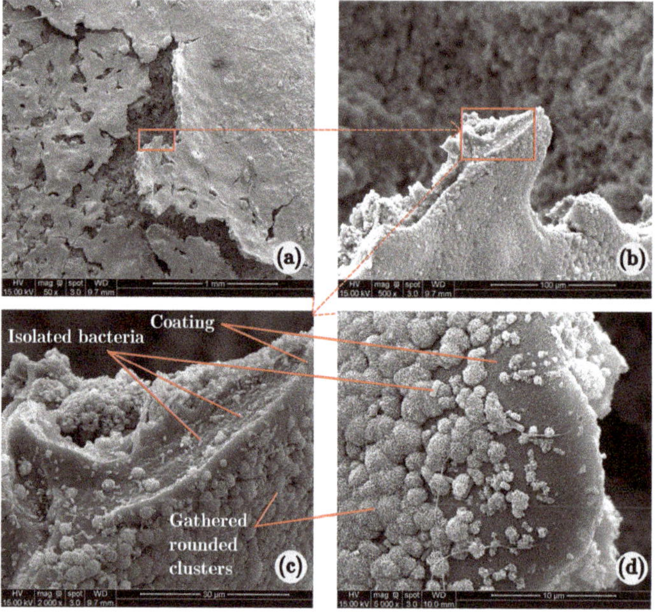

Figure 7. SEM images (secondary electron mode) of TiO$_2$-coated samples irradiated with UV light for 24 h.

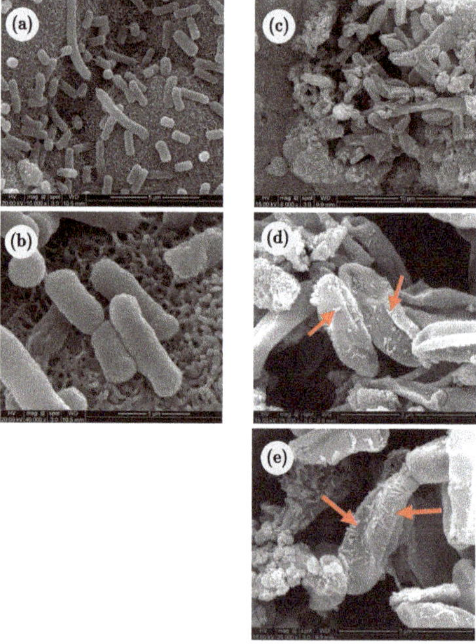

Figure 8. SEM images (secondary electron mode) of intact cells on control-coated samples kept in the dark (**a**,**b**) and of damaged cells observed on TiO$_2$-coated samples irradiated with UV light for 24 h experiment (**c**,**d**,**e**).

4. Conclusions

In this paper, two types of experiments were carried out in order to evaluate the antibacterial properties of a semi-transparent coating incorporating photocatalytic TiO_2 particles. Experimentation focused on the measurement of: (a) the photoinduced antibacterial activity of TiO_2 coating, and (b) its resistance to the formation of bacterial biofilm under accelerated growth conditions.

The tested coating showed significant effects in terms of antibacterial activity and resistance to biofilm formation under low irradiation (2.5 W/m^2). The antibacterial activity was induced by photocatalysis. SEM analyses confirmed the lethal activity of the coating on bacterial cells, even in favorable conditions of biofilm formation considered as a 'worst case' in comparison to the main conditions of use.

The Epifluorescence approach carried out to evaluate biofilm formation was suitable for TiO_2-coated cementitious supports. Fluorescence observations provided good pictures of the colonization patterns on the surface of samples. This work can be seen as a preliminary study exploring the potential of epifluorescence microscopy in a field in which it is little exploited: microbial contamination of building materials. Further tests are necessary to validate the methodology, e.g., quantitative measurements by chemical or molecular methods. Subsequently, the methodology could be used in studies on microbial growth and spread on coatings and/or on building materials.

Regarding the coating, further research should be undertaken to explore the optimal formulation, the distribution of TiO_2 particles and the application of the coating to building materials. More work should also be carried out to evaluate the antimicrobial properties towards gram positive bacteria, including *Staphylococcus aureus* (also recommended by JIS Z 2801) and molds. The molds will be chosen among the most detected in indoor environments such as *Penicillium*, *Aspergillus*, and *Cladosporium* species.

Acknowledgments: The authors would like to thank Université Toulouse III—Paul Sabatier for its financial support.

Author Contributions: T.V. conceived and designed the experiments with the help of C.R.; T.V. performed the experiments; T.V. and B.E. performed the Epifluorescence observations and the corresponding analysis; T.V., A.B. and C.R. analyzed the data; T.V. wrote the paper.

Conflicts of Interest: The authors declare no conflict of interest.

References

1. Gutarowska, B.; Piotrowska, M. Methods of mycological analysis in buildings. *Build. Environ.* **2007**, *42*, 1843–1850.
2. Verdier, T.; Coutand, M.; Bertron, A.; Roques, C. A review of indoor microbial growth across building materials and sampling and analysis methods. *Build. Environ.* **2014**, *80*, 136–149.
3. Bertron, A. Understanding interactions between cementitious materials and microorganisms: A key to sustainable and safe concrete structures in various contexts. *Mater. Struct.* **2014**, *47*, 1787–1806.
4. *Guidelines for Indoor Air Quality*; WHO: Geneva, Switzerland, 2006.
5. Escadeillas, G.; Bertron, A.; Blanc, P.; Dubosc, A. Accelerated testing of biological stain growth on external concrete walls. Part 1: Development of the growth tests. *Mater. Struct.* **2007**, *40*, 1061–1071.
6. Verdier, T.; Bertron, A.; Johansson, P. Overview of Indoor Microbial Growth on Building Materials. In Proceedings of the International RILEM Symposium on the Microorganisms-Cementitious materials Interactions (TC253-MCI), Delft, The Netherlands, 23 June 2016.
7. Johansson, P.; Ekstrand-Tobin, A.; Svensson, T.; Bok, G. Laboratory study to determine the critical moisture level for mould growth on building materials. *Int. Biodeterior. Biodeterior.* **2012**, *73*, 23–32.
8. Pasanen, A.L. A Review: Fungal Exposure Assessment in Indoor Environments. *Indoor Air* **2001**, *11*, 87–98.
9. Warscheid, T.; Braams, J. Biodeterioration of stone: A review. *Int. Biodeterior. Biodeterior.* **2000**, *46*, 343–368.
10. Samson, R.; Flannigan, B.; Flannigan, M.; Verhoeff, A.; Adan, O.; Hoekstra, E. *Health Implications of Fungi in Indoor Environments*; Elsevier: Amsterdam, The Netherlands, 1994.

11. Torvinen, E.; Meklin, T.; Torkko, P.; Suomalainen, S.; Reiman, M.; Katila, M.L.; Paulin, L.; Nevalainen, A. Mycobacteria and Fungi in Moisture-Damaged Building Materials. *Appl. Environ. Microbiol.* **2006**, *72*, 6822–6824.

12. Andersson, M.A.; Nikulin, M.; Köljalg, U.; Andersson, M.C.; Rainey, F.; Reijula, K.; Hintikka, E.L.; Salkinoja-Salonen, M. Bacteria, molds, and toxins in water-damaged building materials. *Appl. Environ. Microbiol.* **1997**, *63*, 387–393.

13. Lacaze, I. Etude des Mécanismes de Colonisation des Produits de Construction Par les Micromycètes. Ph.D. Thesis, Microbiologie Procaryote Et Eucaryote, Université Paris-Diderot, Paris, France, 2016. (in French)

14. Sharpe, R.A.; Cocq, K.L.; Nikolaou, V.; Osborne, N.J.; Thornton, C.R. Identifying risk factors for exposure to culturable allergenic moulds in energy efficient homes by using highly specific monoclonal antibodies. *Environ. Res.* **2016**, *144*, 32–42.

15. Flannigan, B.; Samson, R.A.; Miller, J.D. *Microorganisms in Home and Indoor Work Environments: Diversity, Health Impacts, Investigation and Control,* 2nd ed.; CRC Press: Boca Raton, FL, USA, 2011; p. 529.

16. Parat, S.; Perdrix, A.; Mann, S.; Cochet, C. A study of the relationship between airborne microbiological concentrations and symptoms in office in buildings. In Proceedings of the Healthy Buildings '95: An International Conference on Healthy Buildings in Mild Climate, Milano, Italy, 10–14 September 1995; Volume III, pp. 1481–1486.

17. Spilak, M.P.; Madsen, A.M.; Knudsen, S.M.; Kolarik, B.; Hansen, E.W.; Frederiksen, M.; Gunnarsen, L. Impact of dwelling characteristics on concentrations of bacteria, fungi, endotoxin and total inflammatory potential in settled dust. *Build. Environ.* **2015**, *93*, 64–71.

18. Prescott, L.M.; Harley, J.P.; Klein, D.A.; Willey, J.M.; Sherwood, L.M.; Woolverton, C.J. *Microbiology*, 7th ed.; McGraw-Hill: New York, NY, USA, 2007.

19. Costerton, J.W.; Lewandowski, Z.; DeBeer, D.; Caldwell, D.; Korber, D.; James, G. Biofilms, the customized microniche. *J. Bacteriol.* **1994**, *176*, 2137–2142.

20. Davies, D.G.; Parsek, M.R.; Pearson, J.P.; Iglewski, B.H.; Costerton, J.W.; Greenberg, E.P. The Involvement of Cell-to-Cell Signals in the Development of a Bacterial Biofilm. *Science* **1998**, *280*, 295–298.

21. Shirakawa, M.A.; Tavares, R.G.; Gaylarde, C.C.; Taqueda, M.E.S.; Loh, K.; John, V.M. Climate as the most important factor determining anti-fungal biocide performance in paint films. *Sci. Total Environ.* **2010**, *408*, 5878–5886.

22. Gutarowska, B.; Pietrzak, K.; Machnowski, W.; Danielewicz, D.; Szynkowska, M.; Konca, P.; Surma-Slusarska, B. Application of Silver Nanoparticles for Disinfection of Materials to Protect Historical Objects. *Curr. Nanosci.* **2014**, *10*, 277–286.

23. Shirakawa, M.A.; Gaylarde, C.C.; SahÃ£o, H.D.; Lima, J.R.B. Inhibition of Cladosporium growth on gypsum panels treated with nanosilver particles. *Int. Biodeterior. Biodeterior.* **2013**, *85*, 57–61.

24. De Niederhãusern, S.; Bondi, M.; Bondioli, F. Self-Cleaning and Antibacteric Ceramic Tile Surface. *Int. J. Appl. Ceram. Technol.* **2013**, *10*, 949–956.

25. Valentin, R.; Alignan, M.; Giacinti, G.; Renaud, F.N.R.; Raymond, B.; Mouloungui, Z. Pure short-chain glycerol fatty acid esters and glycerylic cyclocarbonic fatty acid esters as surface active and antimicrobial coagels protecting surfaces by promoting superhydrophilicity. *J. Colloid Interface Sci.* **2012**, *365*, 280–288.

26. Carp, O.; Huisman, C.; Reller, A. Photoinduced reactivity of titanium dioxide. *Prog. Solid State Chem.* **2004**, *32*, 33–177.

27. Chong, M.N.; Jin, B.; Chow, C.W.K.; Saint, C. Recent developments in photocatalytic water treatment technology: A review. *Water Res.* **2010**, *44*, 2997–3027.

28. Dalrymple, O.K.; Stefanakos, E.; Trotz, M.A.; Goswami, D.Y. A review of the mechanisms and modeling of photocatalytic disinfection. *App. Catal. B Environ.* **2010**, *98*, 27–38.

29. Foster, H.A.; Ditta, I.B.; Varghese, S.; Steele, A. Photocatalytic disinfection using titanium dioxide: Spectrum and mechanism of antimicrobial activity. *Appl. Microbiol. Biotechnol.* **2011**, *90*, 1847–1868.

30. Taylor, D.M. *Transmissible Degenerative Encephalopathies: Inactivation of the Unconventional Causal Agents*; John Wiley & Sons: Hoboken, NJ, USA, 2008; p. 324.

31. Fujishima, A.; Rao, T.N.; Tryk, D.A. Titanium dioxide photocatalysis. *J. Photochem. Photobiol. C Photochem. Rev.* **2000**, *1*, 1–21.

32. Maury-Ramirez, A.; Muynck, W.D.; Stevens, R.; Demeestere, K.; Belie, N.D. Titanium dioxide based strategies to prevent algal fouling on cementitious materials. *Cem. Concr. Compos.* **2013**, *36*, 93–100.

33. Guo, M.Z.; Maury-Ramirez, A.; Poon, C.S. Self-cleaning ability of titanium dioxide clear paint coated architectural mortar and its potential in field application. *J. Clean. Prod.* **2016**, *112*, 3583–3588.

34. Maury-Ramirez, A.; Demeestere, K.; Belie, N.D.; Mäntylä, T.; Levänen, E. Titanium dioxide coated cementitious materials for air purifying purposes: Preparation, characterization and toluene removal potential. *Build. Environ.* **2010**, *45*, 832–838.

35. Boonen, E.; Beeldens, A. Recent Photocatalytic Applications for Air Purification in Belgium. *Coatings* **2014**, *4*, 553–573.

36. Martinez, T.; Bertron, A.; Ringot, E.; Escadeillas, G. Degradation of NO using photocatalytic coatings applied to different substrates. *Build. Environ.* **2011**, *46*, 1808–1816.

37. Hot, J.; Martinez, T.; Wayser, B.; Ringo, E.; Bertron, A. Photocatalytic degradation of NO/NO_2 gas discharge in a 10 m^3 experimental chamber. *Environ. Sci. Pollut. Res. Int.* **2016**, *24*, 12562–12570.

38. Maury-Ramirez, A.; Demeestere, K.; Belie, N.D. Photocatalytic activity of titanium dioxide nanoparticle coatings applied on autoclaved aerated concrete: Effect of weathering on coating physical characteristics and gaseous toluene removal. *J. Hazard. Mater.* **2012**, *211–212*, 218–225.

39. Verdier, T. Élaboration de Revêtements Pour Matériaux de Construction Visant à Lutter Contre la Prolifération Microbienne à L'intérieur des Bâtiments: Efficacité et Mode D'action. Ph.D. Thesis, Toulouse III-Paul Sabatier, Toulouse, France, 2015. (in French)

40. Falco, G.D.; Porta, A.; Petrone, A.M.; Gaudio, P.D.; Hassanin, A.E.; Commodo, M.; Minutolo, P.; Squillace, A.; D'Anna, A. Antimicrobial activity of flame-synthesized nano-TiO_2 coatings. *Environ. Sci. Nano* **2017**, *4*, 1095–1107.

41. Wiktor, V.; De Leo, F.; Urzì, C.; Guyonnet, R.; Grosseau, P.; Garcia-Diaz, E. Accelerated laboratory test to study fungal biodeterioration of cementitious matrix. *Int. Biodeterior. Biodeterior.* **2009**, *63*, 1061–1065.

42. Wiktor, V.; Grosseau, P.; Guyonnet, R.; Garcia-Diaz, E.; Lors, C. Accelerated weathering of cementitious matrix for the development of an accelerated laboratory test of biodeterioration. *Mater. Struct.* **2010**, *44*, 623–640.

43. Maness, P.C.; Smolinski, S.; Blake, D.M.; Huang, Z.; Wolfrum, E.J.; Jacoby, W.A. Bactericidal Activity of Photocatalytic TiO_2 Reaction: toward an Understanding of Its Killing Mechanism. *Appl. Environ. Microbiol.* **1999**, *65*, 4094–4098.

44. Sunada, K.; Watanabe, T.; Hashimoto, K. Studies on photokilling of bacteria on TiO_2 thin film. *J. Photochem. Photobiol. A Chem.* **2003**, *156*, 227–233.

45. Blöß, S.P.; Elfenthal, L. Doped titanium dioxide as a photocatalyst for UV and visible light. In Proceedings of the International RILEM Symposium on Photocatalysis, Environment and Construction Materials TDP, Florence, Italy, 8–9 October 2007; pp. 31–38.

46. Fujishima, A.; Zhang, X. Titanium dioxide photocatalysis: present situation and future approaches. *Comptes Rendus Chim.* **2006**, *9*, 750–760.

47. Husken, G.; Hunger, M.; Brouwers, H.J.H. Experimental study of photocatalytic concrete products for air purification. *Build. Environ.* **2009**, *44*, 2463–2474.

48. Martinez, T. Revêtements Photocatalytiques Pour les Matériaux de Construction: Formulation, Évaluation de L'efficacité de la Dépollution de L'air et Écotoxicité. Ph.D. Thesis, Toulouse III-Paul Sabatier, Toulouse, France, 2012. (in French)

49. Ishiguro, H.; Nakano, R.; Yao, Y.; Kajioka, J.; Fujishima, A.; Sunada, K.; Minoshima, M.; Hashimoto, K.; Kubota, Y. Photocatalytic inactivation of bacteriophages by TiO_2-coated glass plates under low-intensity, long-wavelength UV irradiation. *Photochem. Photobiol. Sci.* **2011**, *10*, 1825–1829.

50. *ISO 27447 Fine Ceramics (Advanced Ceramics, Advanced Technical Ceramics)—Test Method for Antibacterial Activity of Semiconducting Photocatalytic Materials*; International Organization for Standardization: Geneva, Switzerland, 2009.

51. Verdier, T.; Coutand, M.; Bertron, A.; Roques, C. Antibacterial Activity of TiO_2 Photocatalyst Alone or in Coatings on *E. coli*: The Influence of Methodological Aspects. *Coatings* **2014**, *4*, 670–686.

52. *JIS Z 2801 Antibacterial Products—Test for Antibacterial Activity And Efficacy*; Biosan Laboratories, Inc.: Warren, MI, USA, 2010.

53. Alasri, A.; Roques, C.; Michel, G.; Cabassud, C.; Aptel, P. Bactericidal properties of peracetic acid and hydrogen peroxide, alone and in combination, and chlorine and formaldehyde against bacterial water strains. *Can. J. Microbiol.* **1992**, *38*, 635–642.

54. Pineau, L.; Roques, C.; Luc, J.; Michel, G. Automatic washer disinfector for flexible endoscopes: A new evaluation process. *Endoscopy* **1997**, *29*, 372–379.

55. Campanac, C.; Pineau, L.; Payard, A.; Baziard-Mouysset, G.; Roques, C. Interactions between Biocide Cationic Agents and Bacterial Biofilms. *Antimicrob. Agents Chemother.* **2002**, *46*, 1469–1474.

56. Ortelli, S.; Blosi, M.; Albonetti, S.; Vaccari, A.; Dondi, M.; Costa, A.L. TiO$_2$ based nano-photocatalysis immobilized on cellulose substrates. *J. Photochem. Photobiol. A Chem.* **2014**, *276*, 58–64.

57. Saito, T.; Iwase, T.; Horie, J.; Morioka, T. Mode of photocatalytic bactericidal action of powdered semiconductor TiO$_2$ on mutans streptococci. *J. Photochem. Photobiol. B Biol.* **1992**, *14*, 369–379.

58. Gogniat, G.; Thyssen, M.; Denis, M.; Pulgarin, C.; Dukan, S. The bactericidal effect of TiO$_2$ photocatalysis involves adsorption onto catalyst and the loss of membrane integrity. *FEMS Microbiol. Lett.* **2006**, *258*, 18–24.

59. Dunlop, P.; Byrne, J.; Manga, N.; Eggins, B. The photocatalytic removal of bacterial pollutants from drinking water. *J. Photochem. Photobiol. A Chem.* **2001**, *148*, 355–363.

60. Rincón, A.G.; Pulgarin, C. Effect of pH, inorganic ions, organic matter and H$_2$O$_2$ on *E. coli* K12 photocatalytic inactivation by TiO$_2$: Implications in solar water disinfection. *Appl. Catal. B Environ.* **2004**, *51*, 283–302.

61. Gogoi, S.K.; Gopinath, P.; Paul, A.; Ramesh, A.; Ghosh, S.S.; Chattopadhyay, A. Green Fluorescent Protein-Expressing Escherichia coli as a Model System for Investigating the Antimicrobial Activities of Silver Nanoparticles. *Langmuir* **2006**, *22*, 9322–9328.

62. Erable, B.; Vandecandelaere, I.; Faimali, M.; Delia, M.L.; Etcheverry, L.; Vandamme, P.; Bergel, A. Marine aerobic biofilm as biocathode catalyst. *Bioelectrochemistry* **2010**, *78*, 51–56.

63. Gregori, G.; Citterio, S.; Ghiani, A.; Labra, M.; Sgorbati, S.; Brown, S.; Denis, M. Resolution of Viable and Membrane-Compromised Bacteria in Freshwater and Marine Waters Based on Analytical Flow Cytometry and Nucleic Acid Double Staining. *Appl. Environ. Microbiol.* **2001**, *67*, 4662–4670.

64. Serra, D.O.; Richter, A.M.; Klauck, G.; Mika, F.; Hengge, R. Microanatomy at Cellular Resolution and Spatial Order of Physiological Differentiation in a Bacterial Biofilm. *mBio* **2013**, *4*, e00103-13.

Article

Simple Coatings to Render Polystyrene Protein Resistant

Marcelle Hecker [1], Matthew Sheng Hao Ting [1,2] and Jenny Malmström [1,2,*]

[1]   Department of Chemical and Materials Engineering, University of Auckland, Auckland 1023, New Zealand; mhec827@aucklanduni.ac.nz (M.H.); mtin749@aucklanduni.ac.nz (M.S.H.T.)
[2]   MacDiarmid Institute for Advanced Materials and Nanotechnology, Wellington 6140, New Zealand
*   Correspondence: j.malmstrom@auckland.ac.nz; Tel.: +64-9-923-5338

Received: 15 December 2017; Accepted: 29 January 2018; Published: 1 February 2018

Abstract: Non-specific protein adsorption is detrimental to the performance of many biomedical devices. Polystyrene is a commonly used material in devices and thin films. Simple reliable surface modification of polystyrene to render it protein resistant is desired in particular for device fabrication and orthogonal functionalisation schemes. This report details modifications carried out on a polystyrene surface to prevent protein adsorption. The trialed surfaces included Pluronic F127 and PLL-*g*-PEG, adsorbed on polystyrene, using a polydopamine-assisted approach. Quartz crystal microbalance with dissipation (QCM-D) results showed only short-term anti-fouling success of the polystyrene surface modified with F127, and the subsequent failure of the polydopamine intermediary layer in improving its stability. In stark contrast, QCM-D analysis proved the success of the polydopamine assisted PLL-*g*-PEG coating in preventing bovine serum albumin adsorption. This modified surface is equally as protein-rejecting after 24 h in buffer, and thus a promising simple coating for long term protein rejection of polystyrene.

Keywords: polystyrene; DOPA; polydopamine; antifouling; polyethylene glycol; Pluronic; QCM-D

---

1. Introduction

Preventing biological fouling is critically important to many industries. Biofouling significantly shortens the lifespan of materials in contact with biological fluids. For example, in the biomedical sector, devices such as biosensors will produce false signals when covered in proteins and filtration membranes are known to suffer, as biomolecular adhesion causes blockage of the membrane, thus preventing flow through the device [1,2]. Bio-fluids, and in particular proteins, cause this fouling phenomenon because they are charged particles that are amphiphilic and amphoteric, and therefore adsorb easily to most materials [3]. Because of the wide-reaching problems biofouling is causing, and because it is effectively halting other advances in science, particularly in biomedical engineering, a significant amount of research has been conducted to create surfaces which can reduce non-specific protein adsorption [2].

On hydrophobic surfaces, such as the commonly used polystyrene, proteins will structurally rearrange to allow the hydrophobic regions of the protein to interact with the hydrophobic surface. The denaturing exposes the hydrophobic segments (originally often on the inside of the protein to be separated from the polar solvent) to increase the hydrophobic bond with the surface, leading to a gain in entropy upon adsorption [3–5]. In the case of hydrophilic surfaces, hydration of the polar surface is very favourable and large entropies must therefore be achieved to ensure that the adsorption energy overcomes the interfering water molecules/water barrier [3–5]. Research has shown that proteins do not denature greatly on hydrophilic surfaces and thus do not spread on the surface of the substrate as readily as on hydrophobic surfaces [6]. Therefore, on hydrophilic surfaces, adsorption is mainly due to coulomb forces between charged proteins and the surface [3–5].

Molecular groups that provide the desired hydrophilic and electrically neutral characteristics, and which have therefore been used to create antifouling polymers in the past, include ethylene glycol, hydroxyl groups, and zwitterionic betaine groups [7]. Polyethyleneglycol (PEG) is one of the most commonly used anti-fouling polymers and it couples water through hydrogen bonding. More recently, zwitterions have become widely researched as they have even stronger, electrostatically induced, hydration [8]. The most common coating technique is to attach polymer 'brushes' on the surface of the substrate. These so-called brushes are polymer chains which have one end bound to the surface, while the other extends into the solution [9]. The brush formation ensures both a steric barrier to protein adsorption and an effective surface for the protective water layer [10]. Adhesion, or the physisorbed grafting-to method, is the least complicated method of surface modification and is often used to attach block copolymers to substrates. However, because this method does not involve the forming of covalent bonds between the anti-fouling polymer and the substrate, adhesion is reversible, generally leading to shorter-term anti-fouling results [11,12].

Pluronics® (BASF trademark name) are amphiphilic triblock copolymers consisting of a central hydrophobic poly(propyleneoxide) (PPO) block surrounded by hydrophilic poly(ethyleneoxide) (PEO) blocks. It is this amphiphilic behaviour that results in the differences in solubility between the blocks and, thus, in the ability of the hydrophobic block to adsorb to a hydrophobic surface (such as polystyrene) through hydrophobic interactions while the hydrophilic PEO chains extend away from the surface as well-solvated brushes, which sterically hinder protein adsorption [11–13]. Pluronics have been tested as anti-adhesion surface modifiers by many research groups [11–15]. The main characteristics of Pluronics proven to affect their anti-fouling properties are the length of their respective hydrophobic and hydrophilic blocks. An early study in 1991, by Bridgett et al. [11], tested 16 Pluronics against the adhesion of bacteria (Staphylococcus epidermidis) to polystyrene. The Pluronics differed in PPO and PEO block length and all proved successful in reducing adhesion levels by up to 97% [11]. The study demonstrated that the Pluronics' ability to adhere to the polystyrene surface via the PPO block (hydrophobic interactions) was directly related to the length of the PPO block and that the PEO chains must be long enough to be sufficiently mobile to reduce protein adsorption, while not so long as to reduce surface adhesion [11]. Although the adsorption mechanism is complex, due to the many different interactions occurring, the method to apply Pluronics to a substrate surface is simple and thus favourable. However, since the physisorption is reversible, the coating may have a limited lifespan [9,16].

Poly(L-lysine)-*graft*-poly(ethylene glycol) (PLL-*g*-PEG) was developed to provide a strong irreversible surface binding of a somewhat similar PEO construct [17,18]. The polymer contains a cationic poly(L-lysine) backbone, with grafted PEG side chains. While Pluronics interact with hydrophobic surfaces through simple adsorption, PLL-*g*-PEG was designed to bind to anionic surfaces through electrostatic interactions between the positively charged amino-terminated side chains on the PLL-backbone and anionic oxide surfaces. PLL-*g*-PEG has excellent anti-fouling properties due to its high PEG grafting density, providing the necessary steric hindrance and volume exclusion effects [17,18]. While PLL-*g*-PEG has been extensively examined as an anti-fouling surface modifier on negatively charged surfaces, it is also known to not repel proteins when deposited on hydrophobic surfaces. For example, Malmström et al. showed that PLL-*g*-PEG could not prevent adsorption of laminin, and was replaced by the adsorbing laminin during QCM-D measurement on hydrophobic gold surfaces [19]. Similarly, a study by Nonckreman et al., reported that a polystyrene surface modified with PLL-*g*-PEG could not resist human serum albumin or fibrinogen adsorption [20]. An earlier paper by Reimhult et al. seemingly contradicted these findings, stating that, while not as promising as on negatively charged surfaces, polystyrene coated in PLL-*g*-PEG showed significant protein resistance [21].

The dopamine molecule is an interesting surface modifier mimicking proteins excreted by mussels, which help them adhere to most surfaces in the marine environment [8,22]. The ability to form an adhering polydopamine film on a variety of surfaces was first demonstrated by Messersmith et al. [22,23]. Since then, a variety of dopamine-assisted grafting techniques and DOPA modified antifouling polymers have been tested to more securely attach antifouling polymers to substrate surfaces [23–35]. The polydopamine layer not only readily adheres to most surfaces, but also provides a surface which allows various secondary reactions to take place and could therefore more easily enable the functionalization of material surfaces [22]. The reactions the catechol group of dopamine undergoes with metals, thiols, and amines has led to many research groups favouring the inclusion of dopamine when treating these surfaces [8,22,24–26,36]. While many DOPA-functionalised grafting-to methods have proved successful on the specific surfaces listed above, researchers have encountered fouling when trying the same approach for hydrophobic surfaces. It has been shown that catechol chain ends cannot sufficiently bind to hydrophobic surfaces due to the high solubility of the anti-fouling polymers to which they are bound (e.g., zwitterions) [37]. The alternative to functionalising the polymer with DOPA and thus overcoming surface adhering limitations, is to coat the entire surface in dopamine. This forms a polydopamine film, to which amine or thiol functionalised polymers can be grafted. This method has been labelled "polydopamine-assisted grafting" [22].

In this study we have chosen to investigate the use of a polydopamine-assisted binding of PLL-*g*-PEG to polystyrene. The use of PLL-*g*-PEG both allows for binding through the amine groups and provides an optimised grafting density controlled by the PLL-*g*-PEG synthetic composition. We are also comparing this approach with Pluronic F127. The results reveal that while a simple Pluronic coating of polystyrene resists protein short-term, the coating is unstable. A polydopamine-assisted PLL-*g*-PEG layer however, show promising longer term stability and excellent protein rejecting properties.

2. Materials and Methods

2.1. Materials

Silicon wafers (100, P-type, B-doped) were purchased from University Wafers (Boston, MA, USA) and divided into samples of approximately 1 cm × 1 cm. QCM-D crystals were purchased from ATA scientific (Sydney, Australia) (301), polystyrene from Polymer Source Inc. (Dorval, QC, Canada) (P3810-S, MW: 25500), PEO-PPO-PEO triblock copolymer (Pluronic F127), TRIZMA (2-Amino-2-(hydroxymethyl)-1,3-propanediol), HEPES (4-(2-Hydroxyethyl)piperazine-1-ethanesulfonic acid) and NaCl was purchased from Sigma Aldrich (Auckland, New Zealand), Dopamine hydrochloride (HPLC grade) from AK Scientific (Union City, CA, USA), bovine serum albumin from MP biomedicals (Auckland, New Zealand) and PLL-*g*-PEG, (20)-[3.5]-(2) from Surface Solutions (Dübendorf, Switzerland).

The buffer used in the majority of experiments was TRIZMA 10 mM, NaCl 150 mM, pH 8.5. The only exception was PLL-*g*-PEG adsorption to bare polystyrene, which was performed in HEPES 10 mM, pH 7.4. All buffers were filtered through a 0.22 μm syringe filter and degassed by ultrasonication for 1 h.

2.2. Sample Preparation

Silicon wafers and AT-cut, quartz QCM-D crystals, were cleaned by ultrasonication in acetone and ethanol (15 min each) and dried under a stream of nitrogen. Previously used QCM-D crystals were also sonicated in toluene prior to acetone and ethanol to remove polystyrene. Clean samples were immediately coated by spin coating a solution of 3 wt % polystyrene in toluene (3000 rpm, 2 min). The polystyrene coated samples were post-baked for 5 min at 90 °C, and stored, submerged in ultrapure water, in sealed petri-dishes, until use. The samples were further ultrasonicated in ultrapure water for five minutes prior to conducting surface modifications.

2.3. Surface Modification with Polydopamine

The samples were exposed to dopamine (2 mg/mL) in TRIZMA buffer, with the samples vertically arranged, to prevent polydopamine sedimentation, and left open to the air, for 5 h at

25 °C. After coating, the samples were rinsed with ultrapure water, dried with N_2 (g) and dried in a vacuum desiccator for 24 h at room temperature.

2.4. Secondary Polymer Layer Deposition on Polydopamine

The desired polymer solutions were prepared; Pluronic F127 (2 mg/mL) and PLL-*g*-PEG (0.1 mg/mL) in TRIZMA buffer (pH 8.5), and subsequently filtered through 0.22 µm syringe filters. The polydopamine coated samples were incubated in the secondary polymer solutions, and were allowed to react with the polydopamine coated samples for 18.5 h at 50 °C. Finally, the samples were thoroughly rinsed with buffer and ultrapure water and dried under a stream of nitrogen.

2.5. QCM-D Experiments

All QCM-D experiments were carried out using the Q-Sense Analyzer from Q-Sense AB (Biolin Scientific, Gothenburg, Sweden). The temperature during all experiments was maintained at 22 °C $\pm$ 0.02 °C, and using a pump speed of 0.5, measured to correspond to approximately 0.23 mL/min. The fundamental, 3rd, 5th, 7th, 11th and 13th overtones of frequency and dissipation were recorded for each test, and at least three tests per surface type were conducted, to allow for statistical analysis.

The frequency and dissipation shifts related to mass adsorption were measured by the Q-Soft 401 software. The QCM-D data was then further assessed using the D-find program. The data from the 7th overtone was used for mass adsorption comparisons between all surfaces.

2.6. X-ray Photoelectron Spectroscopy (XPS)

XPS was performed to identify the elemental composition of the surfaces. The compositional data was obtained using a Kratos Axis Ultra-DLD spectrometer (Kratos Analytical Ltd., Manchester, UK). Spectra excitation on an analysis area of 300 by 700 µm was achieved with a 150 W X-ray power source, emitting A1 Kα X-rays (1486.7 eV) in an ultrahigh vacuum chamber ($\sim 10^{-9}$ torr). Survey scans (0–1300 eV) and core level scans were carried out on each surface, at pass energies of 160 eV and 20 eV, respectively. The resulting spectra were analysed with CasaXPS software (Version 2.3.14). All data was calibrated such that the C(1s) peak (corresponding to C–H/C–C energies) was situated at 285.0 eV. The Shirley model was used for peak fitting, and Gaussian-Lorentzian (30% Lorentzian) peaks were used to fit the core-level peaks.

2.7. Atomic Force Microscopy (AFM)

AFM was carried out using a Cypher ES instrument (Oxford Instruments, Asylum Research, Santa Barbara, CA, USA). Images were acquired in tapping mode with a Tap 150 Al-G cantilever (Nominal f = 150 kHz and k = 5 N/m, Budget sensors, Sofia, Bulgaria) and a free air amplitude of 1 V.

2.8. Contact Angle Goniometry

Wettability testing was carried out with a contact angle goniometer (CAM 101, KSV instruments, Helsinki, Finland). The AttensionTheta2017 software was used for acquisition of images and image analysis. All contact angles were recorded using ultrapure water and using the same drop volume (6 ± 2 µL), impact distance between the drop and surface and time before image was acquired (using automated image acquisition).

2.9. Statistics

One-way ANOVA studies were carried out on QCM-D mass adsorption data and contact angle data. From Post Hoc, Tukey analysis, probability values of <0.05 were considered to indicate significant differences. All results with more than three tests, are expressed as the mean, with standard deviations indicated. The statistical accuracy of results from AFM and XPS testing was not analysed, due to sample sizes for each surface being too small ($n < 3$).

3. Results and Discussion

The results from QCM-D experiments evaluating the protein fouling to the different surface coatings and the sample characterisation using AFM, XPS and contact angles are presented in the following sections.

3.1. QCM-D Analysis

QCM-D uses the piezoelectric properties of quartz and follows the frequency of a mechanical oscillation of a quartz crystal to measure the mass of adsorbed molecules in real time. The technique is also able to determine the viscoelastic properties of the adsorbed layer by recording the dissipation of the oscillation. The change in frequency observed in QCM-D is proportional to the change in mass of the oscillating crystal and adsorbed molecules (as described by the Sauerbrey equation), so long as three conditions are satisfied: the mass is distributed evenly, it obeys a no-slip boundary condition and it is sufficiently rigid, or thin, so as not to be subject to internal friction [38–41]. Most of the QCM-D results reported here show only small differences between overtones, which is why the Sauerbrey equation has been used to calculate the adsorbed mass. The mass measured by QCM-D includes everything adsorbing to the surface, including the water either hydro-dynamically coupled to the protein ad-layer or trapped within the anti-fouling film [38–40]. According to Höök et al., for globular proteins like BSA, the measured mass is around 1.75 times higher than that recorded by 'dry mass' optical methods [39].

QCM-D was carried out to observe the frequency changes associated with BSA adsorption to the different surfaces investigated. The typical frequency change over time curves, for the addition of the protein and the subsequent wash-off step on each tested surface, are shown in Figure 1A.

The majority of the surfaces show an initial rapid frequency decrease upon exposure to the protein and a subsequent short region of slower frequency shift as the system reaches saturation. Different extent of protein dissociation from the surface can be observed after rinsing with protein free buffer (indicated by black arrows in Figure 1A). The size of the frequency shifts, taken as the difference between a point immediately prior to the BSA injection, and a point 10 min into the buffer rinse, have been directly related to the mass of BSA bound to the surface, by use of the Sauerbrey equation [41]. Averages from at least three measurements of each surface type are presented in Figure 1B. From this data it is evident that polystyrene modified by adsorption of Pluronic F127 (PS + F127), and polystyrene modified with polydopamine and PLL-*g*-PEG (PS + DOPA + PLL-*g*-PEG) are the most successful coatings in terms of reducing protein adsorption to polystyrene.

The plain polystyrene coated crystal subjected to 1 mg/mL BSA shows a frequency shift, after rinsing, of around -27 Hz. This translates to 478.3 ng/cm^2 $\pm$ 8.7 ng/cm^2 BSA adsorbed, in good agreement with literature [21].

The polystyrene surface modified with Pluronic F127, presented the expected protein rejecting properties, but the Pluronic coating displayed significant instability on the surface as seen in Figure 1C. In fact, after the buffer rinse, which was done to remove any loosely bound protein, the final frequency change recorded is positive, which indicates an overall reduction of mass. This apparent reduction in mass on the crystals surface, despite BSA addition, can be clearly seen in Figure 1C, as a continuous upward drift, throughout the BSA adsorption stage and the rinse off stage. Jin et al. [42] saw a similar response in their test of Pluronic P-123 modified polystyrene, when subjecting the surface to BSA in QCM-D tests. The group attributed this drift to the release of bound water from the Pluronic brush, as BSA replaces the water molecules [42]. However, we hypothesize that desorption of the Pluronic layer from the polystyrene surface is an additional cause of the continuous upward drift. This drift highlights the issue found in literature regarding the short life-span of this protein resistant coating [14,16], which is why we are also investigating polydopamine assisted coatings.

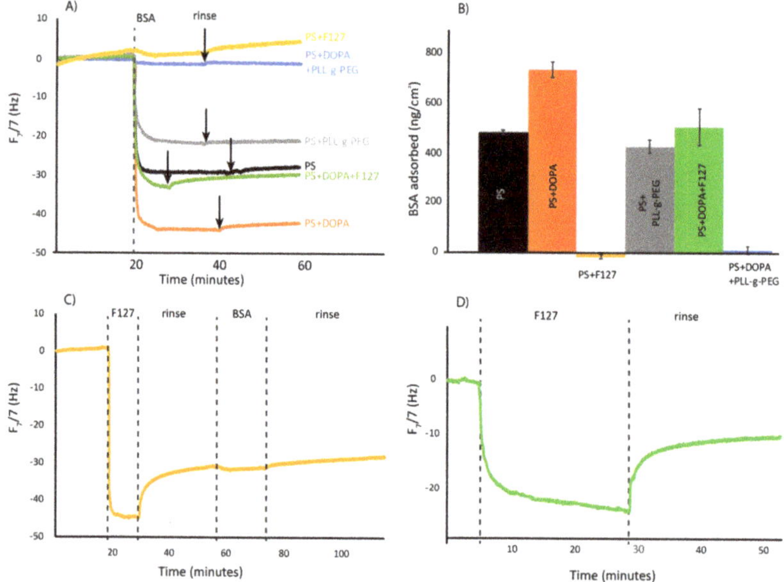

Figure 1. (**A**) Example QCM-D frequency data comparing BSA (bovine serum albumin) adsorption to the different surface chemistries investigated, namely untreated polystyrene (PS) and polystyrene treated with Pluronic F127 (PS + F127), polydopamine (PS + DOPA), PLL-*g*-PEG (PS + PLL-*g*-PEG) or PS + DOPA additionally treated with PLL-*g*-PEG (PS+DOPA + PLL-*g*-PEG) or Pluronic F127 (PS + DOPA + F127). All data has been shifted in time to start the BSA adsorption at the same time. Start of rinsing with buffer is indicated in the graph with arrows for each curve. (**B**) Average adsorbed BSA to the different surface chemistries (error bars represent standard deviation); (**C**) QCM-D frequency graph showing deposition of F127 Pluronic onto a PS coated QCM-D crystal with subsequent rinse-off and BSA adsorption step; (**D**) F127 Pluronic adsorption onto polydopamine coated PS coated crystal showing a lower adsorption of F127 compared to bare PS.

The dopamine molecule has been shown to adhere to many diverse surfaces including polystyrene. It has been found that the initial driving force for adsorption to the hydrophobic polystyrene surface is the displacement of water, followed by the strengthening of the bond through the functional group interactions [43,44]. The interactions between dopamine and polystyrene are believed to occur due to a combination of π–π overlap between the aromatic rings of polystyrene and the catechol rings, and the cation–π interactions between the protonated amine group of dopamine and the π–face of the styrene rings. These bonding theories imply that the DOPA and polystyrene aromatic rings align parallel to each other [43–47]. As there is debate about how exactly the polymer is formed and adheres to surfaces and secondary coatings [48], we tested if Pluronic F127 could in fact bind to such a polydopamine coating and create a protein rejecting interface. Interestingly, a polystyrene surface coated in polydopamine film results in the largest frequency shift of all tested surfaces, possibly due to less surface induced denaturation of BSA on this hydrophilic surface compared to the hy "glue" between polystyrene and F127, the ex situ coating method first described by Messersmith et al. [22] was employed. In Figure 1A,B it can be seen that BSA addition to such a surface resulted in a large negative frequency shift, corresponding to a mass increase of 501.8 ng/cm^2 $\pm$ 73.5 ng/cm^2. It was hoped that introducing the intermediary polydopamine layer may help the protein resistant Pluronic F127 to better adsorb to the surface. However, in light of the high mass adsorption measured, we suggest that the Pluronic either is adsorbed with its hydrophobic backbone facing the solution or that it rinses-off rapidly. Polydopamine and F127 adsorbtion onto the PS film was also explored in situ in the QCM-D, and compared to the protein resistant polystyrene surface modified only with F127 (Figure 1C,D). The DOPA + F127 experiment carried out in the QCM-D (Figure 1D) showed that only around half as much F127 adsorbed onto the polydopamine layer, indicated by a frequency shift of -22.23 Hz $\pm$ 0.81 Hz, compared to that adsorbing on the polystyrene surface, -42.33 Hz $\pm$ 0.6 Hz. The rinsing step also shows a significant desorption of F127 from both surfaces. Such a weak binding of F127 onto polydopamine, suggests that any interactions between the polymer and the substrate, are much weaker than the interactions between the hydrophilic PEO blocks and the solvent, leading in turn to BSA adsorbing to the surface almost to the same level as to the polydopamine coating itself.

Upon failing to produce better anti-fouling results than by simply adsorbing F127 onto polystyrene, PLL-*g*-PEG was trialed, to attach the same protein resisting PEG group to the surface.

PLL-*g*-PEG has been extensively examined as an anti-fouling surface modifier on negatively charged surfaces [17,18]. Accordingly, we had reason to believe it could not attach to polystyrene without assistance. However, due to a publication citing reasonable protein resistance on polystyrene, and for comparative purposes, 0.1 mg/mL PLL-*g*-PEG was adsorbed onto the plain polystyrene surface. Figure 1 shows that upon BSA addition, a frequency drop of -23.9 Hz $\pm$ 1.5 Hz was observed, corresponding to a mass uptake of 423.2 ng/cm^2 $\pm$ 26.4 ng/cm^2. This surface modification provides near 12% protein resistance when compared to the plain polystyrene surface, and statistical analysis proved the difference to be significant ($p < 0.05$), which could explain why one group suggested the modification improved protein resistance. It is worth noting that the apparent reduction in protein adsorption could simply be due to weakly bound PLL-*g*-PEG molecules being replaced by BSA during the protein adsorption phase.

To mediate PLL-g-PEG binding to PS, a polydopamine intermediary layer was formed on the polystyrene coated crystals, upon which PLL-*g*-PEG was allowed to bind. From Figure 1B, a frequency shift of -0.8 Hz and mass of 13.8 ng/cm^2 $\pm$ 13.5 ng/cm^2 were recorded for the subsequent BSA adsorption step. These results prove the PS + DOPA + PLL-*g*-PEG modified surface was as resistant to BSA adsorption, as the well documented PS + F127 coating ($p > 0.05$), and significantly more resistant than all other trialed surfaces ($p < 0.05$). Therefore, this coating can provide a plain polystyrene surface with improved protein resistance of around 97%.

Many reports in literature have shown that thiol and amine terminated functionalized polymers could covalently bond to the dopamine molecule, and several papers also specifically used the polydopamine layer for this bonding process [8,46,49]. Messersmith et al. suggested a possible polymerisation mechanism for dopamine [22], which leads us to believe that at mildly alkaline pH the PLL-*g*-PEG ad-layer could be covalently grafted to the dopamine molecules of the polydopamine film through Michael or Schiff-base reactions. Thus, several different reaction products could be present in the final coating. The extremely low protein adsorption observed in QCM-D experiments proves that the polydopamine coating arranged on polystyrene enables the formation of a dense PLL-*g*-PEG ad-layer with PEG chains extending into the solution. It is worth noting however, that reports of PLL-*g*-PEG on negatively charged surfaces have been stated to produce protein resistance below the detection limit of the OWLS technique (<2 ng/cm^2), and while our measure is from a wet-mass technique, adjusting for the possible coupled water, does not achieve results quite as successful. Therefore, it is likely that the PLL-*g*-PEG layer formed on polydopamine is not as ordered as on negatively charged surfaces. Tsai et al. [31] published a report in 2011, in which it was stated, albeit without any supporting data, that the one-step dip coating method they had trialed, by coating a surface in a mixture of polydopamine and PLL-*g*-PEG, did not provide satisfactory cell resistant results. We show with our QCM-D results that, by first coating the surface in polydopamine and, in a second step, covalently grafting the PLL-*g*-PEG, we are able to achieve very low protein adsorption.

Finally, we investigated whether the PS + DOPA + PLL-*g*-PEG surface could provide better long-term protein resistance than the F127 modified polystyrene surface. To test the longevity of both coatings, the crystals were submerged in buffer for 24 h prior to QCM-D measurements. The response of each soaked surface to protein exposure, along with the original QCM-D data of the successful surfaces prior to soaking, is displayed in Figure 2.

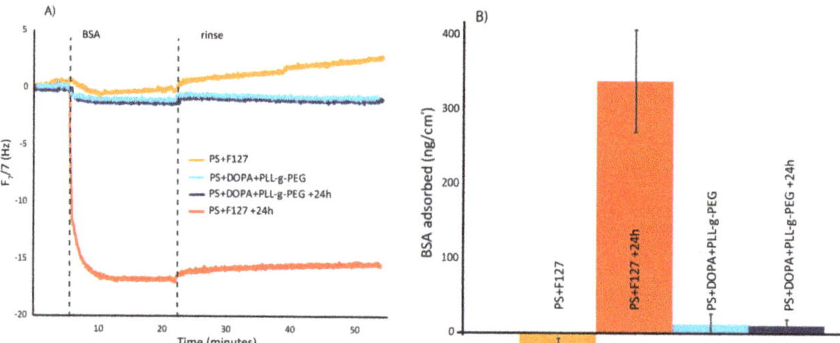

Figure 2. Testing the modified surface's response to BSA with or without pre-soaking in buffer for 24 h. Introduction of BSA and start of rinse is indicated by the dotted lines in (**A**). Average and standard deviation of adsorbed BSA mass (calculated using the Sauerbrey equation) presented in (**B**).

It is clear that F127 is no longer repelling protein after a 24 h soak, while the PLL-*g*-PEG coated polydopamine remains protein rejecting. The trend of F127 to be unstable at the surface was also evident in Figure 1, where the frequency continued to drift upward throughout the measurement. These problems have also been described in literature by several groups. For example, Marsh et al. [14] reported XPS analysis which proved that after 24 h in PBS, the Pluronic coating had washed off, and Su et al. [16] showed that after immersion in water for one week, the polystyrene originally coated in F127 had near plain polystyrene contact angles [26,28].

The frequency shift recorded for the 24 h buffer-soaked Pluronic surface, subject to BSA, was -19.2 Hz, corresponding to 340 ng/cm^2 $\pm$ 68.9 ng/cm^2. This is a significant difference to the coat spared the 24 h treatment ($p < 0.05$). Additionally, it can be seen that the buffer rinse after BSA adsorption does not cause the 24 h soaked F127 (PS+F127 +24 h in Figure 2A) surface to drift as dramatically, as it does the un-soaked coating (PS+F127 in Figure 2A), indicating that the pre-soaked surface has lost much of the anti-fouling Pluronic already. The DOPA + PLL-*g*-PEG coating however, showed no significant difference after 24 h of buffer soak ($p > 0.05$), as is clearly depicted in Figure 2. This shows that the polydopamine assisted grafting of PLL-*g*-PEG will provide the desired protein resistance for a longer time and is thereby a possible solution to overcome the life-span problems encountered by the physisorbed F127 modified surfaces. Judging by the life-span reported in literature for PLL-*g*-PEG on negatively charged surfaces, this coating could last much longer than the 24 h tested. One report shows that flowing HEPES buffer for one week, resulted in a loss of less than 5% mass of the polymer, while another group had reported PLL-*g*-PEG was stable and protein-rejecting for at least three weeks on surface-modified polystyrene microspheres [50].

3.2. Sample Characterisation

XPS allowed us to determine the chemical composition of the modified surfaces and clearly showed the expected increase in both oxygen and nitrogen after modification of the PS surface with polydopamine, in good agreement with literature, atomic percentages from quantified XPS survey spectra are presented in in Table 1 [22,51]. The presence of PLL-*g*-PEG for the polydopamine assisted graft-PLL-*g*-PEG sample is confirmed by a further increase in atomic oxygen and nitrogen percentage, which can be attributed to the protein resisting PEG chains, and the PLL-backbone, respectively [17,18,20,52]. The unmodified polystyrene coating showed unexpected oxygen and nitrogen signals, most likely due to contamination of the PS surface before the XPS was acquired (proteins or microbial contamination suspected as the sample was stored in an aqueous environment), or possibly due to oxidation. It is worth pointing out that while polystyrene is hydrophobic, the binding mechanism of polydopamine to PS may not be the same to other hydrophobic surfaces, and the findings in this study may not apply generically to other hydrophobic materials.

Table 1. The elemental ratios present in the build-up of the PS + DOPA + PLL-*g*-PEG surface.

Surface	XPS Atomic Percent (%)					
	C(1s)	O(1s)	N(1s)	Si(2p)	Na(1s)	S(2p)
PS	94.5 ± 0.5	4.3 ± 0.3	0.8 ± 0.1	0.3 ± 0.1	–	0.1 ± 0.1
PS + DOPA	79.1 ± 0.5	15.3 ± 0.4	5.6 ± 0.1	–	–	–
PS + DOPA + PLL-*g*-PEG	73.8 ± 0.4	19.3 ± 0.3	6.4 ± 0.2	0.3 *	0.2 ± 0.0	–

* This silicon peak was only visible in one of the three scanned spots.

All surface coatings were characterized by recording the contact angles in contact with water. Figure 3, shows clearly that all surfaces containing polydopamine have a lower contact angle, representing a more "wettable" surface. More specifically, contact angle analysis shows the PS + DOPA surface to be more hydrophilic than all other coatings, bar one: PS + DOPA + PLL-*g*-PEG. This coating is the most hydrophilic surface, which helps to explain the significant protein resistance of this surface. Surprisingly, the other successful anti-fouling surface, PS + F127, does not produce the same results in this case. Values found in literature for PS + F127 however show much lower contact angles, of around 64 degrees [16]. The discrepancy may relate to rinsing protocols as F127 is not stably adsorbed on the PS surface. The PS + DOPA + F127 surface, which proved not protein resistant in QCM-D testing, is significantly more hydrophilic than PS + F127 ($p < 0.05$) most likely due to the weakly bound F127 washing off. Such wash-off may also be inferred from AFM images of the surfaces (see Figure S1 and Table S1 in Supplementary Materials). AFM also confirmed a uniform polydopamine coating on polystyrene leading to a modestly increased surface roughness, in agreement with literature [51]. The short-term successful anti-fouling PS + F127 coat appeared largely homogeneous while the dopamine assisted F127 coating, showed large areas of low-lying film which measure around 10 nm deeper than the surrounding surface. These areas could be representative of washed off F127, leaving only the polydopamine coating below. Overall, these F127 surfaces appear rougher than those modified with PLL-*g*-PEG.

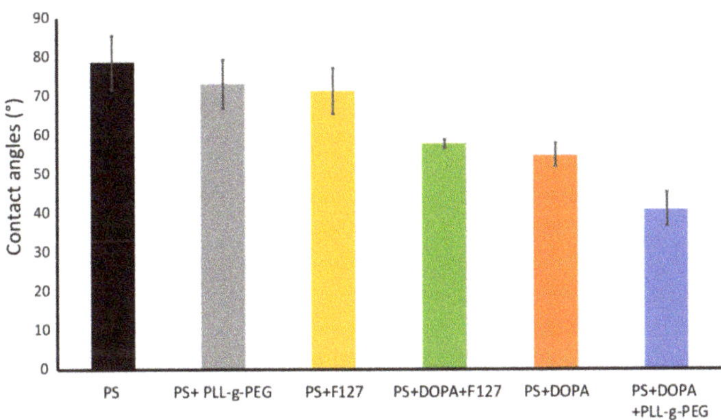

Figure 3. Average contact angles, with standard deviations (error bars), for all surface coatings investigated.

Taken together, these surface characteristics support the conclusions drawn from the QCM-D analysis, and out of these simple coating methods to modify polystyrene to render it protein rejecting, the polydopamine assisted PLL-*g*-PEG modified surface holds particular promise. Future work evaluating this coating with other proteins and in cell culture is needed to fully confirm the long-term stability of the coating.

4. Conclusions

The main conclusions drawn from the present work are:

- F127 significantly reduces BSA adsorption in short-term studies. However, after soaking only 24 h in buffer, the resistance is significantly diminished.
- The polydopamine intermediary film produced, cannot provide the necessary attraction to bind F127 to the surface. Instead, bonds between the Pluronic and the surface are weakened, causing more protein adsorption than on unmodified polystyrene.
- By adsorbing the same "assisting" polydopamine film and subsequently grafting PLL-*g*-PEG, we can successfully prevent 97% of protein adsorption on a polystyrene substrate.
- This modified surface is equally as protein-rejecting after 24 h in buffer, and thus a promising simple coating for long term protein rejection of polystyrene.

Supplementary Materials: The following are available online at www.mdpi.com/2079-6412/8/2/55/s1, Figure S1: AFM tapping mode images; Table S1: Roughness recorded for one sample of each surface.

Acknowledgments: This work is supported by Supported by the Marsden Fund Council and the Rutherford Discovery Fellowship from Government funding, managed by Royal Society Te Apārangi and The MacDiarmid Institute for Advanced Materials and Nanotechnology. Yiran An is acknowledged for his assistance with AFM imaging and Colin Doyle for help with XPS acquisition.

Author Contributions: Jenny Malmström conceived and designed the experiments. Marcelle Hecker and Matthew Sheng Hao Ting performed the experiments, Marcelle Hecker and Jenny Malmström analyzed the data; Marcelle Hecker and Jenny Malmström wrote the paper. All authors have read and approved the manuscript.

Conflicts of Interest: The authors declare no conflict of interest.

References

1. Hatakeyama, E.S.; Ju, H.; Gabriel, C.J.; Lohr, J.L.; Bara, J.E.; Noble, R.D.; Freeman, B.D.; Gin, D.L. New protein-resistant coatings for water filtration membranes based on quaternary ammonium and phosphonium polymers. *J. Membr. Sci.* **2009**, *330*, 104–116. [CrossRef]

2. Lowe, S.; O'Brien-Simpson, N.M.; Connal, L.A. Antibiofouling polymer interfaces: Poly(ethylene glycol) and other promising candidates. *Polym. Chem.* **2015**, *6*, 198–212. [CrossRef]

3. Poncin-Epaillard, F.; Vrlinic, T.; Debarnot, D.; Mozetic, M.; Coudreuse, A.; Legeay, G.; El Moualij, B.; Zorzi, W. Surface treatment of polymeric materials controlling the adhesion of biomolecules. *J. Funct. Biomater.* **2012**, *3*, 528–543. [CrossRef] [PubMed]

4. Malmsten, M. Ellipsometry studies of protein layers adsorbed at hydrophobic surfaces. *J. Colloid Interface Sci.* **1994**, *166*, 333–342. [CrossRef]

5. Roach, P.; Farrar, D.; Perry, C.C. Interpretation of protein adsorption: Surface-induced conformational changes. *J. Am. Chem. Soc.* **2005**, *127*, 8168–8173. [CrossRef] [PubMed]

6. Wertz, C.F.; Santore, M.M. Effect of surface hydrophobicity on adsorption and relaxation kinetics of albumin and fibrinogen: Single-species and competitive behavior. *Langmuir* **2001**, *17*, 3006–3016. [CrossRef]

7. Chou, Y.N.; Wen, T.C.; Chang, Y. Zwitterionic surface grafting of epoxylated sulfobetaine copolymers for the development of stealth biomaterial interfaces. *Acta Biomater.* **2016**, *40*, 78–91. [CrossRef] [PubMed]

8. Sundaram, H.S.; Han, X.; Nowinski, A.K.; Ella-Menye, J.R.; Wimbish, C.; Marek, P.; Senecal, K.; Jiang, S. One-step dip coating of zwitterionic sulfobetaine polymers on hydrophobic and hydrophilic surfaces. *ACS Appl. Mater. Interfaces* **2014**, *6*, 6664–6671. [CrossRef] [PubMed]

9. Raynor, J.E.; Capadona, J.R.; Collard, D.M.; Petrie, T.A.; García, A.J. Polymer brushes and self-assembled monolayers: Versatile platforms to control cell adhesion to biomaterials Review. *Biointerphases* **2009**, *4*, FA3–FA16. [CrossRef] [PubMed]

10. Brittain, W.J.; Minko, S. A structural definition of polymer brushes. *J. Polym. Sci. Part A Polym. Chem.* **2007**, *45*, 3505–3512. [CrossRef]

11. Bridgett, M.J.; Davies, M.C.; Denyer, S.P. Control of staphylococcal adhesion to polystyrene surfaces by polymer surface modification with surfactants. *Biomaterials* **1992**, *13*, 411–416. [CrossRef]

12. Green, R.J.; Tasker, S.; Davies, J.; Davies, M.C.; Roberts, C.J.; Tendler, S.J.B. Adsorption of PEO-PPO-PEO triblock copolymers at the solid/liquid interface: A surface plasmon resonance study. *Langmuir* **1997**, *13*, 6510–6515. [CrossRef]

13. Song, X.; Zhao, S.; Fang, S.; Ma, Y.; Duan, M. Mesoscopic simulations of adsorption and association of PEO-PPO-PEO triblock copolymers on a hydrophobic surface: From mushroom hemisphere to rectangle brush. *Langmuir* **2016**, *32*, 11375–11385. [CrossRef] [PubMed]

14. Marsh, L.H.; Coke, M.; Dettmar, P.W.; Ewen, R.J.; Havler, M.; Nevell, T.G.; Smart, J.D.; Smith, J.R.; Timmins, B.; Tsibouklis, J.; et al. Adsorbed poly(ethyleneoxide)-poly(propyleneoxide) copolymers on synthetic surfaces: Spectroscopy and microscopy of polymer structures and effects on adhesion of skin-borne bacteria. *J. Biomed. Mater. Res.* **2002**, *61*, 641–652. [CrossRef] [PubMed]

15. Zheng, J.; Song, W.; Huang, H.; Chen, H. Protein adsorption and cell adhesion on polyurethane/Pluronic® surface with lotus leaf-like topography. *Colloids Surf. B Biointerfaces* **2010**, *77*, 234–239. [CrossRef] [PubMed]

16. Su, Y.; Wang, L.; Yang, X. A simple swelling and anchoring method for preparing dense and stable poly(ethylene oxide) layers on polystyrene surfaces. *Appl. Surf. Sci.* **2008**, *254*, 4606–4610. [CrossRef]

17. Huang, N.P.; Michel, R.; Voros, J.; Textor, M.; Hofer, R.; Rossi, A.; Elbert, D.L.; Hubbell, J.A.; Spencer, N.D. Poly(L-lysine)-g-poly(ethylene glycol) layers on metal oxide surfaces: Surface-analytical characterization and resistance to serum and fibrinogen adsorption. *Langmuir* **2001**, *17*, 489–498. [CrossRef]

18. Pasche, S.; De Paul, S.M.; Vörös, J.; Spencer, N.D.; Textor, M. Poly(L-lysine)-graft-poly(ethylene glycol) assembled monolayers on niobium oxide surfaces: A quantitative study of the influence of polymer interfacial architecture on resistance to protein adsorption by ToF-SIMS and in situ OWLS. *Langmuir* **2003**, *19*, 9216–9225. [CrossRef]

19. Malmström, J.; Agheli, H.; Kingshott, P.; Sutherland, D.S. Viscoelastic modeling of highly hydrated laminin layers at homogeneous and nanostructured surfaces: Quantification of protein layer properties using QCM-D and SPR. *Langmuir* **2007**, *23*, 9760–9768. [CrossRef] [PubMed]

20. Nonckreman, C.J.; Fleith, S.; Rouxhet, P.G.; Dupont-Gillain, C.C. Competitive adsorption of fibrinogen and albumin and blood platelet adhesion on surfaces modified with nanoparticles and/or PEO. *Colloids Surf. B Biointerfaces* **2010**, *77*, 139–149. [CrossRef] [PubMed]

21. Reimhult, K.; Petersson, K.; Krozer, A. QCM-D analysis of the performance of blocking agents on gold and polystyrene surfaces. *Langmuir* **2008**, *24*, 8695–8700. [CrossRef] [PubMed]

22. Lee, H.; Dellatore, S.M.; Miller, W.M.; Messersmith, P.B. Mussel-inspired surface chemistry for multifunctional coatings. *Science* **2007**, *318*, 426–430. [CrossRef] [PubMed]

23. Dalsin, J.L.; Hu, B.H.; Lee, B.P.; Messersmith, P.B. Mussel adhesive protein mimetic polymers for the preparation of nonfouling surfaces. *J. Am. Chem. Soc.* **2003**, *125*, 4253–4258. [CrossRef] [PubMed]

24. Barclay, T.G.; Hegab, H.M.; Clarke, S.R.; Ginic-Markovic, M. Versatile surface modification using polydopamine and related polycatecholamines: Chemistry, structure, and applications. *Adv. Mater. Interfaces* **2017**, *4*, 1601192. [CrossRef]

25. Liu, M.; Zeng, G.; Wang, K.; Wan, Q.; Tao, L.; Zhang, X.; Wei, Y. Recent developments in polydopamine: An emerging soft matter for surface modification and biomedical applications. *Nanoscale* **2016**, *8*, 16819–16840. [CrossRef] [PubMed]

26. Zhang, X.; Huang, Q.; Deng, F.; Huang, H.; Wan, Q.; Liu, M.; Wei, Y. Mussel-inspired fabrication of functional materials and their environmental applications: Progress and prospects. *Appl. Mater. Today* **2017**, *7*, 222–238. [CrossRef]

27. Chien, C.Y.; Tsai, W.B. Poly(dopamine)-assisted immobilization of Arg-Gly-Asp peptides, hydroxyapatite, and bone morphogenic protein-2 on titanium to improve the osteogenesis of bone marrow stem cells. *ACS Appl. Mater. Interfaces* **2013**, *5*, 6975–6983. [CrossRef] [PubMed]

28. Li, W.; Zheng, Y.; Zhao, X.; Ge, Y.; Chen, T.; Liu, Y.; Zhou, Y. Osteoinductive effects of free and immobilized bone forming peptide-1 on human adipose-derived stem cells. *PLoS ONE* **2016**, *11*, e0150294. [CrossRef] [PubMed]

29. Luo, R.; Wang, X.; Deng, J.; Zhang, H.; Maitz, M.F.; Yang, L.; Wang, J.; Huang, N.; Wang, Y. Dopamine-assisted deposition of poly (ethylene imine) for efficient heparinization. *Colloids Surf. B Biointerfaces* **2016**, *144*, 90–98. [CrossRef] [PubMed]

30. Pan, C.; Chen, L.; Liu, S.; Zhang, Y.; Zhang, C.; Zhu, H.; Wang, Y. Dopamine-assisted immobilization of partially hydrolyzed poly(2-methyl-2-oxazoline) for antifouling and biocompatible coating. *J. Mater. Sci.* **2016**, *51*, 2427–2442. [CrossRef]

31. Tsai, W.B.; Chien, C.Y.; Thissen, H.; Lai, J.Y. Dopamine-assisted immobilization of poly(ethylene imine) based polymers for control of cell-surface interactions. *Acta Biomater.* **2011**, *7*, 2518–2525. [CrossRef] [PubMed]

32. Yuan, S.; Li, Z.; Zhao, J.; Luan, S.; Ma, J.; Song, L.; Shi, H.; Jin, J.; Yin, J. Enhanced biocompatibility of biostable poly(styrene-b-isobutylene-b-styrene) elastomer via poly(dopamine)-assisted chitosan/hyaluronic acid immobilization. *RSC Adv.* **2014**, *4*, 31481–31488. [CrossRef]

33. Gao, C.; Li, G.; Xue, H.; Yang, W.; Zhang, F.; Jiang, S. Functionalizable and ultra-low fouling zwitterionic surfaces via adhesive mussel mimetic linkages. *Biomaterials* **2010**, *31*, 1486–1492. [CrossRef] [PubMed]

34. Kuang, J.; Messersmith, P.B. Universal surface-initiated polymerization of antifouling zwitterionic brushes using a mussel-mimetic peptide initiator. *Langmuir* **2012**, *28*, 7258–7266. [CrossRef] [PubMed]

35. Wu, J.; Zhang, L.; Wang, Y.; Long, Y.; Gao, H.; Zhang, X.; Zhao, N.; Cai, Y.; Xu, J. Mussel-inspired chemistry for robust and surface-modifiable multilayer films. *Langmuir* **2011**, *27*, 13684–13691. [CrossRef] [PubMed]

36. Kord Forooshani, P.; Lee, B.P. Recent approaches in designing bioadhesive materials inspired by mussel adhesive protein. *J. Polym. Sci. Part A Polym. Chem.* **2017**, *55*, 9–33. [CrossRef] [PubMed]

37. Sundaram, H.S.; Han, X.; Nowinski, A.K.; Brault, N.D.; Li, Y.; Ella-Menye, J.R.; Amoaka, K.A.; Cook, K.E.; Marek, P.; Senecal, K.; et al. Achieving one-step surface coating of highly hydrophilic poly(carboxybetaine methacrylate) polymers on hydrophobic and hydrophilic surfaces. *Adv. Mater. Interfaces* **2014**, *1*, 1400071. [CrossRef] [PubMed]

38. Höök, F.; Kasemo, B.; Nylander, T.; Fant, C.; Sott, K.; Elwing, H. Variations in coupled water, viscoelastic properties, and film thickness of a Mefp-1 protein film during adsorption and cross-linking: A quartz crystal microbalance with dissipation monitoring, ellipsometry, and surface plasmon resonance study. *Anal. Chem.* **2001**, *73*, 5796–5804. [CrossRef] [PubMed]

39. Höök, F.; Vörös, J.; Rodahl, M.; Kurrat, R.; Böni, P.; Ramsden, J.J.; Textor, M.; Spencer, N.D.; Tengvall, P.; Gold, J.; et al. A comparative study of protein adsorption on titanium oxide surfaces using in situ ellipsometry, optical waveguide lightmode spectroscopy, and quartz crystal microbalance/dissipation. *Colloids Surf. B Biointerfaces* **2002**, *24*, 155–170. [CrossRef]

40. Höök, F.; Rodahl, M.; Brzezinski, P.; Kasemo, B. Energy dissipation kinetics for protein and antibody-antigen adsorption under shear oscillation on a quartz crystal microbalance. *Langmuir* **1998**, *14*, 729–734. [CrossRef]

41. Sauerbrey, G. Verwendung Von Schwingquarzen Zur Wagung Dunner Schichten Und Zur Mikrowagung. *Zeitschrift Fur Physik* **1959**, *155*, 206–222. (In German) [CrossRef]

42. Jin, J.; Huang, F.; Hu, Y.; Jiang, W.; Ji, X.; Liang, H.; Yin, J. Immobilizing PEO-PPO-PEO triblock copolymers on hydrophobic surfaces and its effect on protein and platelet: A combined study using QCM-D and DPI. *Colloids Surf. B Biointerfaces* **2014**, *123*, 892–899. [CrossRef] [PubMed]

43. Baty, A.M.; Leavitt, P.K.; Siedlecki, C.A.; Tyler, B.J.; Suci, P.A.; Marchant, R.E.; Geesey, G.G. Adsorption of adhesive proteins from the marine mussel, Mytilus edulis, on polymer films in the hydrated state using angle dependent X-ray photoelectron spectroscopy and atomic force microscopy. *Langmuir* **1997**, *13*, 5702–5710. [CrossRef]

44. Baty, A.M.; Suci, P.A.; Tyler, B.J.; Geesey, G.G. Investigation of mussel adhesive protein adsorption on polystyrene and poly(octadecyl methacrylate) using angle dependent XPS, ATR-FTIR, and AFM. *J. Colloid Interface Sci.* **1996**, *177*, 307–315. [CrossRef]

45. Lee, H.; Scherer, N.F.; Messersmith, P.B. Single-molecule mechanics of mussel adhesion. *Proc. Natl. Acad. Sci. USA* **2006**, *103*, 12999–13003. [CrossRef] [PubMed]

46. Moulay, S. Dopa/catechol-tethered polymers: Dioadhesives and biomimetic adhesive materials. *Polym. Rev.* **2014**, *54*, 436–513. [CrossRef]

47. Jiang, J.; Zhu, L.; Zhu, L.; Zhu, B.; Xu, Y. Surface characteristics of a self-polymerized dopamine coating deposited on hydrophobic polymer films. *Langmuir* **2011**, *27*, 14180–14187. [CrossRef] [PubMed]

48. Liebscher, J.; Mrówczyński, R.; Scheidt, H.A.; Filip, C.; Hadade, N.D.; Turcu, R.; Bende, A.; Beck, S. Structure of polydopamine: A never-ending story? *Langmuir* **2013**, *29*, 10539–10548. [CrossRef] [PubMed]

49. Mian, S.A.; Yang, L.M.; Saha, L.C.; Ahmed, E.; Ajmal, M.; Ganz, E. A fundamental understanding of catechol and water adsorption on a hydrophilic silica surface: Exploring the underwater adhesion mechanism of mussels on an atomic scale. *Langmuir* **2014**, *30*, 6906–6914. [CrossRef] [PubMed]

50. Wattendorf, U.; Kreft, O.; Textor, M.; Sukhorukov, G.B.; Merkle, H.P. Stable stealth function for hollow polyelectrolyte microcapsules through a poly(ethylene glycol) grafted polyelectrolyte adlayer. *Biomacromolecules* **2008**, *9*, 100–108. [CrossRef] [PubMed]
51. Zangmeister, R.A.; Morris, T.A.; Tarlov, M.J. Characterization of polydopamine thin films deposited at short times by autoxidation of dopamine. *Langmuir* **2013**, *29*, 8619–8628. [CrossRef] [PubMed]
52. Heuberger, R.; Sukhorukov, G.; Vörös, J.; Textor, M.; Möhwald, H. Biofunctional polyelectrolyte multilayers and microcapsules: Control of non-specific and bio-specific protein adsorption. *Adv. Funct. Mater.* **2005**, *15*, 357–366. [CrossRef]

Article

Metakaolin-Based Geopolymer with Added TiO$_2$ Particles: Physicomechanical Characteristics

Luis A. Guzmán-Aponte [1], Ruby Mejía de Gutiérrez [1,*] and Anibal Maury-Ramírez [2]

[1]   Composites Materials Group (CENM), School of Materials Engineering Universidad del Valle, Cali 76001, Colombia; luis.guzman@correounivalle.edu.co

[2]   Department of Civil and Industrial Engineering, Pontificia Universidad Javeriana, Cali 760031, Colombia; anibal.maury@javerianacali.edu.co

*   Correspondence: ruby.mejia@correounivalle.edu.co; Tel.: +57-2-3302436

Academic Editor: Silvia Gross

Received: 26 October 2017; Accepted: 12 December 2017; Published: 15 December 2017

Abstract: The effect of the TiO$_2$ addition on the physicomechanical properties of a geopolymer system based on metakaolin (MK) and hydroxide and potassium silicate as activators is presented in this article. Three different liquid-solid systems (0.35, 0.40, and 0.45) and two titanium additions were investigated (5% and 10% of the cement content). The flowability, setting time, and mechanical strength of the geopolymer mixtures and their microstructural characteristics were evaluated using techniques such as X-ray diffraction (XRD), Fourier transform infrared spectroscopy (FTIR), and scanning electron microscopy (SEM). It was concluded that a percentage of up to 10% TiO$_2$ does not affect the mechanical properties of the geopolymer, although it does reduce the fluidity and setting times of the mixture.

Keywords: geopolymer; metakaolin; titanium oxide; physical and mechanical properties

---

1. Introduction

Population growth and the consequent demand of more infrastructure have increased the consumption of ordinary Portland cement (OPC), which is the main cementitious material in concrete technology. In spite of the benefits of concrete, there are several factors, such as high energy consumption, high CO$_2$ emissions, and limited limestone reserves that have led to the reconsideration of the use of this material [1]. Owing to the above, measures have been taken both to improve the production processes of Portland cement, via the use of alternative fuels and different sources of raw materials to develop new cementitious materials that equal or exceed the properties of Portland cement and, in turn, exhibit better environmental performance. Geopolymers have been considered as an alternative at a global level; they are considered to be the third generation of cement, after lime and OPC. It is estimated that, depending of the precursors and activators, the production of geopolymers results in ~70% less greenhouse gas emissions than the production of cement, which makes some geopolymers environmentally friendly [2–4].

Geopolymers are obtained from the optimum blend of a material, mineral or industrial by-product based on SiO$_2$ and Al$_2$O$_3$ (precursors) with a chemical agent (alkaline activator), which, through a series of reactions at low temperature (<100 °C), leads to the formation of a product with cementitious characteristics [5]. The main criterion for the stable development of a geopolymer is that the source of the aluminosilicate material is highly amorphous and contains sufficient reactive glass content. The main alkaline activators used are sodium hydroxide (NaOH), potassium hydroxide (KOH), sodium silicate (Na$_2$SiO$_3$), and potassium silicate (K$_2$SiO$_3$). Compared with NaOH, KOH exhibits a higher degree of alkalinity; however, it has been reported that NaOH has a greater ability to separate the monomers of silicates and aluminates in the phase of dissolution of the precursor [5]. Geopolymer cements are considered materials that can replace Portland cement owing to properties such as high resistance at early ages, resistance to chemical attack, low thermal and acoustic conductivity, and high temperature and fire resistance; these properties, depending of the type of raw materials and mix formulation of the geopolymer, can be similar, or even superior, to those of traditional Portland cement.

As in the case of Portland cement, the addition of nanoparticles to geopolymers has focused on improving the mechanical properties. For example, Assaedi et al. added SiO_2 nanoparticles in a proportion of up to 3 wt % to a fly ash-based geopolymer, finding that the addition of these materials improved the mechanical performance of the material by 27%, while decreasing the porosity and water absorption [6]. Riahi and Nazari synthesized geopolymers from rice husk ash and fly ash, to which SiO_2 and Al_2O_3 nanoparticles were added; they also obtained an improvement in the mechanical properties of the material [7]. However, the best performance was for the samples with added nano-silica, owing to their greater amorphous character, unlike the alumina nanoparticles, which, because they were crystalline, did not participate in the process of geopolymerization [7]. Recently, Duan et al. developed geopolymers from fly ash, focusing on the mechanical performance and durability to carbonation when nano-titanium oxide (TiO_2) particles were added; the authors reported that the addition of these up to 5 wt % improves the mechanical properties and durability and attributed this behaviour to pore refinement and densification of the geopolymer microstructure [8]. Similarly, Yang et al. studied the mechanical and physical behaviour of an alkaline activated slag to which TiO_2 nanoparticles were added at 0.5 wt % and reported a decrease in porosity and material shrinkage; in addition, the material was densified and had a greater resistance to compression [9]. On the other hand, Zhang et al. evaluated the effect of adding titanium dioxide and hollow glass microspheres to a geopolymer coating from metakaolin (MK) on the geopolymer's optical and thermal properties [10]. The study found that by adding hollow glass microspheres at 6 wt % and titanium dioxide based on the cementant at 12 wt %, increases of 12% and 90% of the thermal insulation and reflectivity, respectively. The authors attributed this behaviour to the pigment character of the titanium dioxide, for which the high coefficient of refraction enabled it to reflect wavelengths close to infrared and visible light [10].

In other studies that were not particularly focused on geopolymer materials, it has been found that nanoparticles of TiO_2, because they are a semiconductor substance, are able to confer photocatalytic properties to the system to which they are added [11], and this results in self-cleaning and air-purifying properties, which is highly desirable for building materials.

Based on the above, the present article aims to evaluate the effect of the addition of TiO_2 particles on the physical and mechanical behaviour of a geopolymer based on MK as a precursor. The variables considered are the liquid/solid ratio (L/S) of the system and the percentage by weight of TiO_2 added. Properties in the fresh state, such as the setting time and flowability, and properties of the hardened state, such as the compressive strength, density, absorption, and porosity, are studied. The study is complemented with a microstructural analysis of the geopolymers using the X-ray diffraction (XRD), Fourier transform infrared spectroscopy (FTIR), and scanning electron microscopy (SEM) techniques. The contribution of the photocatalytic properties of TiO_2 to the geopolymer, specifically its self-cleaning and algaecide properties, will be the object of future studies.

2. Materials and Experimental Methodology

2.1. Materials

A high-purity commercial MK (Metamax BASF, Florham Park, NJ, USA) was used as the primary source or precursor in the production of the geopolymer material, whose chemical composition is presented in Table 1. The chemical composition was determined via X-ray fluorescence (XRF) using a Phillips MagiX-Pro PW 2440 spectrometer (PANalytical, Tollerton, UK) equipped with a Rhodium tube and a maximum power of 4 kW. The SiO_2/Al_2O_3 molar ratio of the MK was 1.97. High-purity analytical titanium oxide (TiO_2; Merck, reference 1008081000) was used as an additive. Particle size and distribution analysis was performed with a Mastersizer-2000 laser granulometer from Malvern Instruments (Malvern, UK) with a Hydro2000MU dispersion unit; distilled water was used as a dispersant medium. The mean particle sizes D (4,3) of the MK and titanium oxide were 6.57 and 1.59 μm, respectively (Figure 1).

Table 1. Chemical composition of the precursor (MK).

Element/Composite	wt %
SiO_2	52.018
Al_2O_3	44.950
TiO_2	1.730
Fe_2O_3	0.469
Na_2O	0.295
MgO	0.192
K_2O	0.158
P_2O_5	0.058
CaO	0.023
Ce	0.023
V	0.021
S	0.018
Cr	0.014
Zr	0.012
Ga	0.006
Sr	0.005
Zn	0.003
Nb	0.003

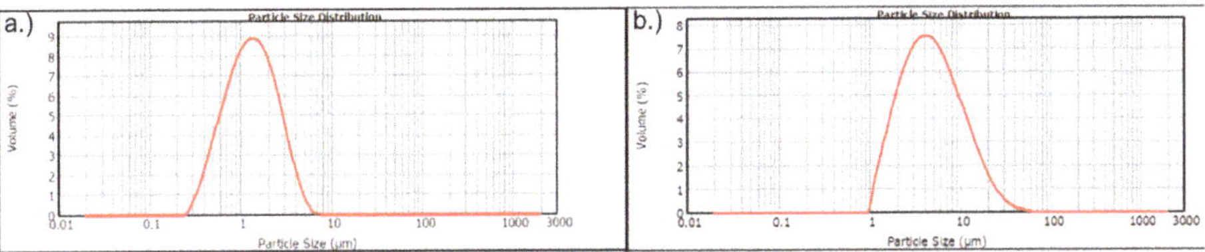

Figure 1. Particle size distribution of (**a**) titanium dioxide, and (**b**) MK.

Figures 2 and 3 show the XRD and FTIR results for MK and TiO_2, respectively. For the XRD analysis, a Bruker diffractometer equipped with a wide-angle goniometer RINT2000 was used, using the Kα1 signal of Cu at 45 kV and 40 mA. A 0.02° pitch was used within a range of 5–70° at a rate of 5°/min. Information processing was performed using the X'pert HighScore Plus software package, version 2.2.5. FTIR spectroscopy was performed using a PerkinElmer Spectrum 100 spectrometer (Perkin Elmer, Shelton, United States) in transmittance mode in the frequency range of 450–4000 cm^{-1}. Samples were prepared using the KBr compressed method. In Figure 2, it can be observed that MK has a high level of amorphicity due to the halo located in the range between 20 and 30° 2θ and presents small traces of a crystalline product identified as anatase (reference pattern 01-078-2486), which corresponds to the observed peaks at approximately 25°, 38°, 48°, 55°, and 63° 2θ. This coincides with the percentage of TiO_2 reported in the chemical composition of MK (Table 1). There is also evidence that the TiO_2 used is in the anatase phase, which is an important phase because of its photocatalytic potential [11]. In the FTIR spectrum (Figure 3) of the MK, a band at approximately 3437 cm^{-1} is observed; this corresponds to the asymmetric vibration of OH–groups. At 1089 cm^{-1}, a Si–O–Al vibrational peak is observed, which corroborates the aluminosilicate character of this raw material. The peaks at 814 and 473 cm^{-1} are attributed to the amorphous Al–O stretching vibration and Si–O–Si flexion, respectively [12]. On the other hand, the FTIR spectrum obtained for TiO_2 indicates the presence of peaks near 3430 and 1632 cm^{-1}, which correspond to vibration via stretching of the –OH bonds and vibration via deformation of the bonds in adsorbed surface water molecules, respectively. The peak located near 687 cm^{-1} is characteristic of the Ti–O–Ti stretching of the anatase phase, thus confirming the presence of these photoactive species [13].

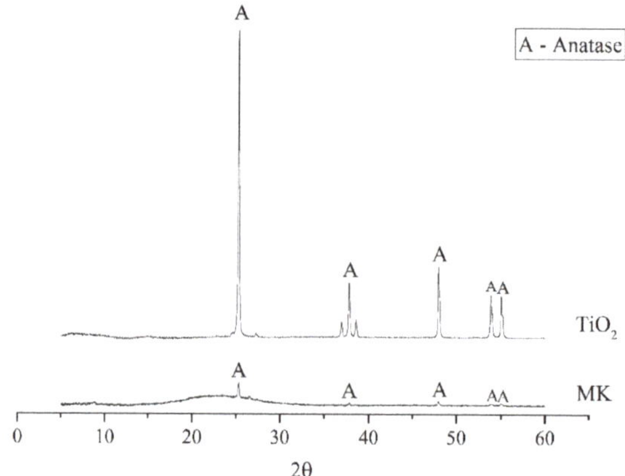

Figure 2. XRD patterns of the raw materials.

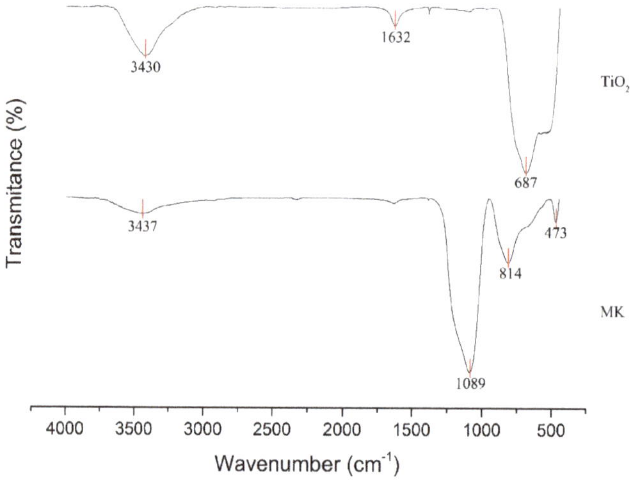

Figure 3. FTIR spectra of the raw materials.

2.2. Preparation of Geopolymeric Mixtures

The geopolymeric materials (GP) were prepared using the molar ratios SiO$_2$/Al$_2$O$_3$ = 2.5 and K$_2$O/SiO$_2$ = 0.28 based on results from previous research [14]. A potassium hydroxide (KOH) analytical-grade reagent and commercial potassium silicate (K$_2$SiO$_3$) distributed by Pan American Chemicals (SiO$_2$ = 26.38%, K$_2$O = 13.06%, H$_2$O = 60.56%) were dosed to obtain the activator solution module (Ms: SiO$_2$/K$_2$O = 0.76) required in the geopolymer mixture. The percentage of addition of TiO$_2$ as a function of the cementant was varied in three levels, 0 wt %, 5 wt % and 10 wt %. The liquid/solid ratio (L/S) was also varied in three levels, 0.35 (dry consistency), 0.40 (mean consistency), and 0.45 (fluid consistency). In total, nine mixtures per system were made, each with two repetitions, for a total of 18 mixtures. Table 2 presents the blend specifications employed in this research.

Table 2. Compositions of the mixtures.

Mixture	TiO$_2$%	L/S Ratio	SiO$_2$/Al$_2$O$_3$	K$_2$O/SiO$_2$
GP	0			
GP 5 Ti	5	0.35, 0.40 and 0.45	2.5	0.28
GP 10 Ti	10	–	–	–

The mixing process of the solid components (MK and titanium dioxide) and liquids (activator) was performed in a HOBART Vulcan 1249 mixer. The solids were homogenized for 15 s, and the activating solution (KOH + K$_2$SiO$_3$) was added and mixed for 3 min at low speed, followed by 2 min at medium speed and, finally, 1 min at low speed until a homogeneous paste was obtained. The geopolymeric paste was then casted into cubic silicone moulds of dimensions 20 mm × 20 mm × 20 mm. The obtained samples were cured at room temperature (25 °C) for 24 h and then demoulded and taken to a humidity chamber (relative humidity >90% and 25 °C) until the test age was reached (seven and 28 days).

2.3. Physical and Mechanical Properties

In fresh conditions, the setting time and fluidity were determined. The setting time was determined using a Vicat needle, following ASTM C191. The fluidity was measured following the procedures described in ASTM C230. A modification to the test was that a cone of smaller volume or minislump with dimensions of 57 mm (high) × 38.20 mm (diameter) was employed on the flow table prior to the 25 strokes standardized in the guidelines. The flow diameters reported were measured using a digital caliper.

In the hardened state, the compressive strength was evaluated at ages of seven and 28 days using an INSTRON 3369 universal test machine with a capacity of 50 kN at a deformation rate of 1 mm/min. In each case, a minimum of three specimens were tested. The density, pore volume, and water absorption capacity of each material were also determined following the procedures detailed in ASTM C642-13. It should be noted that the drying of the sample was performed at 60 °C for 48 h.

2.4. Microstructural Analysis

The following techniques were used for the microstructural study of the blends: Electron microscopy, FTIR, and XRD. A JEOL scanning electron microscope (JSM-6490LV, high vacuum of 3 × 10^{-6} Torr, Peabody, MA, USA) was employed. The equipment has an INCAPentaFETx3 Brand Oxford Instruments model 7573 detector. The samples were metalized with gold in a Denton Vacuum Desk IV tank.

3. Results and Discussion

3.1. Fluidity and Setting Time

Figure 4 shows the diameters measured after subjecting the different systems evaluated to the flow table test. As expected, as the L/S ratio of the system was increased, the fluidity increased, regardless of the TiO$_2$ content. However, in all cases, the samples with no TiO$_2$ addition had a higher fluidity percentage. For a fixed L/S ratio, increased TiO$_2$ content reduced the workability of the samples; using 10 wt % with respect to the cement decreased the fluidity by as much as 30.13%. Duan et al., found similar results when measuring flowability in fly ash geopolymer mortars and reported that the addition of 3% and 5% TiO$_2$ nanoparticles decreased the fluidity of the material by factors of 21.86% and 31.12%, respectively [8].

Figure 5 shows the effect of the TiO$_2$ additions on the setting time. It is evident that both the higher liquid content (L/S ratio) and the addition of TiO$_2$ particles affect the setting time, although their effects have opposite signs. By increasing the percentage of addition up to 10 wt % TiO$_2$ particles, the setting time was decreased by 20.7%, 31.1%, and 27.6% for the dry, medium, and fluid consistency materials, respectively. The reduction of the setting time of the geopolymers with addition of TiO$_2$, at a constant L/S ratio, can be attributed to a nucleation and filling effect that contributes to accelerate the degree of reaction of the geopolymer. It is also evident that, by increasing the mixing water content, the concentration of ions in the aqueous solution decreases, decreasing the gel formation process, in addition to leading to a greater availability of water in the system to be evaporated for the formation of the consolidated structure, which is reflected in an increase in the setting time [15].

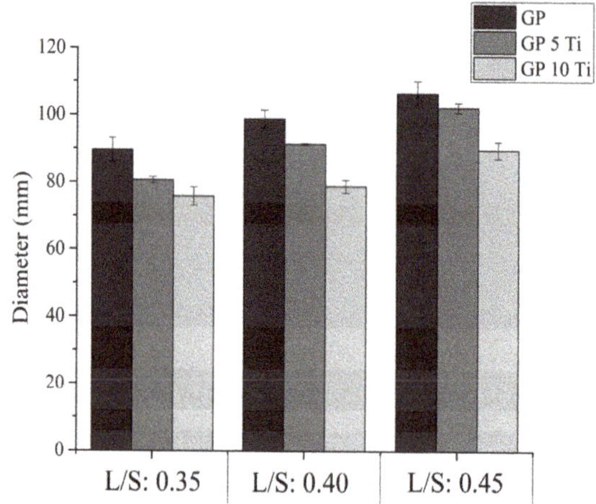

Figure 4. Fluidity of the various systems evaluated.

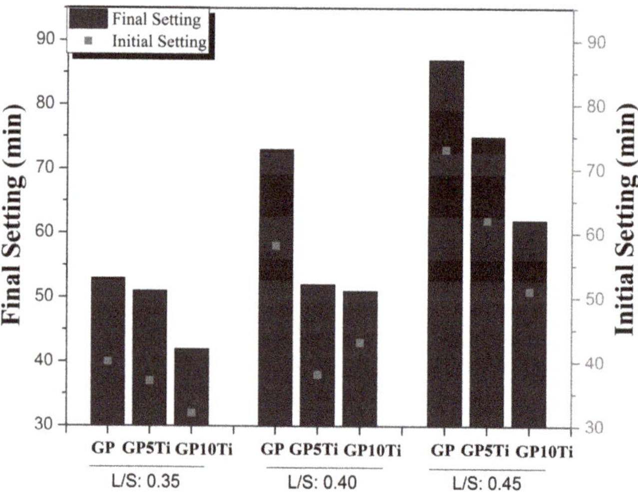

Figure 5. Setting times of the various materials evaluated.

3.2. Physical and Mechanical Properties

The obtained results regarding density, absorption, and porosity for the different mixtures can be observed in Figure 6a,b, and those of mechanical resistance to compression are presented in Figure 7. In general, the density fluctuated between 1230 and 1376 kg/m^3, with no significant dependence on the L/S or the percentage of added TiO$_2$ observed (Figure 6a). The results obtained for the density are in accordance with those reported by Liew et al. who evaluated the effect of using different L/S ratios on the density of geopolymers based on MK and found that the system of medium consistency had the highest density, followed by the more fluid system and, finally, the drier one [16]. The authors concluded that increased liquid content can promote the speed of dissolution of the Al and Si species of the precursor, but it obstructs the processes of polycondensation; in contrast, a small amount of liquid does not favour dissolution and, therefore, generates increased viscosity [16]. Zuhua et al. also evaluated the role of water in the geopolymerization of materials based on MK using the calorimetry technique, noting that increased water reduced the rate of geopolymerization owing to the lower concentration of ions present in the dissolution processes; they also mentioned that water promotes hydrolysis processes, but decreases the rate of polycondensation [15].

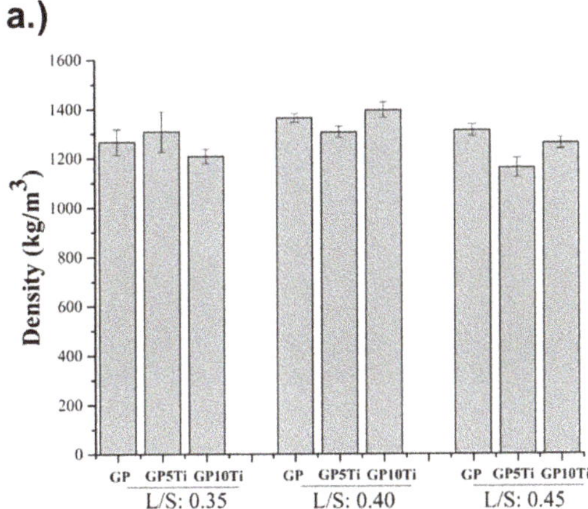

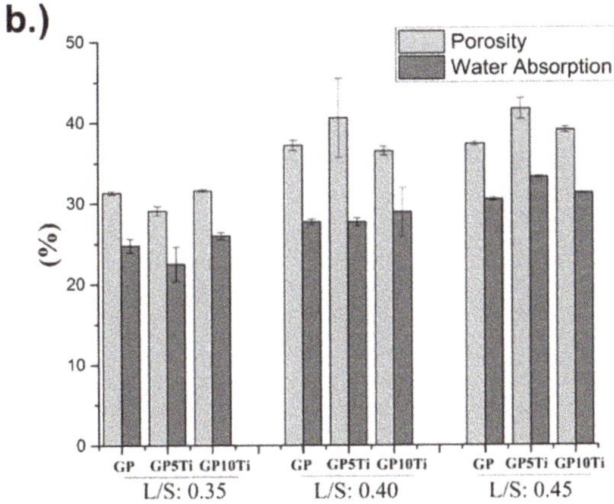

Figure 6. (**a**) Density and (**b**) water absorption and porosity of the geopolymeric materials evaluated.

In general, an increased proportion of liquid adversely affects the permeable pore content of the blends (Figure 6b). This higher pore volume for the higher L/S ratio (0.45) is related to the lower mechanical performance of these mixtures. It was observed that the addition of TiO_2 particles decreased the pore volume for materials with lower water content or dry consistency (L/S = 0.35), particularly when 5 wt % (GP5Ti) was added, coinciding with increased mechanical strength (Figure 7). This result seems to indicate that the addition of TiO_2 particles contributes to a better mechanical performance; however, there is an optimum concentration of TiO_2, so in the presence of an excess of the particles, these will act as filler [17].

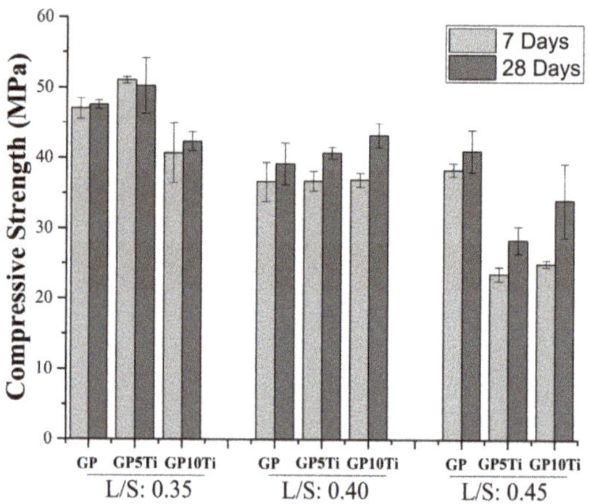

Figure 7. Compressive strength of the various materials evaluated.

Of the different mixtures evaluated, those with the lowest water content (L/S = 0.35) presented higher mechanical performance. Among them, that with 5 wt % TiO_2 added exhibited the most superior resistance, a 5.8% increase (50.27 MPa) after 28 days of curing, relative to the reference (GP); however, adding 10 wt % TiO_2 yielded the opposite effect, possibly due to the low availability of liquid to hydrate the particles and promote the formation of reaction products, taking into account that there is a greater presence of non-reactive TiO_2 particles coexisting in the gel. In contrast, for the materials with L/S = 0.4, it was found that the highest strength was achieved by adding 10 wt % TiO_2 (43.25 MPa), yielding a strength 10.3% greater than that of the reference. This effect is attributed to the TiO_2 particles having a smaller particle size than the precursor and behaving as nucleation points that enhance the formation of reaction products [17].

3.3. Characterization of the Geopolymer Microstructure

Figure 8 shows the XRD patterns obtained for the evaluated geopolymer systems. Compared with the MK precursor diffractogram (Figure 2), an amorphous halo between 25° and 35° can be observed in all cases, which corresponds to the aluminosilicate gel that forms the fundamental binder phase of the geopolymer matrix (KASH) and is responsible for the resistance of the geopolymer [18]. Likewise, the presence of the crystalline phase of the precursor is observed in the GP, which increases with the addition of TiO_2 in the material owing to its contribution.

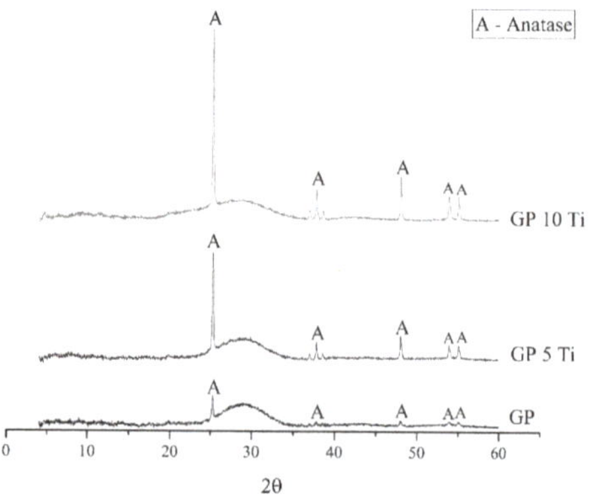

Figure 8. XRD pattern of the synthesized geopolymers.

The FTIR spectra of the synthesized geopolymers are shown in Figure 9, in which no large differences between the non-added and those with 5 wt % and 10 wt % TiO$_2$ added are observed. The presence of bands characteristic of precursor materials (Figure 4), in addition to the displacement of the one located at 1089 cm^{-1} in MK to 1007 cm^{-1} due to the asymmetric stretching of the Si–O–T band (T is a tetrahedron of Al or Si) in KASH gels and a band at 463 cm^{-1} corresponding to a lower Si–O–Si band intensity due to the reaction of MK for KASH gel formation, can be observed. The band located at approximately 703 cm^{-1} verifies the presence of Ti–O–Ti photoactive species, and it also generates overlap with amorphous Al–O bands, in addition to exhibiting a decrease in its intensity due to gel formation. When comparing these spectra with the MK FTIR spectrum (Figure 3), it is important to highlight the displacement at higher frequencies in the 3436 and 1650 cm^{-1} bands, which correspond to OH– and HOH group vibrations, suggesting increased hydrophobicity of the material as a result of the increased TiO$_2$ additive content [19].

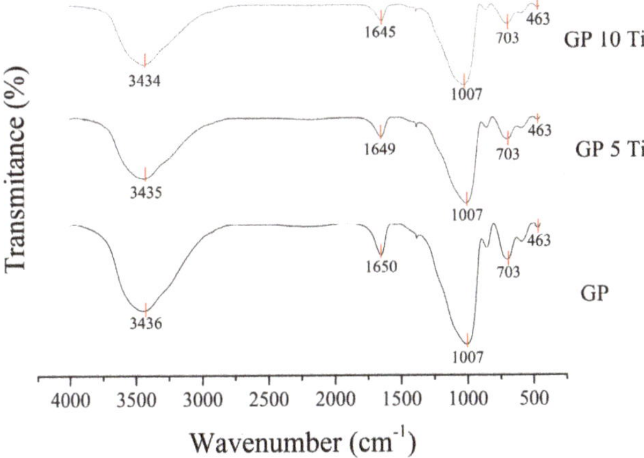

Figure 9. FTIR spectra of geopolymers.

The surface morphologies of the GP and GP10Ti geopolymers at seven and 28 days of curing evaluated by SEM and the corresponding EDS analysis results are shown in Figures 10 and 11, and the compositions of elements present at the evaluated points are presented in Tables 3 and 4. In general, all samples had a compact and continuous structure with small traces of unreacted MK, which is characterized by preservation of its lamellar structure in layers, mainly for the earliest age of curing; this behaviour was verified via EDS when performing the measurement at point 1 of Figure 10a, in which only high contents of O, Al, and Si, and lower proportions of K, were identified. By increasing the curing age, a greater gel formation is observed in the structure that is identified in the micrographs due to its dark tone, in addition to a lower presence of unreacted MK; in addition, the presence of O, Al, Si, and K is evidenced by the measurement of EDS at point 2 of Figure 10b, suggesting adequate KASH gel formation. It is noteworthy that there is no significant difference between the morphologies of the systems with and without TiO$_2$, but the EDS measurements at point 2 of the sample with 10 wt % added (Figure 11) reveals the presence of O, Al, Si, K, and Ti, the latter being present in a smaller proportion, indicating that the formation of the gel was not impeded by the addition of TiO$_2$ particles and that these coexist in the structure of the material [19].

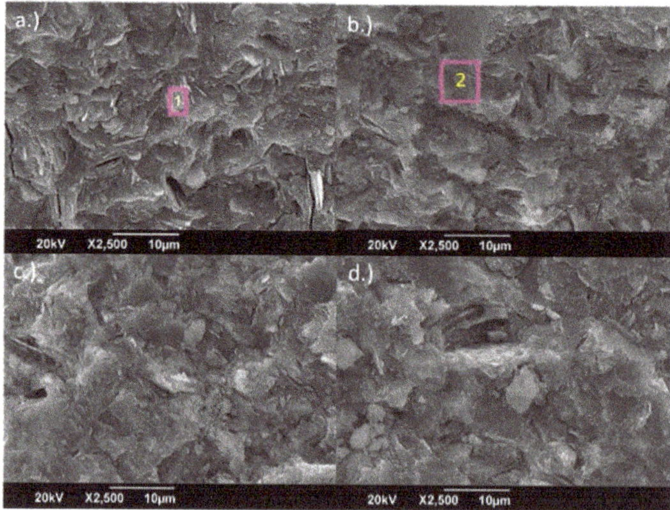

Figure 10. SEM micrographs. (**a**) GP at seven days; (**b**) GP 0 at 28 days; (**c**) GP 10 Ti at seven days, and (**d**) GP 10 Ti at 28 days.

Figure 11. SEM micrographs with EDS for sample to which 10 wt % TiO_2 was added (28 days).

Table 3. Composition of elements measured via EDS as shown in Figure 10.

Composition, %	Spectrum 1	Spectrum 2
O	53.42	36.40
Al	18.59	13.96
Si	23.42	20.89
K	4.56	28.74

Table 4. Composition of elements measured via EDS as shown in Figure 11.

Composition, %	Spectrum 2
O	50.67
Al	13.43
Si	19.23
K	11.02
Ti	5.65

4. Conclusions

The effects of adding titanium dioxide particles and modifying the system L/S ratio on the physicomechanical and structural properties of an MK-based geopolymer were evaluated. The results indicate that the addition of titanium dioxide particles at up to 10 wt % did not affect the formation of the KASH gel and that these particles coexisted within it. The compressive strength was maximized by adding titanium dioxide particles in proportions of 5 and 10 wt % at a dry (L/S = 0.35) and an average (L/S = 0.40) consistency, respectively. No improvement in the compressive strength was found for materials with a fluid consistency (L/S = 0.45). The fluidity and the setting time of the material were affected by the addition of titanium oxide particles and the L/S ratio; adding titanium dioxide particles at 10 wt % decreased the fluidity by up to 20.27%, and the setting time decreased by up to 30.13%. In contrast, an increased L/S ratio resulted in increases in the flowability and setting time of up to 20% and 60%, respectively.

Acknowledgments: This study was funded by the Colombian Institute for the Development of Science, Technology, and Innovation COLCIENCIAS (project, "Construction of prototype at the scale of rural housing using innovative materials with low carbon footprint," contract 096-2016). The authors, who are members of the Composite Materials Group (GMC) from the Centre of Excellence in New Materials (CENM), would like to thanks to Universidad del Valle (Cali, Colombia), in which the experimental work was carried out.

Author Contributions: Mejía de Gutiérrez R. and Maury-Ramírez A. conceived and designed the experiments; Guzmán-Aponte L. performed the experiments; Guzmán-Aponte L., Mejía de Gutiérrez R. and Maury-Ramírez A. analyzed the data; Mejía de Gutiérrez R. contributed reagents/materials/analysis tools; Guzmán-Aponte L., Mejía de Gutiérrez R., and Maury-Ramírez A. wrote the paper.

Conflicts of Interest: The authors declare no conflict of interest.

References

1. Singh, B.; Ishwarya, G.; Gupta, M.; Bhattacharyya, S. Geopolymer concrete: A review of some recent developments. *Constr. Build. Mater.* **2015**, *85*, 78–90. [CrossRef]
2. Robayo-Salazar, R.A.; Mejia, J.; Mejia de Gutierrez, R. Eco-efficient alkali-activated cement based on red clay brick wastes suitable for the manufacturing of building materials. *J. Clean. Prod.* **2017**, *166*, 242–252. [CrossRef]
3. McLellan, C.; Williams, R.P.; Lay, J.; Van Riessen, A.; Corder, G.D. Costs and carbon emissions for geopolymer pastes in comparison to ordinary portland cement. *J. Clean. Prod.* **2011**, *19*, 1080–1090. [CrossRef]
4. Weil, M.; Dombrowski, K.; Buchwald, A. Life-cycle analysis of geopolymers. In *Geopolymers: Structure Processing Properties and Industrial Applications*, 1st ed.; Provis, J.L., van Deventer, J.S.J., Eds.; Woodhead Publishing Limited: Cambridge, UK, 2009; pp. 194–210.
5. Duxon, P.; Fernandez-Jimenez, A.; Provis, J.; Luckey, G.; Palomo, A.; van Deventer, J. Geopolymer technology: The current state of the art. *J. Mater. Sci.* **2007**, *42*, 2917–2933. [CrossRef]
6. Assaedi, H.; Shaikh, F.; Low, I. Influence of mixing methods of nano silica on the microstructural and mechanical properties of flax fabric reinforced geopolymer composites. *Constr. Build. Mater.* **2016**, *123*, 541–552. [CrossRef]
7. Riahi, S.; Nazari, A. The effects of nanoparticles on early age compressive strength of ash-based geopolymers. *Ceram Int.* **2012**, *38*, 4467–4476. [CrossRef]
8. Duan, P.; Yan, C.; Luo, W.; Zhou, W. Effects of adding nano-TiO_2 on compressive strength, drying shrinkage, carbonation and microstructure of fluidized bed fly ash based geopolymer paste. *Constr. Build. Mater.* **2016**, *106*, 115–125. [CrossRef]
9. Yang, L.; Jia, Z.; Zhang, Y.; Dai, J. Effects of nano-TiO_2 on strength, shrinkage and microstructure of alkali activated slag pastes. *Cem. Concr. Compos.* **2015**, *57*, 1–7. [CrossRef]
10. Zhang, Z.; Wang, K.; Mo, B.; Li, X.; Cui, X. Preparation and characterization of a reflective and heat insulative coating based on geopolymers. *Energy Build.* **2015**, *87*, 220–225. [CrossRef]
11. Ohama, Y.; Van Gemert, D. Introduction. In *Application of Titanium Dioxide Photocatalysis to Construction Materials*, 1st ed.; Ohama, Y., Van Gemert, D., Eds.; State of the Art Report of the RILEM Technical committee 194-TDP; Springer: Dordrecht, The Netherlands, 2011; Volume 5, pp. 1–4, ISBN 978-94-007-1296-6.

12. Wan, Q.; Rao, F.; Song, S.; Garcia, R.; Estrella, R.; Patiño, C.; Zhang, Y. Geopolymerization reaction, microstructure and simulation of metakaolin-based geopolymers at extended Si/Al ratios. *Cem. Concr. Compos.* **2017**, *79*, 45–52. [CrossRef]
13. Wang, W.; Chen, J.; Gao, M.; Huang, Y.; Zhang, X.; Yu, H. Photocatalytic degradation of atrazine by boron-doped TiO$_2$ with a turnable rutile/anatase ratio. *Appl. Catal. B* **2016**, *195*, 69–76. [CrossRef]
14. Villaquirán-Caicedo, M.; Mejía de Gutiérrez, R.; Sulekar, S.; Davis, C.; Nino, J. Thermal properties of novel binary geopolymers based on metakaolin and alternative silica sources. *Appl. Clay Sci.* **2015**, *118*, 276–282. [CrossRef]
15. Zuhua, Z.; Xiao, Y.; Huajun, Z.; Yue, C. Role of water in the synthesis of calcined kaolin-based geopolymer. *Appl. Clay Sci.* **2009**, *43*, 218–223. [CrossRef]
16. Liew, Y.; Kamarudin, H.; Mustafa, A.; Bnhussain, M.; Luqman, M.; Nizar, I.; Ruzaidi, C.; Heah, C. Optimization of solids-to-liquid and alkali activator ratios of calcined kaolin geopolymeric powder. *Constr. Build. Mater.* **2012**, *37*, 440–451. [CrossRef]
17. Essawy, A.; Aleem, S. Physico-mechanical properties, potent adsorptive and photocatalytic efficacies of sulfate resisting cement blends containing micro silica and nano-TiO$_2$. *Constr. Build. Mater.* **2014**, *52*, 1–8. [CrossRef]
18. Yun, L.; Cheng, H.; Mustafa, M.; Hussin, K. Structure and properties of clay-based geopolymer cements: A review. *Prog. Mater. Sci.* **2016**, *83*, 595–629. [CrossRef]
19. Falah, M.; Mackenzie, K. Synthesis and properties of novel photoactive composites of P25 titanium dioxide and copper (I) oxide with inorganic polymers. *Ceram. Int.* **2015**, *41*, 13702–13708. [CrossRef]

 *coatings*

Review

State-of-the-Art Green Roofs: Technical Performance and Certifications for Sustainable Construction

Alejandra Naranjo [1], Andrés Colonia [2], Jaime Mesa [3], Heriberto Maury [4] and Aníbal Maury-Ramírez [1,5,*

[1] Department of Civil and Industrial Engineering, Pontificia Universidad Javeriana Cali, Cali 760031, Colombia; alejanaranjo@gmail.com

[2] Gerencia, Dos Mundos, Cali 760050, Colombia; gerencia@dosmundos.org

[3] Department of Mechanical Engineering, Universidad Tecnológica de Bolívar, Cartagena de Indias 130001, Colombia; jmesa@utb.edu.co

[4] Department of Mechanical Engineering, Universidad del Norte, Barranquilla 081007, Colombia; hmaury@uninorte.edu.co

[5] Engineering Faculty, Universidad De La Salle, Bogotá 111711, Colombia

* Correspondence: amaury@lasalle.edu.co; Tel.: +57-1-3417900

Received: 30 November 2019; Accepted: 4 January 2020; Published: 13 January 2020

Abstract: Green roof systems, a technology which was used in major ancient buildings, are currently becoming an interesting strategy to reduce the negative impact of traditional urban development caused by ground impermeabilization. Only regarding the environmental impact, the application of these biological coatings on buildings has the potential of acting as a thermal, moisture, noise, and electromagnetic barrier. At the urban scale, they might reduce the heat island effect and sewage system load, improve runoff water and air quality, and reconstruct natural landscapes including wildlife. In spite of these significant benefits, the current design and construction methods are not completely regulated by law because there is a lack of knowledge of their technical performance. Hence, this review of the current state of the art presents a proper green roof classification based on their components and vegetation layer. Similarly, a detailed description from the key factors that control the hydraulic and thermal performance of green roofs is given. Based on these factors, an estimation of the impact of green roof systems on sustainable construction certifications is included (i.e., LEED—Leadership in Energy and Environment Design, BREEAM—Building Research Establishment Environmental Assessment Method, CASBEE—Comprehensive Assessment System for Built Environment Efficiency, BEAM—Building Environmental Assessment Method, ESGB—Evaluation Standard for Green Building). Finally, conclusions and future research challenges for the correct implementation of green roofs are addressed.

Keywords: green roofs; biological coatings; hydraulic performance; thermal performance; sustainable construction certification; LEED; BREEAM; CASBEE; BEAM; ESGB

---

1. Introduction

Due to its multiple benefits, biological coatings on buildings (i.e., green roofs and walls) were used by different human civilizations [1,2]. According to the scientific community, there is testimonial evidence of the use of biological coatings on Babylon's Hanging Gardens and Babel Tower [2–4]. Similarly, religious buildings called "Ziggurats" were constructed in Mesopotamia using plant growth on their building surfaces most probably to reduce heat interchange with the environment [3]. Biological coatings were also part of vernacular architecture in Scandinavia. For instance, in Norway, green roof systems are still used as a thermal isolation system [5].

Even when green roofs were present on the major buildings of the ancient world, their use on contemporaneous architecture reappeared only in the 20th century with the Swiss architect Le Corbusier who included them among modern architecture principal elements [6]. In fact, he stated that "the garden terrace will be the reunion place preferred by citizens and it will also mean the recovery of the built surface from the city". In this sentence, the preoccupation that exists since then about the accelerated urban development is synthesized. The truth is that the construction of urban infrastructure with non-permeable materials such as traditional concrete and asphalt significantly reduces rainwater flow to the soil layers and water table [7]. In a basin, the water cycle alteration can produce floods, river and lake disappearances, and consequently ecosystem extinction. Biological coatings for buildings such as green roofs and walls are used to mitigate urban threat and other environmental problems. In general, the benefits of green roofs and walls can be classified into three categories: environmental, economic, and social. Aspects such as thermal, moisture, noise, and electromagnetic protection in buildings, heat island effect reduction, sewage system load reduction, runoff water quality and air quality improvement, habitat development, and natural landscape reconstruction are among the major environmental benefits from green roofs and walls. From an economic viewpoint, green roofs and walls on buildings increase the property's commercial value, increase waterproof membrane lifespan, increase fire resistance, and reduce energy consumption. Regarding social aspects, these biological coatings improve the occupant's health and wellbeing, while they also help to generate employment for maintenance and even for food production at the urban scale (urban agriculture).

Recently, the use of green roofs on buildings increased significantly around the world due to different government incentives and due to pressure exerted by a new market that seeks products and processes that are more environmentally friendly. However, the lack of a detailed characterization and guidelines for their design resulted in the implementation of these biological coatings on buildings being currently based on experience, making them inefficient and even risky. In order to improve this situation, this review of the state of the art presents a proper green roof classification, which is based on their components. Similarly, a detailed description of the factors that control the hydraulic and thermal performance of green roofs is given. Based on these, an estimation of the impact of green roof systems in sustainable construction certifications is included. Finally, research challenges for the correct implementation of these biological coatings on buildings are addressed.

2. Components and Classification

2.1. Components

Although the green roof's most visible part is the vegetation layer, green roofs are multilayer systems in which each layer has different functions, impact on the complete life cycle (life-cycle analysis (LCA), and certifications for sustainable construction. This natural-based solution simulates the natural soil's characteristics, as presented in Figure 1. The multi-layer components of green roofs are described below.

(a) Vegetation layer: This can be composed of different biological species, usually plants or even trees. The respective species selection is based on availability, weather, maintenance conditions, and the substratum's depth, which depends on the structural building capacity [9]. Experience shows that green roof plants need to be chosen with care, as not all plants are suitable for growing in this way. For example, when choosing plants for a green roof, they need to be able to withstand wind and frost, be drought-resistant, tolerate living in poor soil, and require low maintenance. Also, green roof plants should also be attractive and offer food and shelter for wildlife. Although, many plants were proposed for green roofs in the literature, there is a need for comparisons among various types of plants to provide design guidelines for selecting the most appropriate vegetative layer for a given green roof [10]. Interesting research is now being conducted to analyze CO_2 reduction and O_2 production from different plant species; hence, this factor should also be included in the plant selection process for increasing the positive environmental impact of the biological coatings in life-cycle analysis.

(b) Substratum: The function of this layer is to physically support the vegetation and to supply the required nutrients for their development. Also, this layer should have the ability to store and gradually release excess rainwater, keeping enough moisture to later reduce maintenance activities. This layer is usually based on a combination of minerals with organic matter and other nutrients such as nitrogen, phosphorus, potassium, and magnesium [11,12]. Physical properties of a typical substratum are presented in Table 1.

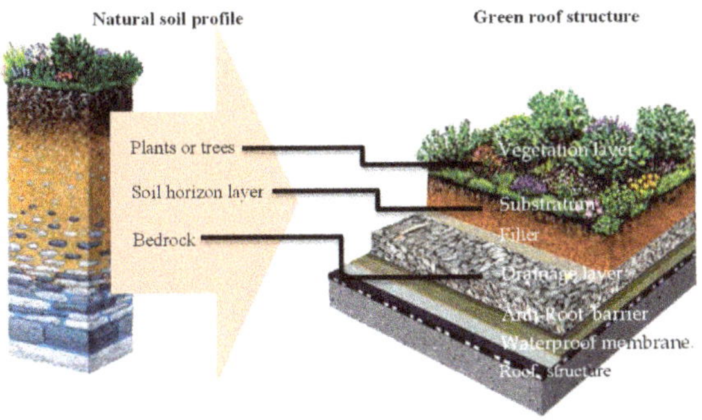

Figure 1. Green roof's basic components and their similarity with the natural soil. Modified from Ovacen (2014) [8].

Table 1. Physical properties of a typical substratum used in green roofs [13].

Physical Property	One-Layer System	Multi-Layered System
Water retention	Minimum 20%	Minimum 35%
Water permeability	Minimum 60 mm/min	Minimum 0.6 mm/min
Air content (fully saturated)	Minimum 10%	Minimum 10%
pH	6.5 to 9.5	6.5 to 8.0

(c) Filter: This building protective layer is usually a geotextile; its main function is to block the substratum's material flow to the drainage layer and to keep it in the right position. This geotextile, chemically neutral, has to be resistant to acid and alkaline attacks [3].

(d) Drainage layer: Also known as a drainage system, this layer is composed of granular-based material layers and pipelines. This is the key to appropriate vegetation growth and control of building water-associated problems such as filtrations. In particular, the drainage layer is responsible for the equilibrium between water excess and scarcity. It is normally formed by gravel, which is a natural crushed rock [14]. Research is now being conducted to reduce the use of natural aggregates to minimize the negative environmental impact of using nonrenewable resources in green roofs [15].

(e) Anti-root barrier: This high-density polyethylene layer protects the waterproof membrane from possible tearing caused by roots. It also functions as a second protection from the substratum to the building [3].

(f) Waterproof membrane: This layer is used to block the water flow to the building slab. Some of the most used waterproof membranes are thermo-polymer elastomers such as EPDM (ethylene propylene diene) or thermoplastic polyolefin (TPO), which are also good as root barriers [14].

2.2. Classification

Although there are different criteria to classify green roofs, the most common classification at the international level is based on the characteristics of the vegetation layer, as presented below [16].

(a) Extensive: This green roof type uses plants with low-moisture needs (Figure 2). As a consequence, it uses a substratum with thickness between 5 and 15 cm. This biological coating is usually implemented on places with difficult access for pedestrians, can be moistened with rainwater, and can be placed on existing structures when the building has a proper design (i.e., roof live load considered as indicated in design codes). Its weight varies from 60 to 140 kg/m^2 [17].

Figure 2. Extensive green roofs: (a) Icelandic house [18]; (b) Nanyang Technological University, Singapore [19].

(b) Intensive: The green roofs belonging to this type have the possibility of using a great variety of plants and even tree species (Figure 3). Hence, they generally require substratum of great depth, usually superior to 15 cm. Based on the last aspect, these coatings can allow pedestrian access, require artificial water irrigation systems, and are placed on structures specifically designed to bear these additional loads. Their weight varies from 250 to 400 kg/m^2 [20].

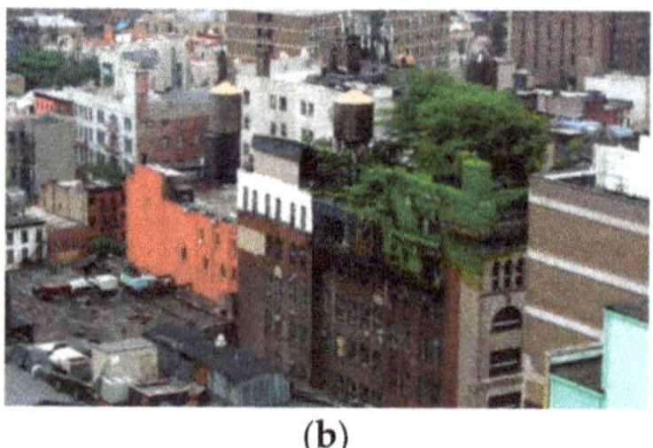

Figure 3. Intensive green roofs: (a) Chicago City Hall, Illinois [21]; (b) apartment building, New York [22].

(c) Semi-intensive: This type of green roof considers an intermediate system between intensive and extensive systems with species that grow over a 10 to 30-cm-deep substratum (Figure 4). It allows partial pedestrian access and requires artificial irrigation systems. Its weight varies from 60 to 140 kg/m^2 [23].

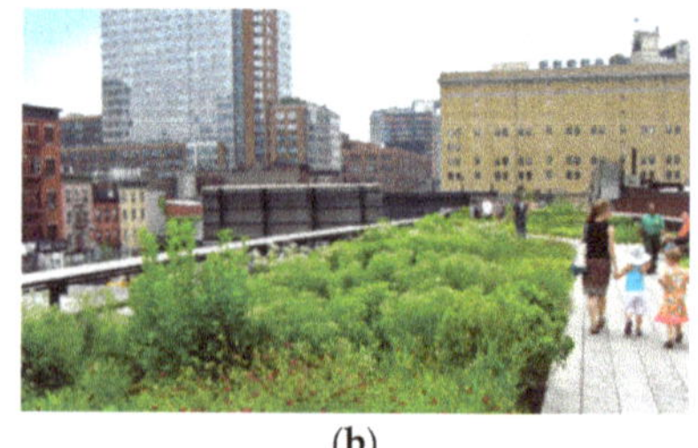

Figure 4. Semi-intensive green roofs: (a) Buenos Aires, Argentina [24]; (b) High Line park building, New York [25].

3. Hydraulic Performance

Based on the non-permeable large areas generated by conventional urban development, green roofs might reduce the negative environmental impact, particularly by partially recovering the natural water cycle in a basin. These biological coatings on buildings have the ability to store water in the substratum, in which a fraction is absorbed by plants and returned to the atmosphere via evaporation and transpiration processes [26,27]. In addition to the efficient management of the evapotranspiration water fraction, another water fraction is absorbed, infiltrated, and stored inside the system layers. Depending on the precipitation intensity, substratum type, and depth, studies showed that green roofs have the capacity to store between 40% and 80% of annual precipitation that falls on them. Additionally, it was reported that a 12-cm layer takes 12 h to start releasing the stored water during a rain event, and it continues releasing it during the next 21 h, approximately [28]. Based on Kok et al. (2016) [29] and Masseroni and Cislagi (2016) [30], the discharge peak from a storm can be reduced by 26% to 35%. Hence, it is deduced that green roofs, in addition to reducing the magnitude of water volume captured by rain drainage systems, also delay the water discharge time, because the substratum and other layers require time to saturate and discharge the stored water, reducing potential flood impacts when used intensively at the urban scale [26]. In Germany, Spain, Holland, England, and the United States, the performance of different green roof types was evaluated toward the quantity and quality of urban runoff water. For example, van Woert et al. (2005) [14] indicated that the implementation of green roofs reduces the excess of runoff water, and Stovin (2009) [17] found 34% rainwater retention in the 11 precipitation events monitored. In this article, the retention capacity, measured as the water percentage stored in different green roofs compared with the water that passes through them, is presented in Table 2. From the results, it can be observed that retention capacities can range from 45% to 78% for different green roof systems. The variation in the results is associated with the substratum depth, initial water content, vegetation age and type, and precipitation intensity and distribution.

Table 2. Retention capacity from different green roof systems evaluated worldwide [31].

Rainwater Retention (%), Average during Study Period	Rainwater Retention (%), Range for Studied Events	Monitoring Time	Short Description and References Taken from Reference [31]
46	-	17 months	Bengtsson et al. (2005) [32] studied a hydrological function of a *Sedum* moss thin vegetated roof from Malmö, Sweden (roof slope 2.6%, substrate thickness 30 mm). The real rain events (mid July 2001 through December 2002) and artificial storms were investigated both on the study in real scale and on the smaller model of the green roof, designed identically to the original roof but with the possibility of changing the slope and the drainage material.
61	-	15 months	Van Woert et al. (2005) [14] investigated vegetated roof water retention and its dependence on roof surface type, slope, and media depth. Three roof platforms were constructed in a model scale with a slope of 2%: gravel, vegetated, and un-vegetated (study period: August 2002 through October 2003). Vegetated roofs consisted of the following layers: drainage (15 mm), water retention fabric (7.5 mm), additional retention fabric as vegetation carrier (7.5 mm), and 25 mm of growing media. Twelve additional roof platforms were used to examine roof slope (6 with slope 2% and 6 with slope 6.5%) and media depth (for 2% slope 25 mm and 40 mm; for 6% 40 mm and 60 mm).

Table 2. *Cont.*

Rainwater Retention (%), Average during Study Period	Rainwater Retention (%), Range for Studied Events	Monitoring Time	Short Description and References Taken from Reference [31]
45	19–98	2 months	De Nardo et al. (2005) [33] investigated stormwater retention by three green roofs located in Pennsylvania, United States of America (USA). The roofs consisted of the following layers: waterproof membrane, drainage layer of 12 mm, growing medium 89 mm, plant support medium 25 mm. Rainfall and runoff data were collected for seven rain events during October–November 2002.
63—Roof 1	-	18 months	Moran et al. (2005) [34] monitored two green roofs installed in North Carolina, USA (in Goldsboro: soil depth 75 mm, flat, and in Raleigh: soil depth 100 mm, slope of 7%) to estimate water retention and peak flow reduction. Two different commercially available drainage systems were used: one in Goldsboro with negligible storage and one in Raleigh with water storage capacity of 2.4 L/m^2. The runoff data were collected at Goldsboro during April 2003–September 2004 and in Raleigh during July 2003–September 2004. To investigate water quality (P-tot, N-tot) runoff water samples were collected during 11 rain events from the Goldsboro green roof (soil mix consisting of 55% expanded slate, 30% sand, and 15% compost).
55—Roof 2	-	15 months	
78	39–100	13 months	Carter and Rasmussen (2006) [35] investigated water retention of a newly constructed green roof plot in Athens, Georgia, USA. The construction followed a design of a commercial product. In total, 31 rainfall–runoff events were registered during a study period of 13 months (November 2003–November 2004). The roof layers included a root protection sheet of negligible thickness, 4.8-mm-thick moisture retention mat with water retention capacity of 5 L/m^2, 38.1-mm-thick synthetic drainage panel with water retention capacity of 4 L/m^2 (both layers provided about 9-mm retention). The growing medium was 76.2 mm thick.
49	-	4 storm events	Monterusso et al. (2004) [36] performed a pilot investigation of water retention (calculated for individual rainfall events) and water quality of runoff among combinations of various commercial growing systems and vegetation types in Michigan, USA. Twelve roof platforms were installed, each divided into 3 parts with different vegetation. All platforms duplicated typical commercial green roof construction; 4 commercial drainage systems were installed. Platforms were set at a slope of 2%. The substrate depth was 100 mm for three types of drainage systems and 20 mm (*Sedum*) or 60 mm (natives) for the fourth drainage system. The soil mix consisted of 60% heat-expanded slate, 25% grade sand, 5% aged compost, and 10% peat. Three groups of vegetation were used, the first consisting of seven *Sedum* spp. propagated from seed, the second from two *Sedum* spp. planted from plugs, and the third consisting of 18 species of region (Michigan) native plants planted from plugs.

Table 2. *Cont.*

Rainwater Retention (%), Average during Study Period	Rainwater Retention (%), Range for Studied Events	Monitoring Time	Short Description and References Taken from Reference [31]
-	5–70	6 months	Bliss et al. (2009) [37] monitored a prototype green roof in Pittsburgh, Pennsylvania, USA. The 1150-m^2 extensive green roof consisted of a bitumen built-up membrane with root barrier, drainage, and filter fabric layer beneath 140-mm-thick synthetic growing medium made of expanded shale, perlite, and coconut husk. Regarding water quantity investigation, the data were obtained for 13 storms (August 2006–January 2007). Water quality tests (phosphorus, sulfate, nitrogen, chemical oxygen demand (COD) for unfiltered and filtered samples, pH and turbidity for unfiltered samples and lead, cadmium, and zinc for filtered samples) were reported for one of the two selected storms (17 October 2006 or 1 December 2006).

On the other hand, as mentioned previously for the drainage layer, recent research is evaluating new material sources to replace natural aggregates traditionally used on green roofs. In this case, drainage systems composed of PEAD plates (recycled high-density polyethylene), reused PET bottles (polyethylene terephthalate), and vulcanized recycled rubber particles were evaluated and compared with basal gravel, the conventional system used as a drainage system (Figure 5). The results indicated that, for low-intensity precipitations (simulations of 1 and 4 mm) and duration of a couple of hours, granular drainage systems (rubber and gravel) were very efficient, as they kept almost all precipitation water, and retention coefficients were about 100%. For the same conditions, reused PET bottles and PEAD plates retained half of the simulated precipitation, while retention coefficients were about 50%. For higher-intensity precipitations (simulations around 50 mm), the retention coefficients were near 30% for all green roof systems, except for the system composed of PET bottles, which had a retention capacity near to 0% [15].

Using the same experimental methodology but including a simulation for the Singapore precipitation regime, extensive green roofs of 12 cm were not enough to obtain a significant retention coefficient. The authors concluded that a proper green roof design has to be done for each precipitation condition [38]. Hence, more research has to be conducted to improve the retention coefficients in green roofs.

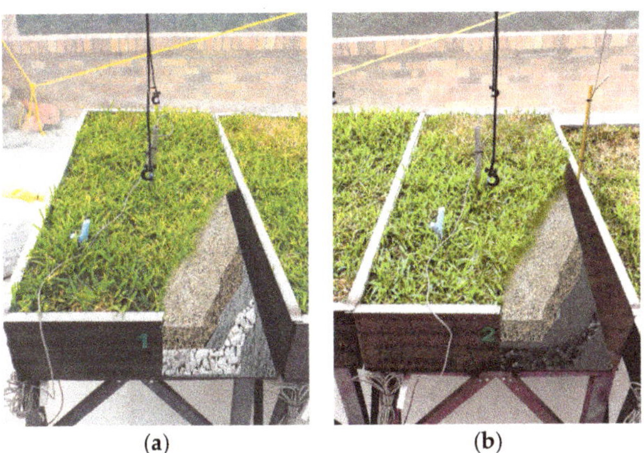

(a) (b)

Figure 5. *Cont.*

(c)   (d)

Figure 5. Cross-transversal sections of green roof prototypes evaluated using reused and recycled materials as their drainage systems: (**a**) roof with basal gravel (reference); (**b**) roof with vulcanized recycled rubber particles; (**c**) roof with PET (polyethylene terephthalate) bottles; (**d**) roof with PEAD (recycled high-density polyethylene) plates [15].

4. Thermal Performance

In general, green roof systems produce a high thermal isolation effect in buildings by increasing thermal inertia. This interesting property is mainly due to the green roof components which work as thermal isolation chambers, preventing and delaying the temperature amplitudes due to their high thermal capacities in hot regions, as well as heat winter loss reduction in cold regions. Indeed, a green roof can reduce energy consumption from air conditioning in buildings to 50% [39]. In addition to the good thermal performance of green roofs, they also reduce the surrounding environment's temperature via vegetative physiological processes such as evapotranspiration, photosynthesis, and ability to store water [39]. Scale experiments in outdoor conditions on different green roof systems developed in Cali (Colombia) showed that, during days where the environmental temperature was high (over 35° C), there was a temperature reduction between 10.6 and 11.7 °C below the green roof prototypes [15]. These results are similar to those reported in the literature for tropical climates. More details about the prototype evaluation can be seen in Figure 6.

Figure 6. Green roof prototypes evaluated using reused and recycled materials in their drainage systems [15].

Table 3 summarizes results from several researchers about the thermal performance of different green roof systems. From these results, a trend to evaluate the thermal behavior of green roof systems during warm and cold seasons can be identified. Although some researchers showed a more efficient behavior in hot weather compared to cold weather, some others showed almost the same behavior. Based on the review, more research should be conducted when the substratum is in saturated conditions during cold weather; then, the isolating effect of green roofs seems to be significantly reduced [40].

Table 3. Relevant work related to the thermal performance of green roofs.

Approach Proposed	Major Findings	Reference
Long-term experimental analysis to compare thermal performance from a conventional roof respect to a green roof.	Under typical Mediterranean climate conditions, the green roof system provides different behavior according to the season. The green roof performance is meaningful during warm seasons, while this technology does not show a significant difference with a conventional roof during cold seasons.	[41]
Model based on energy balanced equations expressed for foliage soil media and simulations.	During the exposure to warm environmental conditions, the evapotranspiration provides evaporative cooling that increases the thermal resistance of a green roof system.	[42]
Analysis of transmittance and heat flux.	During the winter season, the green roof provides further isolation even in saturated conditions. In the summer, the green roof decreases incoming heat fluxes and ceiling temperatures.	[43]
Study during warm and cold periods with 3 different roof conditions.	During warm periods, the evaluated green roof reduced heat gain over 90%. During the cold period, the evaluated green roof system reduced heat loss between 70% and 84%. These results are comparable to those obtained with conventional ceramic and metallic roofs.	[44]
Numerical model and experimental validation for energy savings (comparison approach).	After evaluating extensive, semi-intensive, and intensive green roofs, the extensive one involved higher cooling energy demand than semi-intensive (2.8-fold) and intensive ones (5.9-fold).	[45]
Mathematical model and experimental validation.	Cooling potential of green roofs can be around 3.02 kWh per day for an LAI (leaf area index) of 4.5. This is enough to maintain an average room temperature of 25.7 °C.	[46]
Experimental study for measuring energy savings in cooling for Mediterranean continental climate.	Energy savings (16.7%) for cooling and an increase in energy consumption for heating (11.1%) were observed compared to conventional flat roofs. These results correspond to a 20% to >85% area covered by vegetation.	[47,48]

5. Certifications for Sustainable Construction

In order to evaluate the impact of green roof system implementation on major sustainable construction certifications, a preliminary qualitative impact analysis must be developed to assess the decision process. A suggested impact of the green roof system that varies from not related to high is presented in Tables 4 and 5, which summarize the impact of green roofs on the overall certification systems based on the degree of relevance of green roof benefits concerning the main criteria for each system. Thus, criteria regarding energy and water involve a strong (high impact) relationship (efficiency, cost reduction), while indoor and outdoor environment improvements involve indirect benefits (medium impact). Finally, innovation is considered as a less relevant criterion since its measurement in buildings is still subjective in terms of green roofs, and it can be interpreted as an indirect benefit or consequence.

Table 4. Suggested impact of green roof systems in sustainable construction certifications.

Impact	Description
High	In this level, a green roof has a direct influence on the evaluation criteria providing points in relation to conventional roofs. There is a strong relation to energy and water efficient use criteria.
Medium	Although, in this level, the implementation of green roof systems has an impact on the criteria, complementary technologies are required to be explicitly valued.
Low	At this level, the green roof implementation is related to the criteria, but it does not necessarily mean a performance improvement.
Not related	The implementation of green roof systems has no impact on the evaluation criteria in this level.

Table 5. Impact of green roof systems on different sustainable construction certifications.

BREEAM—Building Research Establishment Environmental Assessment Method	LEED—Leadership in Energy and Environment Design	CASBEE—Comprehensive Assessment System for Built Environment Efficiency	BEAM Plus—Building Environmental Assessment Method	ESGB—Evaluation Standard for Green Building
Management	Sustainable sites	Indoor environment	Site aspect	Sustainable site and outdoor environment
Health and wellbeing	Water efficiency	Quality of service	Material aspect	Energy use
Energy	Energy and atmosphere	Outdoor environment	Energy use	Water use
Transport	Material and resources	Energy resources and materials	Water use	Material use
Water	Indoor environmental quality	Off-site environment	Indoor environmental quality	Indoor environmental quality
Materials	Innovation in design		Innovation and performance enhance	Operation management
Waste	Regional priority			
Land use and ecology				
Pollution				
Innovation				

High impact: ■; medium impact: ■; low impact: ■; not related: □.

From the analysis of the impact of green roofs on construction certifications, it is clear that the British Accreditation, BREEAM (Building Research Establishment Environmental Assessment Method), considers the use of this technology more relevant. BREEAM assumes a high impact of green roofs in energy, water, pollution, land use, and ecology. Contrary to BREEAM, the Japanese Accreditation, CASBEE (Comprehensive Assessment System for Built Environment Efficiency), does not consider the use of green roofs crucial for sustainable construction. However, this accreditation system gives a reasonable importance to green roofs regarding energy resources and materials and in outdoor environments. On the other hand, LEED (Leadership in Energy and Environment Design), BEAM Plus (Building Environmental Assessment Method), and ESGB (Evaluation Standard for Green Building) give a medium impact of using green roofs in their accreditation systems.

In general, for the accreditation systems, green roofs are considered as a low-level innovation, unless they are integrated with other technologies such as renewable energies, cogeneration, and so on. That is due to the fact that green roof systems were implemented in many projects worldwide in the past. However, the real fact is that the impact of the implementation was never properly quantified for the hydraulic and thermal performances. Thus, other environmental benefits such as reduction of the heat island effect and sewage system load, improvement of runoff water and air quality, and reconstruction of natural landscapes remain to be estimated for each green roof case.

As a reference for the reader, a description of major construction certifications is included in Table 6.

Table 6. Description of major sustainable construction certifications based on Park et al. (2017) [49] and Lee (2013) [50].

Certification Systems	Characteristics	Measuring Method
BREEAM—Building Research Establishment Environmental Assessment Method	Most used certification used worldwide to measure, organize hierarchically, and certify a building's sustainability. More than 250,000 buildings in more than 70 countries have this certification. Origin: United Kingdom.	Hierarchical criteria credit system in the following categories: fulfilling, good, very good, excellent, and outstanding.
LEED—Leadership in Energy and Environment Design	This certification includes measuring and hierarchical organizing systems for design, construction, maintenance, and operation of green buildings that use some type of related technologies. More than 80,000 buildings worldwide have this certification. Origin: USA.	Certification system based on points as follows: platinum: more than 80 gold: between 60 and 79 silver: between 50 and 59 certified: between 40 and 49
CASBEE—Comprehensive Assessment System for Built Environment Efficiency	This certification system was designed to measure the impact on people's life quality, resource consumption, and environmental loads caused by buildings. This certification system is supported by the national government in Japan. Origin: Japan.	Valuation scale from 1 to 5. The minimal condition required by law is 3.
BEAM Plus—Hong Kong Building Environmental Assessment Method	This certification covers a wide variety of building impacts on local, global, and indoor scales. Origin: China.	The evaluation system has four levels: bronze: over 40% credits silver: over 55% credits gold: over 65% credits platinum: over 75% credits
ESGB—Evaluation Standard for Green Building	This certification was designed to evaluate new and existing buildings during the design and construction stages. Origin: China.	This certification has 3 levels: 1 start: over 33% marks 2 starts: over 67% marks 3 starts: over 80% marks

6. Conclusions and Research Challenges

Although green roofs were used in major ancient buildings, their implementation in modern infrastructure was restricted until Le Corbusier included them among the main building conceptual design points. However, green roofs became more a landscaping action rather than a technical solution with significant environmental, technical, economic, and social benefits. This is mainly due to the fact that there was no proper design, construction, and maintenance of these biological coatings. Intensive, semi-intensive, and extensive green roofs require a proper conceptualization that allows identification of each component and their functions for the overall performance.

Currently, the economic benefits from reducing building energy consumption and sewage system load attracted the attention of urban developers for implementing green roofs on buildings. This review showed that, only regarding the hydraulic performance, retention capacities vary from 45% to 78% for different green roof systems reported in the literature. Thus, more research has to be conducted to estimate the real impact of substratum depth, initial water content, vegetation age and type, and precipitation regime on the hydraulic performance. Similarly, although thermal gradients up to

$10°$ were reported using green roofs in tropical climates, more research on the application of green roofs in cold climates has to be done. In this case, when the substratum is saturated, the isolating effect is significantly reduced. Therefore, combined models that integrate hydraulic and thermal performances should be developed. Thermal and hydraulic performances mostly control the complete green roof system.

Even though most accreditation systems for sustainable constructions do not give significant importance to green roof implementation, an accurate quantification of the environmental, technical, economic, and social benefits of green roofs might help to improve this situation, as well as address the lack of legislation for this technology in most countries.

Finally, from this review of the state of the art, research challenges that can be undertaken in the short or medium term toward an effective implementation of green roofs are presented in Table 7.

Table 7. Research challenges facing green roofs reported in the literature.

Component	Research Topics
Vegetation layer	(a) Study the vegetation behavior under different climate conditions. (b) Generate robust databases for plant analysis and its selection. Include CO_2 sequestration performance. (c) Study of influence of vegetation on green roof thermal and hydraulic performances. (d) Analyze the relationship and effect on wildlife.
Substratum	(a) Develop growth media mixtures able to reduce erosion and increase water content for low-depth substratum. (b) Develop growth media using renewable local materials.
Materials	(a) Improve root resistance from waterproof membrane and anti-root barrier. (b) Increase water retention capacity from the drainage layer materials. (c) Include reused, reduced, and recycled materials in all components of green roofs. (d) Durability strategies to increase life span.
Others	(a) Design and implementation of robust policies to promote massive use. (b) Evaluate green roof implementation using the life-cycle analysis (LCA) methodology. (c) Analyze the overall performance from a multidisciplinary perspective. (d) Develop local design and construction guidelines.

Author Contributions: Conceptualization, methodology, and editing, A.M.-R. and H.M.; components, classification, and technical performance, A.N. and A.C.; certification for sustainable construction and research challenges, J.M.

Funding: This research received no external funding.

Acknowledgments: The authors thank Dos Mundos (Cali, Colombia) and Pontificia Universidad Javeriana Cali (Colombia) for the technical support given during the experiments reported in this article.

References

1. Peri, G.; Traverso, M.; Finkbeiner, M.; Rizzo, G. The cost of green roofs disposal in a life cycle perspective: Covering the gap. *Energy* **2012**, *48*, 406–414. [CrossRef]

2. Williams, N.; Rayner, J.; Raynor, K. Green roofs for a wide brown land: Opportunities and barriers for rooftop greening in Australia. *Urban For. Urban Green.* **2010**, *9*, 245–251. [CrossRef]

3. Bliss, D.J. Stormwater Runoff Mitigation and Water Quality Improvements through the Use of a Green Roof in Pittsburgh, Pennsylvania. Master's Thesis, Faculty of the School of Engineering, University of Pittsburgh, Pittsburgh, PA, USA, 2010.

4. Getter, K.; Rowe, D. The role of extensive green roofs in sustainable development. *HortScience* **2006**, *41*, 1276–1285. [CrossRef]

5. Coutts, A.; Daly, E.; Beringer, J.; Tapper, N. Assessing practical measures to reduce urban heat: Green and cool roofs. *Build. Environ.* **2013**, *70*, 226–276. [CrossRef]

6.   Eisenman, T. Raising the bar on green roof design. How does the ASLAs unconventional green roof perform? *Landsc. Archit.* **2006**, *2006*, 22–29.

7.   González, Z. Agricultura. *Rev. Agropecu.* **1995**, *71*, 58–61.

8.   Ovacen. Manuales o Guías Sobre Cubiertas Vegetales. Available online: http://ovacen.com/como-construir-cubiertas-vegetales-o-verdes-manuales-guias/ (accessed on 17 February 2014).

9.   Oberndorfer, E.; Lundholm, J.; Bass, B.; Coffman, R.R.; Doshi, H.; Dunnett, N.; Gaffin, S.; Köhler, M.; Liu, K.K.; Rowe, B. Green roofs as urban ecosystems: Ecological Structures, Functions, and Services. *BioScience* **2007**, *57*, 823–833. [CrossRef]

10.   Berardi, U.; GhaffarianHoseini, A.H.; GaffarianHoseini, A. State-of-the-art analysis of the environmental benefits of green roofs. *Appl. Energy* **2014**, *115*, 411–428. [CrossRef]

11.   Villareal, E. Runoff detention effect of sedum green roof. *Hydrol. Res.* **2007**, *38*, 99–105. [CrossRef]

12.   Wachtel, J. Storm water management, green roof style. *Biocycle* **2007**, *48*, 42–46.

13.   Landscape Development and Landscaping Research Society–FLL. *Green Roof Guidelines-Guidelines for the Planning, Construction and Maintenance of Green Roofs*, 6th ed.; Landscape Development and Landscaping Research Society–FLL: Bonn, Germany, 2018.

14.   Van Woert, N.D.; Rowe, D.B.; Andresen, J.A.; Rugh, C.L.; Fernández, R.T.; Xiao, L. Green roof stormwater retention: Effects of roof surface, slope, and media depth. *J. Environ. Qual.* **2005**, *34*, 1036–1044. [CrossRef] [PubMed]

15.   Naranjo, A. Evaluación de Cubiertas Verdes Semi-Intensivas con Capas de Drenaje Elaboradas con Materiales Reciclados y Reutilizados. Master's Thesis, Pontificia Universidad Javeriana Cali, Cali, Colombia, 2017.

16.   University of Cincinnati. Classification system proposed for green roofs. *ScienceDaily*. Available online: www.sciencedaily.com/releases/2013/10/131022113547.htm (accessed on 12 January 2020).

17.   Stovin, V.; Duannet, N.; Hallam, A. Green Roof. Getting sustainable drainage off the ground. In Proceedings of the 6th International Conference of Sustainable Techniques and Strategies in Urban Water Mangement, Lyon, France, 25–28 June 2007; pp. 11–18.

18.   Ecofield. Tejados de Cesped. Available online: http://ecofield.com.ar/blog/tejados-de-csped-y-muros-de-paja/ (accessed on 20 February 2018).

19.   ArchiExpo. A Swirling Green Roof Tops the Gorgeous Nanyang Technical University in Singapore. Available online: https://projects.archiexpo.com/project-23031.html (accessed on 20 March 2015).

20.   López, C. Un acercamiento a las Cubiertas Verdes. *Medellín FBPSA* **2010**, *2010*, 22–23.

21.   Youtech & Associates. Urban Matter. Chicago City Hall. Available online: http://urbanmatter.com/chicago/chicago-city-hall/ (accessed on 10 May 2016).

22.   EcoInventos. Azotea verde en Manhattan. Available online: http://ecoinventos.com/azotea-verde-en-manhattan/ (accessed on 11 April 2011).

23.   García, I. Beneficios de Los Sistemas de Naturación en Las Edificaciones. In Proceedings of the XXXIV Semana Nacional de Energía Solar, Guanajuato, Mexico, 4–9 October 2010; pp. 1–5.

24.   MAyEP/GCBA. Buenos Aires Ciudad. Techos verdes. [Figura]. Obtenido de. Available online: http://www.buenosaires.gob.ar/noticias/techos-verdes-en-buenos-aires (accessed on 13 April 2010).

25.   Zinco. Referencias Cubiertas Ajardinadas. 2011. Available online: http://www.zinco-greenroof.com.ar/referencias/internacional/cubierta_jardin.html (accessed on 12 October 2011).

26.   Carter, T.; Keeler, A. Life-cycle cost-benefit analysis of extensive vegetated roof systems. *J. Environ. Manag.* **2008**, *87*, 350–363. [CrossRef] [PubMed]

27.   Wong, N.H.; Cheong, D.; Yan, H.; Soh, J.; Ong, C.L.; Sia, A. The effect of rooftop garden on energy consumption of a commercial building in Singapore. *Energy Build.* **2003**, *35*, 353–364. [CrossRef]

28.   Scholz-Barth, K.; Tanner, S. *Federal Technology Alert: Green Roofs. Energy Efficiency and Renewable Energy (EERE). U.S.*; Energy Efficiency and Renewable Energy: Washington, DC, USA, 2004.

29.   Kok, K.H.; Mohd Sidek, L.; Chow, M.F.; Zainal Abidin, M.R.; Basri, H.; Hayder, G. Evaluation of green roof performances for urban stormwater quantity and quality controls. *Int. J. River Basin Manag.* **2016**, *14*, 1–7. [CrossRef]

30.   Masseroni, D.; Cislaghi, A. Green roof benefits for reducing flood risk at the catchment scale. *Environ. Earth Sci.* **2016**, *75*, 1–11. [CrossRef]

31.   Czemiel Berndtsson, J. Green roof performance towards management of runoff water quantity and quality: A review. *Ecol. Eng.* **2010**, *36*, 351–360. [CrossRef]

32. Bengtsson, L.; Grahn, L.; Olsson, J. Hydrological function of a thin extensive green roof in southern Sweden. *Nord. Hydrol.* **2005**, *36*, 259–268. [CrossRef]

33. DeNardo, J.C.; Jarrett, A.R.; Manbeck, H.B.; Beattie, D.J.; Berghage, R.D. Stormwater mitigation and surface temperature reduction by green roofs. *Trans. ASAE* **2005**, *48*, 1491–1496. [CrossRef]

34. Moran, A.; Hunt, B.; Smith, J. Hydrological and water quality performance from green roofs in Goldsboro and Raleigh, North Carolina. In Proceedings of the Green Roofs for Healthy Cities Conference, Washington, DC, USA, 4–6 May 2005.

35. Carter, T.L.; Rasmussen, T.C. Hydrologic behavior of vegetated roofs. *J. Am. Water Resour. Assoc.* **2006**, *42*, 1261–1274. [CrossRef]

36. Monterusso, M.A.; Rowe, D.B.; Rugh, C.L.; Russell, D.K. Runoff water quantity and quality from green roof systems. *Acta Hortic.* **2004**, *639*, 369–376. [CrossRef]

37. Bliss, D.J.; Neufeld, R.D.; Ries, R.J. Storm water runoff mitigation using a green roof. *Environ. Eng. Sci.* **2009**, *26*, 407–417. [CrossRef]

38. Van Spengen, J. The Effect of Large-Scale Green Roof Implementation on the Rainfall-Runoff in a Tropical Urbanized Subcatchment: A Singapore Case Study. Master's Thesis, Faculty of Civil Engineering and Geosciences, TUDelft, Dordrecht, The Netherlands, 2010.

39. Minke, G. *Techos Verdes Planificación, Ejecución, Consejos Prácticos*; Editorial Fin de siglo: Montevideo, Uruguay, 2004.

40. Collins, S.; Kuoppamäki, K.; Kotze, D.J.; Xiaoshu, L. Thermal behavior of green roofs under Nordic winter conditions. *Build. Environ.* **2017**, *122*, 206–214. [CrossRef]

41. Theodosiou, T.; Aravantinos, D.; Tsikaloudaki, K. Thermal behaviour of a green vs. a conventional roof under Mediterranean climate conditions. *Int. J. Sustain. Energy* **2014**, *33*, 227–241. [CrossRef]

42. Ouldboukhitine, S.-E.; Belarbi, R.; Djedjig, R. Characterization of green roof components: Measurements of thermal and hydrological properties. *Build. Environ.* **2012**, *56*, 78–85. [CrossRef]

43. D'Orazio, M.; Di Perna, C.; Di Giuseppe, E. Green roof yearly performance: A case study in a highly insulated building under temperate climate. *Energy Build.* **2012**, *55*, 439–451. [CrossRef]

44. Parizzotto, S.; Lamberts, R. Investigation of green roof thermal performance in temperate climate: A case study of an experimental building in Florianopolis city, Southern Brazil. *Energy Build.* **2011**, *43*, 1717–1722. [CrossRef]

45. Silva, C.M.; Gomes, M.G.; Silva, M. Green roofs energy performance in Mediterranean climate. *Energy Build.* **2016**, *116*, 318–325. [CrossRef]

46. Kumar, R.; Kaushik, S.C. Performance evaluation of green roof and shading for thermal protection of buildings. *Build. Environ.* **2005**, *40*, 1505–1511. [CrossRef]

47. Pérez, G.; Vila, A.; Rincón, L.; Solé, C.; Cabeza, L.F. Use of rubber crumbs as drainage layer in green roofs as potential energy improvement material. *Appl. Energy* **2012**, *97*, 347–354. [CrossRef]

48. Coma, J.; Pérez, G.; Solé, C.; Castell, A.; Cabeza, L.F. Thermal assessment of extensive green roofs as passive tool for energy savings in buildings. *Renew. Energy* **2016**, *85*, 1106–1115. [CrossRef]

49. Park, J.; Yoon, J.; Kim, K.-H. Critical Review of the Material Criteria of Building Sustainability Assessment Tools. *Sustainability* **2017**, *9*, 186. [CrossRef]

50. Lee, W. A comprehensive review of metrics for building environmental assessment schemes. *Energy Build.* **2013**, *62*, 403–413. [CrossRef]

 *coatings*

Review

Cellulose Aerogels for Thermal Insulation in Buildings: Trends and Challenges

Danny Illera *, Jaime Mesa, Humberto Gomez and Heriberto Maury

Department of Mechanical Engineering, Universidad del Norte, Barranquilla 081007, Colombia; jamesa@uninorte.edu.co (J.M.); humgomez@uninorte.edu.co (H.G.); hmaury@uninorte.edu.co (H.M.)
* Correspondence: dillera@uninorte.edu.co; Tel.: +57-5-350-9272

Received: 17 August 2018; Accepted: 27 September 2018; Published: 28 September 2018

Abstract: Cellulose-based aerogels hold the potential to become a cost-effective bio-based solution for thermal insulation in buildings. Low thermal conductivities (<0.025 W·m^{-1}·K^{-1}) are achieved through a decrease in gaseous phase contribution, exploiting the Knudsen effect. However, several challenges need to be overcome: production energy demand and cost, moisture sensitivity, flammability, and thermal stability. Herein, a description and discussion of current trends and challenges in cellulose aerogel research for thermal insulation are presented, gathered from studies reported within the last five years. The text is divided into three main sections: (i) an overview of thermal performance of cellulose aerogels, (ii) an identification of challenges and possible solutions for cellulose aerogel thermal insulation, and (iii) a brief description of cellulose/silica aerogels.

Keywords: cellulose; aerogel; thermal insulation; building envelope; silica

---

1. Introduction

Indoor space heating/cooling accounts for over 30% of all the energy consumed in buildings and around 10% of the world's energy consumption [1]. The building envelope thermal performance is a prime factor dictating the amount of energy required for environmental comfort. The building envelope is composed of different elements such as walls and roofs, which separate indoor and outdoor spaces. In this regard, several studies conclude an improvement in energy consumption by the incorporation of thermal insulation materials in walls and roofs as a passive strategy. For example, reductions in annual cooling energy and peak cooling load of up to 20% and 30% were achieved for a building located in a tropical region via implementation of the aforementioned strategy [2]. A thermal insulation material could be defined as one that slows down the heat flow into or out of the building. For its selection, thermal conductivity is the main property to look for. Table 1 lists the typical thermal conductivities of selected commercial building insulation materials for reference. These commercial materials are either derived from petrochemical sources or demand energy-intensive production processes, imparting questions about their role in a future sustainable economy, thus motivating the development of a family of renewable insulation materials and their corresponding low-energy demand processes [3].

Table 1. Thermal conductivity of selected building insulation materials [4].

Material	Thermal Conductivity (W·m^{-1}·K^{-1})
Expanded polystyrene	0.030–0.040
Polyurethane	0.020–0.030
Fiberglass	0.033–0.044
Mineral wool	0.030–0.040

Cellulose is the most abundant bio-polymer on earth, recognized for its renewability, bio-compatibility, and bio-degradability. Cellulose is a linear polysaccharide (β (1−4) linked D-glucose units) generally found as a structural component of cell walls in plants and algae or as a biofilm secretion of some bacteria species. Secondary interactions (van der Waals and hydrogen bonding) between polymer chains promote parallel stacking and subsequent formation of fibrils with a diameter around 5–50 nm containing crystalline and amorphous regions [5]. These fibrils have a density around 1.5 g·cm^{-3}, an average Young modulus of 125 GPa, an average tensile strength of 2.5 GPa, and a surface area close to 800 m^2·g^{-1}. Therefore, cellulose nanofibrils are suitable for the formation of bio-based aerogels for thermal insulation [6].

According to the International Union of Pure and Applied Chemistry (IUPAC), an aerogel refers to a gel comprising a microporous solid in which the dispersed phase is a gas [7]. An aerogel comprises a solid-smoke look alike and highly porous solid (porosity >90% typically). The bulk density of aerogels typically lies between 0.003–0.200 g·cm^{-3} and the surface area between 800–1000 m^2·g^{-1} [8]. A distinction must be made between foams and aerogels. According to Sakai et al. [9], foams tend to have microscale pores with a wall-like solid phase, whereas aerogels have nanoscale pores formed with a network-like skeleton of interconnected nanoparticles (refer to Figure 1).

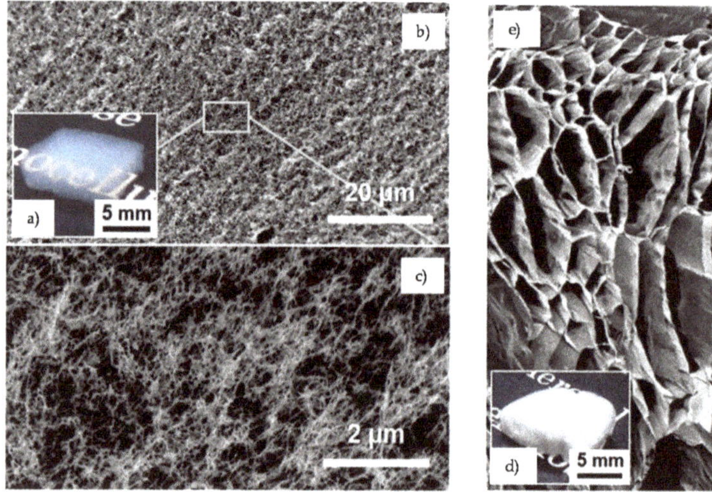

Figure 1. (**a**) Physical appearance of a cellulose aerogel; (**b**,**c**) Scanning electron microscopy images of a cellulose aerogel at different magnifications; (**d**) Physical appearance of a cellulose foam; (**e**) Scanning electron microscopy images of a cellulose foam. Reproduced from Sakai et al. [9], licensed under a Creative Commons Attribution 4.0 International License.

Cellulose aerogels are produced principally by a multistep sol−gel process involving dissolution of cellulose into nanofibrils, solvent exchange, and solvent removal by either supercritical-drying (SD) or freeze-drying (FD) [10–12]. The main technical challenge associated with this approach is to preserve the distribution of the dispersed solids upon solvent removal in order to get a porous material. During supercritical-drying, the solvent is substituted by a fluid at supercritical conditions (CO_2 is the common choice) to prevent the formation of a liquid/vapor interface, and thus, capillary pressure-induced stresses during removal that could lead to collapse of the porous structure. Freeze-drying involves rapid solidification of the dispersion typically by immersion in liquid nitrogen. The solidified solvent (water is the common choice) is then removed by sublimation to prevent the formation of a liquid/vapor interface (refer to Figure 2).

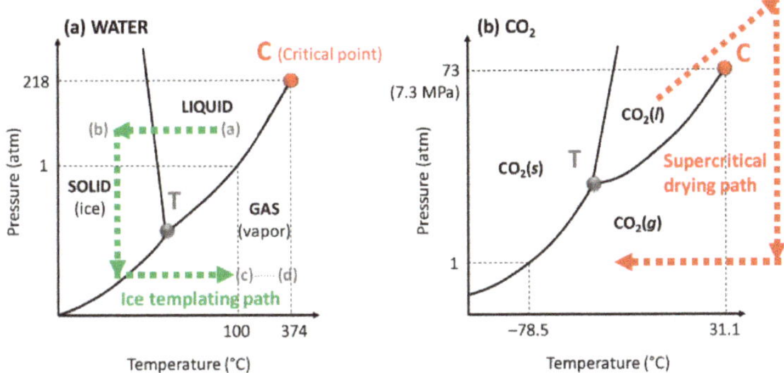

Figure 2. Phase diagrams of freeze-drying (**a**) and supercritical-drying (**b**) superposed on water and CO_2, respectively. Reproduced from from Lavoine & Bergström [10], licensed under a Creative Commons Attribution Non-Commercial 3.0 Unported License.

The thermal conductivity of a porous material like an aerogel could be divided into three contributions: conduction of the solid network, conduction of the gaseous phase, and radiation through or within pores [13]. The contribution of the solid network decreases as the porosity increases because a large quantity of pores restricts the propagation of phonons in the aerogel backbone [14]. The contribution of the gaseous phase is due to elastic collisions between gas molecules. The thermal conductivity of the gaseous phase depends on the mean free path of the gas molecules enclosed in the pores and the pore size. In principle, for the aerogel to be effective designed, its thermal conductivity should be lower than the thermal conductivity of "free" atmospheric air (0.025 $W \cdot m^{-1} \cdot K^{-1}$ at 300 K and 1 atm [15]). The aforementioned is achieved by the presence of pore sizes below the mean free path of air molecules (70 nm at 300 K and 1 atm), thus inhibiting the thermal diffusion of the gas (often call the Knudsen effect [16]). The ratio of the mean free path of gas molecules and the diameter of the pore is known as the Knudsen number. The nanometric size of the pores suppresses gaseous convective thermal transport and radiation.

2. Thermal Performance of Cellulose Aerogels

Cellulose nanofibrils, the building block of cellulose aerogels, present a natural variability as a function of the feedstock and extraction process. Chen et al. [17] measured the thermal properties of cellulose aerogels formed from fibrils extracted by four isolation methods: high-intensity ultra-sonication, hydrochloric acid hydrolysis, (2,2,6,6-tetramethylpiperidin-1-yl) oxidanyl (TEMPO)-mediated oxidation, and sulfuric acid hydrolysis. The highest thermal degradation temperature (342 °C) was reported for the case of hydrochloric-acid hydrolysis-mediated isolation. The lowest thermal degradation temperature (120 °C) corresponded to the case of sulfuric-acid hydrolysis-mediated isolation, attributed to the presence of sulfate groups on the fibril surface. All the aerogels reported thermal conductivities below 0.016 $W \cdot m^{-1} \cdot K^{-1}$ as a consequence of their low densities (0.005 $g \cdot cm^{-3}$) and high porosities.

Surface properties of cellulose nanofibrils also affect the aerogel formation process and final properties. For example, consistent gel formation with sulfated cellulose nanofibrils tends to be more difficult than with carboxylated cellulose nanofibrils [18]. Therefore, aiming to mimic the optical transparency and linear elasticity of silica aerogels, Kobayashi et al. [19] prepared aerogels from surface-carboxylated cellulose nanofibrils dispersed in water in a nematic liquid-crystal order. The lowest thermal conductivity reported was 0.018 $W \cdot m^{-1} \cdot K^{-1}$ for the case of an aerogel with a density of 0.017 $g \cdot cm^{-3}$ and a pore size of approximately 30 nm. The key for obtaining such properties was the preservation of the liquid-crystalline arrangements during the formation of the aerogels.

Regarding the role of morphology on the thermal performance of aerogels, Sakai et al. [9] compared cellulose foams and aerogels with a solid volume fraction ranging from 0.3% to 2.7% made of crystalline cellulose nanofibers. For the case of the foam, the thermal conductivity decreased as the solid volume fraction increased. In principle, the conductivity of air within a microscale pore is due to convection, and thus, has the same conductivity as atmospheric air. However, heat transfer between air and the solid phase plays a major role in the overall conductivity due to the non-interconnected arrangement of the porous structure. Therefore, the tendency between thermal conductivity and solid volume fraction is explained by an increase in the contribution of the interfacial heat transfer as the pore size decreases with the rise in solid volume fraction. For the case of the aerogels, the thermal conductivity increased as the solid volume fraction did. For this case, heat transfer between the gas and solid phases is negligible due to the presence of open pores. Jiménez-Saelices et al. [20] also evaluated the thermal properties of cellulose aerogels as a function of different pore organizations. The aerogels were formed from cellulose nanofibril suspensions by freeze-drying, using two different molds subjected to different temperature gradients. Two cases were evaluated: one aerogel with oriented pore channels and an aerogel with no pore alignment. At a fixed density, aerogels with no pore alignment had more efficient thermal insulating properties than aerogels prepared with oriented pore distribution, and attained a minimal thermal conductivity of 0.024 $W \cdot m^{-1} \cdot K^{-1}$. Although, for the former case, the pores are microscopic; the reduction in thermal conductivity was associated with a decrease in heat transfer by radiation due to the absence of interconnecting channels between pores. This effect is further evidenced in the difference between the thermal conductivity in the transverse (0.030 $W \cdot m^{-1} \cdot K^{-1}$) and axial directions (0.060 $W \cdot m^{-1} \cdot K^{-1}$) of hierarchal aligned cellulose nanofibril foams [21].

Bendahou et al. [22] obtained aerogels with a foam-like morphology; i.e., pores with a wall-like solid phase. The team used nanozeolites and cellulose fibrils to form composite aerogels, expecting a synergy between components. The thermal conductivity decreased 0.028 $W \cdot m^{-1} \cdot K^{-1}$ to 0.018 $W \cdot m^{-1} \cdot K^{-1}$ as the mass fraction of nanoparticles increased. The trend in thermal conductivity was associated with the co-existence of nanozeolite pores and interfibrillar cellulose pores. The porous zeolites were embedded in the mesoporous nanofibrillated structure, thus reducing the gas molecule movement as the mass fraction of the former increased. It is worth noting that the thermal conductivity is similar to the all-cellulose aerogels previously discussed. This behavior could be attributed to the fact that microscopy observation reveaeds that nanozeolites were deposited on cellulose nanofibril films and that the thermal conductivity was relatively high because of the solid conduction between nanofibril films.

Seantier et al. [23] formed aerogels based on combinations of cellulose fiber and cellulose nanofibrils aiming toward tuning the thermal and mechanical properties through aspect ratio and mass proportion modification. The structure of the aerogels could be described as nanofibril films coated upon cellulose fibers. The lowest thermal conductivity reported was 0.023 $W \cdot m^{-1} \cdot K^{-1}$. The insulation performance was described as an interplay between meso- and nanopores. The nanopores were more effective in confining air through the Knudsen effect.

Jiménez-Saelices et al. [24] claimed that the reason for limited reports of aerogels prepared through freeze-drying showing thermal conductivities below 0.020 $W \cdot m^{-1} \cdot K^{-1}$ was due to the presence of macropores in the aerogels prepared by such a method. The team tested a spray freeze-drying technique in order to overcome the aforementioned issue. Aerogels showed a porous interconnected morphology with a pore size ranging from a few tenths of nanometers to a few microns. The lowest thermal conductivity reported was 0.018 $W \cdot m^{-1} \cdot K^{-1}$.

Coquard & Baillis [25] developed a numerical model for estimating the phonon thermal conductivity of cellulose-based aerogels in order to gain a deeper understanding of its thermal behavior. The model used a Kelvin cell representation of the porous structure whose size varied between 57 and 500 nm. At a fixed density, the aerogels with small cells reduced the phonon transport, and consequently, the heat flux due to interactions between phonons and the boundary of the solid phase.

Beyond 100 nm, phonon–phonon scattering events were predominant and the macroscopic laws of heat conduction were then applicable.

Jiménez-Saelices et al. [26] also prepared aerogels from freeze-dried cellulose-stabilized Pickering emulsions as gels (refer to Figure 3). It was claimed that the resistance of these emulsions to shrinkage during sublimation was responsible for the formation of an alveolar morphology and a closed porosity in the walls of such alveolar cells. The aerogels showed low thermal conductivities (0.018 $W \cdot m^{-1} \cdot K^{-1}$) in the "super-insulating" range.

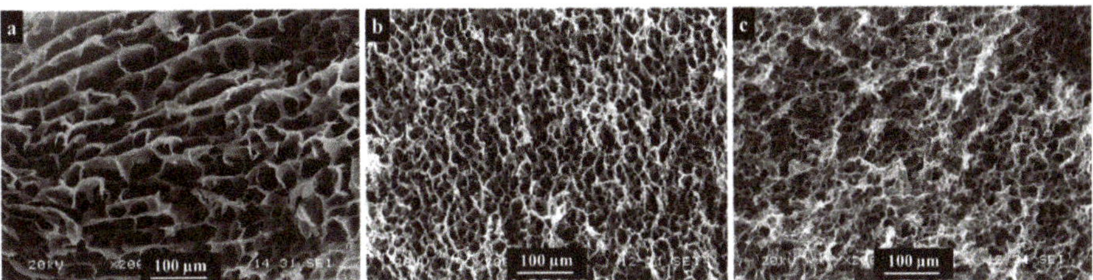

Figure 3. SEM images of aerogels prepared from cellulose-stabilized emulsions at different concentrations: (**a**) 1 wt %, (**b**) 2 wt %, and (**c**) 3 wt %. Reprinted with permission from [26]. Copyright 2018 American Chemical Society.

3. Challenges for Practical Application of Cellulose Aerogels in Thermal Insulation

In addition to thermal performance-related requirements strictly associated with the main function of thermal insulator materials, there are several challenges that need to be overcome for the practical application of cellulose aerogels for such a function.

3.1. Moisture Sensitivity

The hydrophilic characteristic of cellulose is detrimental for the thermal performance of the aerogels because it induces an eventual increase in thermal conductivity of the material through densification of the matrix as air moisture is retained. In a study carried out by Apostolopoulou-Kalkavoura et al. [27], the thermal conductivity of cellulose foams increased more than threefold as the relative humidity (RH) did from 2% to 80%. The main strategy aiming to solve the moisture sensitivity of cellulose involves modification of the surface of the aerogel.

Shi et al. [28,29] performed a hydrophobic modification of cellulose/silica aerogels using CCl_4 as plasma. The aerogels were prepared using NaOH/thiourea as a solvent system and the freeze-drying method. The modification turned the surface of aerogels from hydrophilic to hydrophobic, reporting a water contact angle of 132° for the cellulose/silica system.

Zhai et al. [30] formed poly(vinyl alcohol)/cellulose aerogels using a unidirectional freeze-drying process followed by chemical vapor deposition of methyltrichlorosilane. The silane-coated aerogels became hydrophobic, reporting a water contact angle of 141.8°.

Li et al. [31] fabricated cellulose/silica aerogels surface-treated by immersion in a trimethylchlorosilane/n-hexane solution. The aerogels reported a water contact angle of 139.6°. Trimethylchlorosilane reacts with water and Si–OH groups on the surface of the aerogels, substituting the hydrophilic –OH groups with hydrophobic $-CH_3$ groups.

Jiang et al. [32] aimed to simultaneously improve the mechanical properties and increase hydrophobicity of cellulose nanofibril aerogels by cross-linking cellulose at their surface hydroxyls with diisocyanate. Cross-linked aerogels reported a higher modulus (209 kPa) than their counterparts (94 kPa).

3.2. Flame-Retardant Performance and Thermal Stability

Cellulose by itself is a highly flammable biopolymer with a limiting oxygen index (LOI) near 18 [33]. Furthermore, the large surface area of cellulose aerogels facilitates the ignition process. LOI is a measure of the limiting concentration of oxygen in an oxygen/nitrogen mixture necessary for sustaining combustion once a material is ignited. Therefore, a material with an LOI value above 21 should not burn in air (at room temperature) unless there is a constant transfer of heat from a source. The general strategy explored to overcome this issue is to create cellulose-composite aerogels either by addition or in situ formation of flame-retardant agents. In this regard, inorganic nanoparticles are the main choice.

Fan et al. [34] prepared an aerogel based on cellulose nanofibrils and aluminum oxyhydroxide. Aluminum oxyhydroxide is a widely used inorganic flame-retardant additive because of its non-toxic and non-volatile nature. When exposed to a flame, the aerogel did not burn, and kept an integrated shape even after 60 s of continuous exposure. A control sample aerogel made only of cellulose nanofibrils burned immediately upon contact with the flame.

Yang et al. [35] prepared an aerogel consisting of a cellulose nanofibril backbone encapsulated by a layer of ultrathin MoS_2 nanosheets. The nanocomposite aerogel reported an LOI of 34.7 and a thermal conductivity of 0.028 $W \cdot m^{-1} \cdot K^{-1}$. A vertical burning test further demonstrated a self-extinguishing capability of the aerogel. Evidence suggested the presence of bonding between MoS_2 nanosheets and the cellulose nanofibril as a result of chemical cross-linking between Mo^{4+} cations, and carboxyl (–COOH) and hydroxyl (–OH) groups in the cellulose.

Han et al. [36] proposed an alternative approach to the conventional mixing of flame-retardant additives with cellulose nanofibrils, aiming to avoid additive agglomeration. The team used a three-dimensional nanoporous cellulose gel prepared by dissolution and coagulation of cellulose from an aqueous NaOH/urea solution as template for the growth of magnesium-hydroxide nanoparticles. Once ignited, the flame propagation slowed and extinguished within 40 s for the cellulose/MgOH aerogel. A control pure cellulose aerogel completely burned after 10 s.

Köklükaya et al. [37] deposited nanometric films of cationic chitosan, anionic poly(vinylphosphonic acid), and anionic montmorillonite clay on cellulose nanofibril aerogels through a layer-by-layer approach. The aerogels showed a non-ignition behavior when exposed to heat flux, whereas an unmodified sample ignited and burned completely. It was suggested that chitosan, poly(vinylphosphonic acid), and montmorillonite promote char formation and reduce the release of flammable volatiles.

Kaya [38] prepared a cellulose nanofibril aerogel cross-linked with citric acid. The non-cross-linked aerogel, once ignited, burned 25% of the sample after 34 s. The cross-linked aerogel was more slowly ignited and 25% of the sample burned after 51 s.

The thermal degradation of cellulose in nitrogen is a one-step process involving a competition between depolymerization and dehydration. In air, cellulose undergoes a two-step thermo-oxidation process: (i) aliphatic char and flammable volatile formation (at 300–400 °C) and (ii) oxidation of aliphatic char (at 400–600 °C) and release of carbon monoxide and dioxide [39]. Smith et al. [40] coated cellulose nanofibril aerogel with aluminum oxide through atomic layer deposition aiming toward improving the oxidation resistance of cellulose. The temperature of onset decomposition increased by 120 °C.

Silica layers were grown on the surface of bacterial cellulose nanofibrils in order to prepare freeze-dried aerogels by Liu et al. [41]. The silica-coated cellulose nanofibrils were further exposed to >400 °C, removing the cellulose cores and leading to silica nanotubes with internal dimensions equivalent to the dimensions of the nanofibrils. The decomposition temperature shifted 70 °C, reaching 335 °C under nitrogen atmosphere. The coverage of silica on the cellulose restricted the accessibility of oxygen and limited the degassing of degradation products.

Hydroxyapatite is an inorganic component found in bones making it an abundant and non-toxic choice as a flame-retardant additive. This motivated Guo et al. [42] to fabricate a series of cellulose/hydroxyapatite aerogels through a freeze-drying approach. Although the thermal conductivity was relatively high (0.038 W·m^{-1}·K^{-1}), the addition of the inorganic phase provided the material a low peak heat release rate (20.4 kW·m^{-2}) and total heat release (1.21 MJ·m^{-2}) according to cone calorimetry tests. It was believed that the layer of hydroxyapatite covering cellulose nanofibers inhibited oxygen diffusion into the fibers, restricting the escape of volatile products.

3.3. Processing Energy Demand and Cost of Cellulose Aerogels

"Super-insulating" cellulose aerogels were reported promising half the material thickness for thermal insulation panels compared to expanded polystyrene, polyurethane, fiberglass, or mineral wool. However, the cost of an aerogel panel (silica for example) is up to 10 times that of those made of traditional insulation materials [43]. Although cellulose aerogels are only developed and tested at the lab scale so far, they are expected to have the same cost tendency as conventional aerogel panels. The cost of the panel is partially dictated by raw materials and their processing toward aerogel formation. In this regard, the main drawbacks of both supercritical-drying and freeze-drying approaches are their high capital investment, high energy demand, and long processing times.

In principle, breakthroughs in cost reduction of cellulose aerogel production could be achieved by performing the process at pressures and temperatures near atmospheric reference. Therefore, ambient pressure drying (APD) was proposed as a suitable alternative to FD and SD. For this approach to be effective, the collapse of the porous structure during drying due to capillary forces must be avoided. The general strategy to accomplish the aforementioned goal involves surface modification of cellulose into a hydrophobic nature. The techniques used to transform cellulose surfaces into a hydrophobic nature during APD differ from those presented in Section 3.1 because the former must be performed prior drying of the hydrogel. Surface modifications using trityl chloride [44], anhydride [45], and octylamine [46] were reported. In addition to surface hydrophobization, enhancing the hydrogel mechanical strength by means of cellulose nanofibril entanglement allowed the formation of low-density (0.018 g·cm^{-3}) APD aerogels [47].

APD requires solvent exchange prior to drying; examples in the literature include the use of n-hexane [48], 2-propanol [47], ethanol [49,50], and 2-propanol followed by octane [51]. This represents another opportunity for cost reduction: finding ways to incorporate solvents with low vapor pressure, such as ionic liquids [52]. Upon finding a suitable ionic solvent for cellulose aerogel processing, the main advantage will be the opportunity to reuse these solvents after the drying step.

De France et al. [53] identified other processing methods that could reduce the cost of aerogel preparation by targeting scalability, including spray drying, freeze spray drying, and pressurized gas-expanded drying.

4. Cellulose/Silica Aerogels for Thermal Insulation

Silica aerogels are recognized due to their low thermal conductivity (–0.012 W·m^{-1}·K^{-1}) and low flammability [54]. Production costs and brittle mechanical behavior hinder their application for thermal insulation. To overcome this, several strategies were reported involving cellulose nanofibrils as cross-linkers or as a scaffold/template for aerogel formation. However, the improvement in mechanical properties generally comes at the expense of an increase in thermal conductivity of the resulting composite due to densification of the aerogel by the addition of a reinforcement phase. For example, Sai et al. [55] used a freeze-dried bacterial cellulose fibrous mat impregnated with silica-based sol to obtain a composite achieving a thermal conductivity of 0.037 W·m^{-1}·K^{-1} compared to 0.030 W·m^{-1}·K^{-1} for cellulose aerogel alone.

Zhao et al. [56] formed an interpenetrating network of mesoporous silica within a silylated-cellulose nanofibril template. The cellulose/silica system reported increases (relative to a control silica aerogel) in the Young modulus and ultimate stress under compression of 55% and 126%, respectively, while maintaining a thermal conductivity of 0.017 W·m^{-1}·K^{-1}.

Wong et al. [57] prepared a cellulose/silica aerogel by dispersing cellulose nanofibrils on a poly(ethoxydisiloxane) sols prior to gelation. No significant enhancements in elastic modulus, compressive strength, or fracture strain could be identified or separated from the effects of increasing density. However, the cellulose/silica aerogel reported tensile strengths 25%–40% higher than that for a silica aerogel of comparable density. The thermal conductivity of the cellulose/silica system increased by -11% relative to the silica aerogel.

Silica aerogels are formed through a two-step acid–base-catalyzed sol–gel route from tetraalkoxysilane precursors. Therefore, Fu et al. [58] evaluated the effect of cellulose nanofibril concentration, tetraethyl orthosilicate concentration, pH of the condensation process, and immersion time on the physical and mechanical properties of cellulose/silica aerogels. Aerogels were prepared by the impregnation of a cellulose aerogel with a solution of silica, and subsequent hydrolysis and condensation at different pH values [59]. Optimal process parameters were determined through response surface methodology based on a Box–Behnken experimental design. The compressive Young modulus and the ultimate strength were 13–36 times higher and 8–30 times higher, respectively, when compared with silica aerogel.

Demilecamps et al. [60] aimed to decrease the thermal conductivity of cellulose/silica aerogels by "filling" their pores with silica and using cellulose as the backbone. The thermal conductivity decreased from 0.033 $W \cdot m^{-1} \cdot K^{-1}$ for an all-cellulose aerogel to 0.027 $W \cdot m^{-1} \cdot K^{-1}$ for the cellulose/silica system. The thermal conductivity remained above the "super-insulating" range possibly due to the increase in skeletal heat conduction of the composite with the increasing load on silica.

Laskowski et al. [61] approached a strategy to get cellulose/silica aerogel composites in which silica particles were added during the formation of a cellulose aerogel without disturbing the process of gel conformation. The thermal conductivity lay between 0.040 and 0.052 $W \cdot m^{-1} \cdot K^{-1}$ due to the fibrillary three-dimensional network continuously connected that allowed heat conduction through the backbone.

5. Outlook

Cellulose aerogels with "super-insulating" performance (thermal conductivity <0.025 $W \cdot m^{-1} \cdot K^{-1}$) can be produced through supercritical-drying or freeze-drying. However, the nature of such processes lags the commercialization of the aerogels. The main reasons associated with the aforementioned issue involves challenges in the industrial scalability of the processes, the slow-batch nature of the methods, and the need for specialized equipment due to the temperatures and pressures required. Furthermore, these processes are intensive in chemical usage (*N*-methylmorpholine *N*-oxide, alkali hydroxide/urea, lithium chloride/dimethylacetamine, and lithium chloride/dimethyl sulfoxide to name a few [62]) imparting questions about their sustainability. Therefore, there is a need for novel sustainable cost-effective processes like APD. In addition to APD, other techniques were also reported, although they were either less or not explored for thermal insulation, such as vacuum filtration [51] and freeze-casting [63]. Other techniques used for porous scaffold synthesis for tissue engineering could be explored in order to assess their feasibility in the fabrication of cellulose aerogels, including electrospinning [64], additive manufacturing [65], gas-foaming [66], and compression-molding [67], among others.

Cellulose by itself is an attractive choice as a raw material for aerogel construction because of its natural abundance. However, cellulose nanofibrils, with a diameter of 5–20 nm, constitute the backbone of reported aerogels. These nanofibrils are not widely commercially available because their production is time-consuming and their yield is usually low [68]. Furthermore, depending on the cellulose source and isolation technique, the resulting nanofibril can vary in crystal structure, degree of crystallinity, morphology, aspect ratio, and surface chemistry [69]. Despite the concern, no detailed studies could be found regarding the effect of these parameters on aerogel thermal performance.

Table 2 is a summary of the properties of cellulose aerogels reported within the last five years. Direct comparison among aerogel properties is interfered with by the lack of standardized characterization protocols. Furthermore, most studies focus on the characteristics of the aerogels themselves. Inspection of Table 2 suggests that SD offers lower thermal conductivities than FD.

This could be explained because the growth of ice crystals during FD leads to larger pores compared to SD. A minimum in thermal conductivity is reached for a specific aerogel density. This is because the contribution of solid-phase heat conduction increases with density, whereas, at really low densities, pore size goes beyond 70 nm.

Table 2. Properties of interest for selected cellulose aerogels gathered from scientific articles published within the last five years.

Material	Aerogel Formation Method	Density $(g \cdot cm^{-3})$	Young Modulus (MPa)	Ultimate Strength (MPa)	Thermal Conductivity $(W \cdot m^{-1} \cdot K^{-1})$	T_{Onset} (°C)	Water Contact Angle (°)	LOI (%)
Cellulose (HCl hydrolysis) [17]	FD	0.005	–	–	0.016	342	–	–
Cellulose (surface-carboxylated) [18]	SD	0.017	0.40	–	0.018	–	–	–
Cellulose [18]	–	–	–	13.000	0.030	–	–	–
Cellulose (TEMPO-mediated oxidation) [22]	FD	–	0.10	–	0.018	–	–	–
Cellulose (TEMPO-mediated oxidation) [9]	FD	–	–	–	0.022	–	–	–
Cellulose (TEMPO-mediated oxidation) [9]	SD	–	–	–	0.017	–	–	–
Cellulose (TEMPO-mediated oxidation) [23]	FD	0.035	0.18	–	0.023	–	–	–
Cellulose (TEMPO-mediated oxidation) [24]	Spray FD	0.022	–	–	0.018	–	–	–
Cellulose (TEMPO-mediated oxidation) [20]	FD	0.020	–	–	0.024	–	–	–
Cellulose/silica + cold plasma modification [28,29]	FD	0.023	–	–	0.026	–	132.0	–
Cellulose [26]	FD	0.020	0.60	–	0.018	–	–	–
Poly(vinyl alcohol)/ cellulose (TEMPO-mediated oxidation) + chemical vapor deposition of methyltrichlorosilane [30]	Unidirectional FD	0.018	10.4	–	–	495 at N_2	141.8	–
Cellulose(sulfuric acid hydrolysis)/silica [31]	APD	0.137	–	–	0.035	253 at air	139.6	–
Cellulose cross-linked with diisocyanate [32]	Freezing–thawing	–	0.209	0.066	–	200 at N_2	–	–
Cellulose/MoS$_2$ [35]	FD	0.005	0.326	–	0.028	300 at air	–	34.7
Cellulose/MgOH [36]	FD	0.400	–	0.467	0.081	225 at N_2	–	–
Cellulose/poly(vinyl alcohol)/chitosan/ montmorillonite [37]	FD	–	–	–	–	273 at air	–	–
Cellulose/aluminum oxide [40]	SD	–	–	–	–	295 at air	–	–
Bacterial cellulose/silica [41]	FD	–	–	–	–	335 at N_2	–	–
Cellulose/silica [56]	SD	0.122	1.750	0.375	0.017	–	–	–
Cellulose/silica [57]	SD	0.143	2.200	1.500	0.015	–	·	–
Cellulose/silica [58]	APD	0.100	6.570	1.37	0.023	–	152.1	–
Cellulose/silica [60]	SD	0.225	11.500	–	0.027	–	–	–
Cellulose/silica [61]	SD	0.150	2.000	–	0.045	–	–	–
Cellulose/hydroxipatite [42]	FD	–	0.253	0.060	0.038	–	–	–

FD: freeze-drying; SD: supercritical-drying; APD: ambient pressure drying; T_{Onset}: initial decomposition temperature; LOI: limiting oxygen index; TEMPO: (2,2,6,6-tetramethylpiperidin-1-yl)oxidanyl.

Regarding the composite-based strategies for overcoming moisture sensitivity, flammability, and thermal stability, the main challenge is associated with the solution-based approach for composite-cellulose aerogel formation. In particular, the question to be answered is how to get a fine dispersion of the cellulose/host component into the viscous cellulose hydrogel without disrupting the distribution of the dispersed solids during solvent removal. Similarly, the improvement in mechanical properties of silica/cellulose aerogels comes at the expense of high loading of cellulose in the system. However, increasing the cellulose content may lead to an increase in density and a subsequent decrease in pore size, which will negatively impact thermal insulation performance.

6. Conclusions

Cellulose-based aerogels can have thermal conductivities as low as –0.015 $W \cdot m^{-1} \cdot K^{-1}$ falling into the category of "super-insulating" materials. This range of thermal conductivities seems to be a consequence of an interplay between different factors, such as a non-interconnected porous morphology with an average pore size bellow 70 nm. Low density (<0.025 $g \cdot cm^{-3}$) is also required to decrease the contribution of heat conduction through the solid matrix. However, a continuous decrease in density will eventually lead to an increase in pore size beyond 70 nm, thereby increasing the thermal conduction of the porous solid.

The following topics could be drawn as points of interest for further development of cellulose aerogels as thermal insulators in buildings:

- In addition to cost reduction, the improvement of cellulose aerogel production techniques should also point toward sustainability. Otherwise, the benefits from cellulose as a raw material will be swept away by processing.

- Improvements in moisture sensitivity, flammability, and thermal stability of aerogels following the current approaches will come at the expense of a densification of the porous matrix, increasing the contribution of solid conduction.

- The effects of crystal structure, degree of crystallinity, morphology, aspect ratio, and surface chemistry of cellulose nanofibrils on the cellulose aerogel thermal performance are partially understood.

- Defining and adopting standardized aerogel characterization protocols is required to allow proper comparison among research outcomes. In particular, for the determination of thermal conductivity, the guarded hot-plate technique is considered the most accurate and reliable for such a matter, although other techniques such as transient hot-wire/hot disk and laser flash analysis represent suitable alternatives.

Author Contributions: Conceptualization, D.I.; Investigation, D.I. and J.M.; Resources, D.I. and J.M.; Writing-Original Draft Preparation, D.I. and J.M.; Writing-Review & Editing, D.I., J.M., H.M. and H.G.; Supervision, H.M. and H.G.; Project Administration, H.M. and H.G.

Acknowledgments: The authors would like to acknowledge the Universidad Del Norte and the Departamento Administrativo de Ciencia, Tecnología e Innovación (Colciencias) for the support given within the framework of the PhD National Scholarship Colciencias N° 617-2014 (Contract Identification Number UN-OJ-2014-26159 and UN-OJ-2014-24072).

References

1. LaFrance, M. *Technology Roadmap: Energy Efficient Building Envelopes*; IEA: Paris, France, 2013.
2. Mirrahimi, S.; Mohamed, M.F.; Haw, L.C.; Ibrahim, N.L.N.; Yusoff, W.F.M.; Aflaki, A. The effect of building envelope on the thermal comfort and energy saving for high-rise buildings in hot–humid climate. *Renew. Sustain. Energy Rev.* **2016**, *53*, 1508–1519. [CrossRef]
3. Asdrubali, F.; D'Alessandro, F.; Schiavoni, S. A review of unconventional sustainable building insulation materials. *Sustain. Mater. Technol.* **2015**, *4*, 1–17. [CrossRef]
4. Jelle, B.P. Traditional, state-of-the-art and future thermal building insulation materials and solutions—Properties, requirements and possibilities. *Energy Build.* **2011**, *43*, 2549–2563. [CrossRef]
5. Moon, R.J.; Martini, A.; Nairn, J.; Simonsen, J.; Youngblood, J. Cellulose nanomaterials review: Structure, properties and nanocomposites. *Chem. Soc. Rev.* **2011**, *40*, 3941–3994. [CrossRef] [PubMed]
6. Long, L.-Y.; Weng, Y.-X.; Wang, Y.-Z. Cellulose Aerogels: Synthesis, Applications, and Prospects. *Polymers* **2018**, *10*, 623. [CrossRef]
7. Nič, M.; Jirát, J.; Košata, B.; Jenkins, A.; McNaught, A. *IUPAC Compendium of Chemical Terminology*, 2nd ed.; IUPAC: Research Triagle Park, NC, USA, 2009.

8. Pierre, A.C. History of Aerogels. In *Aerogels Handbook*; Aegerter, M.A., Leventis, N., Koebel, M.M., Eds.; Springer: New York, NY, USA, 2011; pp. 3–18.

9. Sakai, K.; Kobayashi, Y.; Saito, T.; Isogai, A. Partitioned airs at microscale and nanoscale: Thermal diffusivity in ultrahigh porosity solids of nanocellulose. *Sci. Rep.* **2016**, *6*, 20434. [CrossRef] [PubMed]

10. Lavoine, N.; Bergström, L. Nanocellulose-based foams and aerogels: Processing, properties, and applications. *J. Mater. Chem. A* **2017**, *5*, 16105–16117. [CrossRef]

11. Şahin, İ.; Özbakır, Y.; İnönü, Z.; Ulker, Z.; Erkey, C. Kinetics of Supercritical Drying of Gels. *Gels* **2017**, *4*, 3. [CrossRef]

12. García-González, C.A.; Camino-Rey, M.C.; Alnaief, M.; Zetzl, C.; Smirnova, I. Supercritical drying of aerogels using CO_2: Effect of extraction time on the end material textural properties. *J. Supercrit. Fluids* **2012**, *66*, 297–306. [CrossRef]

13. Fricke, J. Thermal transport in porous superinsulations. In *Aerogels*; Springer: New York, NY, USA, 1986; pp. 94–103.

14. Fricke, J.; Emmerling, A. Aerogels. *J. Am. Ceram. Soc.* **1992**, *75*, 2027–2035. [CrossRef]

15. Vargaftik, N.B. *Handbook of Thermal Conductivity of Liquids and Gases*; CRC Press: Boca Raton, FL, USA, 1993.

16. Smith, D.M.; Maskara, A.; Boes, U. Aerogel-based thermal insulation. *J. Non-Cryst. Solids* **1998**, *225*, 254–259. [CrossRef]

17. Chen, W.; Li, Q.; Wang, Y.; Yi, X.; Zeng, J.; Yu, H.; Liu, Y.; Li, J. Comparative Study of Aerogels Obtained from Differently Prepared Nanocellulose Fibers. *ChemSusChem* **2014**, *7*, 154–161. [CrossRef] [PubMed]

18. Buesch, C.; Smith, S.W.; Eschbach, P.; Conley, J.F.; Simonsen, J. The Microstructure of Cellulose Nanocrystal Aerogels as Revealed by Transmission Electron Microscope Tomography. *Biomacromolecules* **2016**, *17*, 2956–2962. [CrossRef] [PubMed]

19. Kobayashi, Y.; Saito, T.; Isogai, A. Aerogels with 3D Ordered Nanofiber Skeletons of Liquid-Crystalline Nanocellulose Derivatives as Tough and Transparent Insulators. *Angew. Chem. Int. Ed.* **2014**, *53*, 10394–10397. [CrossRef] [PubMed]

20. Jiménez-Saelices, C.; Seantier, B.; Cathala, B.; Grohens, Y. Effect of freeze-drying parameters on the microstructure and thermal insulating properties of nanofibrillated cellulose aerogels. *J. Sol-Gel Sci. Technol.* **2017**, *84*, 475–485. [CrossRef]

21. Li, T.; Song, J.; Zhao, X.; Yang, Z.; Pastel, G.; Xu, S.; Jia, C.; Dai, J.; Chen, C.; Gong, A.; et al. Anisotropic, lightweight, strong, and super thermally insulating nanowood with naturally aligned nanocellulose. *Sci. Adv.* **2018**, *4*, eaar3724. [CrossRef] [PubMed]

22. Bendahou, D.; Bendahou, A.; Seantier, B.; Grohens, Y.; Kaddami, H. Nano-fibrillated cellulose-zeolites based new hybrid composites aerogels with super thermal insulating properties. *Ind. Crops Prod.* **2015**, *65*, 374–382. [CrossRef]

23. Seantier, B.; Bendahou, D.; Bendahou, A.; Grohens, Y.; Kaddami, H. Multi-scale cellulose based new bio-aerogel composites with thermal super-insulating and tunable mechanical properties. *Carbohydr. Polym.* **2016**, *138*, 335–348. [CrossRef] [PubMed]

24. Jiménez-Saelices, C.; Seantier, B.; Cathala, B.; Grohens, Y. Spray freeze-dried nanofibrillated cellulose aerogels with thermal superinsulating properties. *Carbohydr. Polym.* **2017**, *157*, 105–113. [CrossRef] [PubMed]

25. Coquard, R.; Baillis, D. Thermal conductivity of Kelvin cell cellulosic aerogels: Analytical and Monte Carlo approaches. *J. Mater. Sci.* **2017**, *52*, 11135–11145. [CrossRef]

26. Jiménez-Saelices, C.; Seantier, B.; Grohens, Y.; Capron, I. Thermal Superinsulating Materials Made from Nanofibrillated Cellulose-Stabilized Pickering Emulsions. *ACS Appl. Mater. Interfaces* **2018**, *10*, 16193–16202. [CrossRef] [PubMed]

27. Apostolopoulou-Kalkavoura, V.; Gordeyeva, K.; Lavoine, N.; Bergström, L. Thermal conductivity of hygroscopic foams based on cellulose nanofibrils and a nonionic polyoxamer. *Cellulose* **2018**, *25*, 1117–1126. [CrossRef]

28. Shi, J.; Lu, L.; Guo, W.; Sun, Y.; Cao, Y. An environment-friendly thermal insulation material from cellulose and plasma modification. *J. Appl. Polym. Sci.* **2013**, *130*, 3652–3658. [CrossRef]

29. Shi, J.; Lu, L.; Guo, W.; Zhang, J.; Cao, Y. Heat insulation performance, mechanics and hydrophobic modification of cellulose–SiO_2 composite aerogels. *Carbohydr. Polym.* **2013**, *98*, 282–289. [CrossRef] [PubMed]

30. Zhai, T.; Zheng, Q.; Cai, Z.; Turng, L.-S.; Xia, H.; Gong, S. Poly(vinyl alcohol)/Cellulose Nanofibril Hybrid Aerogels with an Aligned Microtubular Porous Structure and Their Composites with Polydimethylsiloxane. *ACS Appl. Mater. Interfaces* **2015**, *7*, 7436–7444. [CrossRef] [PubMed]

31. Li, M.; Jiang, H.; Xu, D.; Yang, Y. A facile method to prepare cellulose whiskers–silica aerogel composites. *J. Sol-Gel Sci. Technol.* **2017**, *83*, 72–80. [CrossRef]

32. Jiang, F.; Hsieh, Y.-L. Cellulose Nanofibril Aerogels: Synergistic Improvement of Hydrophobicity, Strength, and Thermal Stability via Cross-Linking with Diisocyanate. *ACS Appl. Mater. Interfaces* **2017**, *9*, 2825–2834. [CrossRef] [PubMed]

33. Tian, C.M.; Shi, Z.H.; Zhang, H.Y.; Xu, J.Z.; Shi, J.R.; Guo, H.Z. Thermal degradation of cotton cellulose. *J. Therm. Anal. Calorim.* **1999**, *55*, 93–98. [CrossRef]

34. Fan, B.; Chen, S.; Yao, Q.; Sun, Q.; Jin, C. Fabrication of cellulose nanofiber/AlOOH aerogel for flame retardant and thermal insulation. *Materials* **2017**, *10*, 311. [CrossRef] [PubMed]

35. Yang, L.; Mukhopadhyay, A.; Jiao, Y.; Yong, Q.; Chen, L.; Xing, Y.; Hamel, J.; Zhu, H. Ultralight, highly thermally insulating and fire resistant aerogel by encapsulating cellulose nanofibers with two-dimensional MoS$_2$. *Nanoscale* **2017**, *9*, 11452–11462. [CrossRef] [PubMed]

36. Han, Y.; Zhang, X.; Wu, X.; Lu, C. Flame Retardant, Heat Insulating Cellulose Aerogels from Waste Cotton Fabrics by in Situ Formation of Magnesium Hydroxide Nanoparticles in Cellulose Gel Nanostructures. *ACS Sustain. Chem. Eng.* **2015**, *3*, 1853–1859. [CrossRef]

37. Köklükaya, O.; Carosio, F.; Wågberg, L. Superior Flame-Resistant Cellulose Nanofibril Aerogels Modified with Hybrid Layer-by-Layer Coatings. *ACS Appl. Mater. Interfaces* **2017**, *9*, 29082–29092. [CrossRef] [PubMed]

38. Kaya, M. Super absorbent, light, and highly flame retardant cellulose-based aerogel crosslinked with citric acid. *J. Appl. Polym. Sci.* **2017**, *134*, 45315. [CrossRef]

39. Soares, S.; Camino, G.; Levchik, S. Comparative study of the thermal decomposition of pure cellulose and pulp paper. *Polym. Degrad. Stab.* **1995**, *49*, 275–283. [CrossRef]

40. Smith, S.W.; Buesch, C.; Matthews, D.J.; Simonsen, J.; Conley, J.F. Improved oxidation resistance of organic/inorganic composite atomic layer deposition coated cellulose nanocrystal aerogels. *J. Vac. Sci. Technol. A Vac. Surf. Film.* **2014**, *32*, 041508. [CrossRef]

41. Liu, D.; Wu, Q.; Andersson, R.L.; Hedenqvist, M.S.; Farris, S.; Olsson, R.T. Cellulose nanofibril core–shell silica coatings and their conversion into thermally stable nanotube aerogels. *J. Mater. Chem. A* **2015**, *3*, 15745–15754. [CrossRef]

42. Guo, W.; Wang, X.; Zhang, P.; Liu, J.; Song, L.; Hu, Y. Nano-fibrillated cellulose-hydroxyapatite based composite foams with excellent fire resistance. *Carbohydr. Polym.* **2018**, *195*, 71–78. [CrossRef] [PubMed]

43. Koebel, M.; Rigacci, A.; Achard, P. Aerogel-based thermal superinsulation: An overview. *J. Sol-Gel Sci. Technol.* **2012**, *63*, 315–339. [CrossRef]

44. Pour, G.; Beauger, C.; Rigacci, A.; Budtova, T. Xerocellulose: Lightweight, porous and hydrophobic cellulose prepared via ambient drying. *J. Mater. Sci.* **2015**, *50*, 4526–4535. [CrossRef]

45. Sehaqui, H.; Zimmermann, T.; Tingaut, P. Hydrophobic cellulose nanopaper through a mild esterification procedure. *Cellulose* **2014**, *21*, 367–382. [CrossRef]

46. Cervin, N.T.; Andersson, L.; Ng, J.B.S.; Olin, P.; Bergström, L.; Wågberg, L. Lightweight and Strong Cellulose Materials Made from Aqueous Foams Stabilized by Nanofibrillated Cellulose. *Biomacromolecules* **2013**, *14*, 503–511. [CrossRef] [PubMed]

47. Li, Y.; Tanna, V.A.; Zhou, Y.; Winter, H.H.; Watkins, J.J.; Carter, K.R. Nanocellulose Aerogels Inspired by Frozen Tofu. *ACS Sustain. Chem. Eng.* **2017**, *5*, 6387–6391. [CrossRef]

48. Fu, J.; He, C.; Wang, S.; Chen, Y. A thermally stable and hydrophobic composite aerogel made from cellulose nanofibril aerogel impregnated with silica particles. *J. Mater. Sci.* **2018**, *53*, 7072–7082. [CrossRef]

49. Zhang, X.; Liu, P.; Duan, Y.; Jiang, M.; Zhang, J. Graphene/cellulose nanocrystals hybrid aerogel with tunable mechanical strength and hydrophilicity fabricated by ambient pressure drying technique. *RSC Adv.* **2017**, *7*, 16467–16473. [CrossRef]

50. Markevicius, G.; Ladj, R.; Niemeyer, P.; Budtova, T.; Rigacci, A. Ambient-dried thermal superinsulating monolithic silica-based aerogels with short cellulosic fibers. *J. Mater. Sci.* **2017**, *52*, 2210–2221. [CrossRef]

51. Toivonen, M.S.; Kaskela, A.; Rojas, O.J.; Kauppinen, E.I.; Ikkala, O. Ambient-Dried Cellulose Nanofibril Aerogel Membranes with High Tensile Strength and Their Use for Aerosol Collection and Templates for Transparent, Flexible Devices. *Adv. Funct. Mater.* **2015**, *25*, 6618–6626. [CrossRef]

52. Shukla, N.; Fallahi, A.; Kosny, J. Aerogel Thermal Insulation-Technology Review and Cost Study for Building Enclosure Applications. *ASHRAE Trans.* **2014**, *120*, 294–307.

53. De France, K.J.; Hoare, T.; Cranston, E.D. Review of Hydrogels and Aerogels Containing Nanocellulose. *Chem. Mater.* **2017**, *29*, 4609–4631. [CrossRef]

54. Pierre, A.C.; Rigacci, A. SiO$_2$ Aerogels. In *Aerogels Handbook*; Springer: New York, NY, USA, 2011; pp. 21–45.

55. Sai, H.; Xing, L.; Xiang, J.; Cui, L.; Jiao, J.; Zhao, C.; Li, Z.; Li, F.; Zhang, T. Flexible aerogels with interpenetrating network structure of bacterial cellulose–silica composite from sodium silicate precursor via freeze drying process. *RSC Adv.* **2014**, *4*, 30453. [CrossRef]

56. Zhao, S.; Zhang, Z.; Sèbe, G.; Wu, R.; Rivera Virtudazo, R.V.; Tingaut, P.; Koebel, M.M. Multiscale assembly of superinsulating silica aerogels within silylated nanocellulosic scaffolds: Improved mechanical properties promoted by nanoscale chemical compatibilization. *Adv. Funct. Mater.* **2015**, *25*, 2326–2334. [CrossRef]

57. Wong, J.C.H.; Kaymak, H.; Tingaut, P.; Brunner, S.; Koebel, M.M. Mechanical and thermal properties of nanofibrillated cellulose reinforced silica aerogel composites. *Microporous Mesoporous Mater.* **2015**, *217*, 150–158. [CrossRef]

58. Fu, J.; He, C.; Huang, J.; Chen, Z.; Wang, S. Cellulose nanofibril reinforced silica aerogels: Optimization of the preparation process evaluated by a response surface methodology. *RSC Adv.* **2016**, *6*, 100326–100333. [CrossRef]

59. Fu, J.; Wang, S.; He, C.; Lu, Z.; Huang, J.; Chen, Z. Facilitated fabrication of high strength silica aerogels using cellulose nanofibrils as scaffold. *Carbohydr. Polym.* **2016**, *147*, 89–96. [CrossRef] [PubMed]

60. Demilecamps, A.; Beauger, C.; Hildenbrand, C.; Rigacci, A.; Budtova, T. Cellulose–silica aerogels. *Carbohydr. Polym.* **2015**, *122*, 293–300. [CrossRef] [PubMed]

61. Laskowski, J.; Milow, B.; Ratke, L. The effect of embedding highly insulating granular aerogel in cellulosic aerogel. *J. Supercrit. Fluids* **2015**, *106*, 93–99. [CrossRef]

62. Shen, X.; Shamshina, J.L.; Berton, P.; Gurau, G.; Rogers, R.D. Hydrogels based on cellulose and chitin: Fabrication, properties, and applications. *Green Chem.* **2016**, *18*, 53–75. [CrossRef]

63. Munier, P.; Gordeyeva, K.; Bergström, L.; Fall, A.B. Directional Freezing of Nanocellulose Dispersions Aligns the Rod-Like Particles and Produces Low-Density and Robust Particle Networks. *Biomacromolecules* **2016**, *17*, 1875–1881. [CrossRef] [PubMed]

64. Peng, F.; Shaw, M.T.; Olson, J.R.; Wei, M. Hydroxyapatite Needle-Shaped Particles/Poly(I-lactic acid) Electrospun Scaffolds with Perfect Particle-along-Nanofiber Orientation and Significantly Enhanced Mechanical Properties. *J. Phys. Chem. C* **2011**, *115*, 15743–15751. [CrossRef]

65. Yun, H.; Kim, S.; Hyun, Y.; Heo, S.; Shin, J. Three-Dimensional Mesoporous–Giantporous Inorganic/Organic Composite Scaffolds for Tissue Engineering. *Chem. Mater.* **2007**, *19*, 6363–6366. [CrossRef]

66. Chen, W.; Zhou, H.; Tang, M.; Weir, M.D.; Bao, C.; Xu, H.H.K. Gas-Foaming Calcium Phosphate Cement Scaffold Encapsulating Human Umbilical Cord Stem Cells. *Tissue Eng. Part A* **2012**, *18*, 816–827. [CrossRef] [PubMed]

67. Jing, D.; Wu, L.; Ding, J. Solvent-Assisted Room-Temperature Compression Molding Approach to Fabricate Porous Scaffolds for Tissue Engineering. *Macromol. Biosci.* **2006**, *6*, 747–757. [CrossRef] [PubMed]

68. Brinchi, L.; Cotana, F.; Fortunati, E.; Kenny, J.M. Production of nanocrystalline cellulose from lignocellulosic biomass: Technology and applications. *Carbohydr. Polym.* **2013**, *94*, 154–169. [CrossRef] [PubMed]

69. Sacui, I.A.; Nieuwendaal, R.C.; Burnett, D.J.; Stranick, S.J.; Jorfi, M.; Weder, C.; Foster, E.J.; Olsson, R.T.; Gilman, J.W. Comparison of the Properties of Cellulose Nanocrystals and Cellulose Nanofibrils Isolated from Bacteria, Tunicate, and Wood Processed Using Acid, Enzymatic, Mechanical, and Oxidative Methods. *ACS Appl. Mater. Interfaces* **2014**, *6*, 6127–6138. [CrossRef] [PubMed]

MDPI

St. Alban-Anlage 66

4052 Basel

Switzerland

Tel. +41 61 683 77 34

Fax +41 61 302 89 18

www.mdpi.com

Coatings Editorial Office

E-mail: coatings@mdpi.com

www.mdpi.com/journal/coatings